Moments of Inertia
of Simple Areas

Polar Moment of Inertia

Shape	Figure	$\bar{I}_x$	$\bar{I}_y$	$\bar{J}$
Rectangle		$\dfrac{bh^3}{12}$	$\dfrac{hb^3}{12}$	$\dfrac{bh}{12}(h^2 + b^2)$
Triangle		$\dfrac{bh^3}{36}$		
Circle		$\dfrac{\pi d^4}{64}$ $\dfrac{\pi R^4}{4}$	$\dfrac{\pi d^4}{64}$ $\dfrac{\pi R^4}{4}$	$\dfrac{\pi d^4}{32}$ $\dfrac{\pi R^4}{2}$
Circular ring		$\dfrac{\pi(d_0^4 - d_i^4)}{64}$ $\dfrac{\pi(R_0^4 - R_i^4)}{4}$	$\dfrac{\pi(d_0^4 - d_i^4)}{64}$ $\dfrac{\pi(R_0^4 - R_i^4)}{4}$	$\dfrac{\pi(d_0^4 - d_i^4)}{32}$ $\dfrac{\pi(R_0^4 - R_i^4)}{2}$
Semicircle		$\left(\dfrac{\pi}{8} - \dfrac{8}{9\pi}\right)R^4$ $= 0.1098R^4$	$\dfrac{\pi R^4}{8}$	$\left(\dfrac{\pi}{4} - \dfrac{8}{9\pi}\right)R^4$ $= 0.5025R^4$
Quarter circle		$\left(\dfrac{\pi}{16} - \dfrac{4}{9\pi}\right)R^4$ $= 0.0549R^4$	$\left(\dfrac{\pi}{16} - \dfrac{4}{9\pi}\right)R^4$ $= 0.0549R^4$	$\left(\dfrac{\pi}{8} - \dfrac{8}{9\pi}\right)R^4$ $= 0.1098R^4$

Applied Strength of Materials

Applied Strength of Materials

Fa-Hwa Cheng, Ph.D., P.E.

Professor of Engineering and General Technology
Virginia Western Community College

Macmillan Publishing Company
New York
Collier Macmillan Publishers
London

A portion of this book is reprinted from *Statics and Strength of Materials* copyright © 1985 by Macmillan Publishing Company.

Macmillan Publishing Company
866 Third Avenue, New York, New York 10022

Collier Macmillan Canada, Inc.

Library of Congress Cataloging in Publication Data

Cheng, Fa-Hwa.
  Applied strength of materials.

  ''A portion of this book is reprinted from Statics and strength of materials. c1985''—T.p. verso.
  Includes index.
  1. Strength of materials.   I. Title.
TA405.C418   1986   620.1'12   85-4933
ISBN 0-02-322320-0

Printing: 1 2 3 4 5 6 7 8   Year: 6 7 8 9 0 1 2 3

ISBN 0-02-322320-0

*To the beloved memory of
my grandparents*

Preface

This book is written primarily for students in four-year engineering technology programs and architectural schools to provide a clear, practical, and easy to understand textbook for a course in Applied Strength of Materials. A workable knowledge of algebra, geometry, and trigonometry is expected of the students. Calculus is used in a few sections for formula derivations and in only one section (beam deflection by the double integration method) for problem solutions, but a lack of background in calculus will not cause a major problem in using the book. Students are assumed to have completed a course in statics. The brief review of statics in Chapter 1 will refresh their memory on the subject.

Both U.S. customary units and SI units are used side by side throughout the book. The U.S. customary units are slightly favored in design topics because most design code, section property tables, and design aids are available only in U.S. customary units.

Basic concepts and fundamental principles are emphasized throughout the book. Topics are presented in a logically organized sequence, generally with increasing complexity. Each topic is carefully developed and clearly explained. An ample amount of example problems with detailed solutions are provided to illustrate each particular phase of the topic under consideration. Extra care has been taken to make the logic of the solution process easy for students to follow.

The book contains an extensive and well-developed coverage of design topics. These topics are designed to familiarize students with the general procedure involved in the design process and to provide them with some perception of the design work performed by engineers. The design of members is introduced as early as Chapter 2 and is scattered throughout the book. Since design is so closely related to analysis, it is logical to discuss the two types of problems at the same time. Furthermore, the early introduction of design problems gives students a greater incentive to study the subject.

A large number of problems of various levels of difficulty are provided at the end of each section within the chapter. In this way students immediately relate the problems to the section covered. The problems in each group are graded according to increasing difficulty, beginning with relatively simple, uncomplicated problems to help students gain confidence and develop technique. In many instances the given data are carefully arranged in order to simplify the numerical solution so that students can concentrate more on the logic and procedure of the solution. The last few problems in the group are usually more involved than the others. These problems provide students with enough challenge to maintain their interest. Answers to two-thirds of the problems are given at the end of the book.

A solutions manual, which provides detailed solutions to all the problems, is available to instructors.

Computer program assignments are included at the ends of Chapters 8, 11, and 12, where computer programming can be used advantageously to handle general problems in beam deflections, transformation of plane stress, and column buckling. These programs can be assigned to students as projects. Some instructors may prefer to load the FORTRAN programs listed in the solutions manual into the school computer and let the students input data and run the programs.

The book starts with a review of the fundamentals in Chapter 1, in which systems of units, differential and integral calculus, and statics are reviewed; followed by discussions on simple stresses caused by simple loading conditions in Chapter 2 and the relationship between stress and strain in Chapter 3. Shaft and beam topics are covered in the subsequent chapters. Before the shear force and bending moment diagrams are discussed in Chapter 6, axial force diagrams and torque diagrams are introduced in Chapters 2 and 5, respectively. The author believes that this arrangement helps students understand better the meaning and significance of the shear and moment diagrams.

All the material pertaining to statically indeterminate members is treated collectively in Chapter 9. This is done for two reasons: First, because of the similarity in the method of solution of all statically indeterminate problems, the solutions to one type of problem will help students to understand the solutions to the other type of problem. Second, most students find statically indeterminate problems more difficult. Therefore, the author feels strongly that none of the statically indeterminate problems should be discussed in the first few chapters.

The subject of combined stresses is discussed in Chapter 10. This chapter can be regarded as an overall review of the material studied previously. Transformation of plane stress and Mohr's circle are introduced in Chapter 11. The geometry of Mohr's circle is used for the derivation of the formulas for principal stresses; thereby more involved mathematics is avoided. The buckling of columns is discussed in Chapter 12. This discussion is mainly based on the Euler formula for long columns and the J. B. Johnson formula for intermediate columns. Finally, structural connections, including riveted connections, high-strength-bolt connections, and welded connections are covered in Chapter 13.

The book contains more material than can be covered in the first course in Applied Strength of Materials. The extra material is included to provide flexibility in the selection of topics for instructors and to serve as a valuable reference book for students. Topics identified with an asterisk on the section title can be omitted without affecting the continuity of the text.

The author is confident that this book will prove to be an effective textbook for its intended purpose. The initial lecture notes on which this book is based have been field-tested for many years and have been extremely well received by students.

To the many reviewers whose valuable suggestions made this book a better work, the author expresses deep appreciation. The author wants also to thank his wife, Rosa, and his two sons, Lincoln and Lindsay, for their loving support.

Fa-Hwa Cheng

Contents

List of Tables

Applied Strength of Materials

CHAPTER 1

Review of the Fundamentals

1–1
INTRODUCTION

Strength of materials deals with the internal forces in a body and the changes of shape and size of the body, particularly in the relationship of the internal force to the external forces that act on the body. The body is usually a structural or machine member, such as a shaft, a beam, or a column. The external forces acting on a body consist of loads and reactions. The body reacts to the external forces by developing internal resisting forces. The intensities of the internal resisting forces are called *stresses*. The changes in the dimensions of the body are called *deformations*.

The subject of strength of materials involves analytical methods for determining the *strength* (load-carrying capacity based on stresses inside a member), *stiffness* (deformation characteristics), and *stability* (the ability of a thin or slender member to maintain its initial configuration without buckling while being subjected to compressive loading). The sizes of all structural or machine members must be properly designed according to the requirements for strength, stiffness, and/or stability. For example, the wall of a pressure vessel must be of adequate strength to withstand the internal pressure for which the vessel is designed. On the other hand, if a thin-walled vessel is subjected to partial vacuum that causes compressive stress in the wall, then the safe level of vacuum at which the stability of the thin wall can be maintained must be determined. The floor of a building must be strong enough to carry the design load, while being stiff or rigid enough so that it will not deflect excessively under the applied load.

Strength of materials is one of the most fundamental subjects in the engineering curriculum. Its methods are needed by structural engineers in the design of bridges, buildings, and aircraft; by mechanical engineers in the design of machines, tools, and pressure vessels; and by mining engineers, chemical engineers, and electrical engineers in those phases of their jobs that involve the analysis and design of structural or machine members.

Keep in mind that in relating internal resisting forces to external forces, the methods developed in statics still apply because the body or part of the body under consideration is only slightly deformed and the small deformations have a negli-

1

gible effect on equilibrium conditions. Therefore, free-body diagrams and application of the static equilibrium equations are essential to the determination of both the external reactions and the internal resisting forces in a body.

Statics is reviewed briefly in Section 1–7. Differential and integral calculus are discussed briefly in Section 1–8.

1–2
SYSTEMS OF UNITS

Currently, there are two systems of units used in engineering practice in the United States. They are the U.S. customary system of units and the International System of units, or SI units (from the French "Système International d'Unités"). The SI units have now been widely adopted throughout the world. In industrial and commercial applications in the United States, U.S. customary units are gradually being replaced by SI units. During the transition years, engineers in this country must be familiar with both systems. For this reason, both systems of units are presented in this book. The U.S. customary system is slightly favored in design problems because most design codes, section property tables, and design aids are available only in U.S. customary units.

1–3
U.S. CUSTOMARY UNITS

The U.S. Customary system of units is commonly used in engineering practice in the United States, especially in civil, architectural, and mechanical engineering. The base units in this system are

length: foot   (ft)

force: pound (lb)

time: second (s)

Because the base unit for force, pound, is dependent on the gravitational attraction of the earth, this system is referred to as the *gravitational system* of units.

The unit of mass in this system is the slug, which is a derived unit. Newton's second law states that

force = mass $\times$ acceleration

or

$$F = ma \tag{1-1}$$

Thus

$$m = \frac{F}{a} = \frac{\text{lb}}{\text{ft/s}^2} = \text{lb}-\text{s}^2/\text{ft} = \text{slug}$$

The unit "slug" is rarely used in strength of materials.

Other U.S. customary units frequently encountered in mechanics are

$$\text{mile (mi)} = 5280 \text{ ft}$$
$$\text{yard (yd)} = 3 \text{ ft}$$
$$\text{inch (in.)} = \tfrac{1}{12} \text{ ft}$$
$$\text{kilopound (kip)} = 1000 \text{ lb}$$
$$\text{U.S. ton (ton)} = 2000 \text{ lb}$$
$$\text{minute (min)} = 60 \text{ s}$$
$$\text{hour (h)} = 60 \text{ min} = 3600 \text{ s}$$

1–4
SI UNITS

The three base SI units are

$$\text{length: meter (m)}$$
$$\text{mass:} \quad \text{kilogram (kg)}$$
$$\text{time:} \quad \text{second (s)}$$

The SI units are called an *absolute system* of units, since the three base units chosen are independent of the location where the measurement is made.

The unit of force, called the newton (N), is a derived unit expressed in terms of the three base units. One newton is defined as the force that produces an acceleration of 1 m/s² (read "meters per second squared" or "meters per second per second") when applied to a mass of 1 kg. From Eq. (1–1),

$$F = ma$$

or

$$1 \text{ N} = (1 \text{ kg})(1 \text{ m/s}^2) = 1 \text{ kg} \cdot \text{m/s}^2$$

Thus the newton is equivalent to kg · m/s².

The acceleration of a freely falling body under the action of its own weight (which is the force exerted on the mass by gravity) is approximately 9.81 m/s² on the surface of the earth. This quantity is usually denoted by g and is called the *gravitational acceleration*. From Eq. (1–1), for a freely falling body on the surface of the earth, we have

$$W = mg = (1 \text{ kg})(9.81 \text{ m/s}^2) = 9.81 \text{ kg} \cdot \text{m/s}^2 = 9.81 \text{ N}$$

which means that the weight of 1-kg mass is 9.81 N on the surface of the earth.

Multiples of the SI units are abbreviated by use of the prefixes shown in Table 1–1.

TABLE 1–1   Recommended SI Prefixes

	Exponential Form	Prefix	SI Symbol
1 000 000 000*	10^9	giga	G
1 000 000	10^6	mega	M
1 000	10^3	kilo	k
0.001	10^{-3}	milli	m
0.000 001	10^{-6}	micro	μ

*A space rather than a comma is used to separate numbers in groups of three, counting from the decimal point in both directions. Space may be omitted for four-digit numbers.

The following are typical examples of the use of prefixes:

$$10^6 \text{ g} = 10^3 \text{ kg} = 1 \text{ Mg}$$

$$10^3 \text{ m} = 1 \text{ km}$$

$$10^3 \text{ N} = 1 \text{ kN}$$

$$10^{-3} \text{ kg} = 1 \text{ g}$$

$$10^{-3} \text{ m} = 1 \text{ mm}$$

1–5
CONVERSION OF UNITS

In this book, problems are solved in the system of units used in the data given. There is no need to convert units from one system to the other. In actual engineering applications, however, there are many occasions when it is necessary to convert units. For this purpose, the following unit conversion factors are useful:

$$1 \text{ ft} = 0.3048 \text{ m}$$

$$1 \text{ slug} = 14.59 \text{ kg}$$

$$1 \text{ lb} = 4.448 \text{ N}$$

The following examples illustrate the conversion of units.

━━━━ **EXAMPLE 1–1** ━━━━━━━━━━━━━━━━━━━━━━━━━━━━━━━━━━━━━━

Convert a moment of 1 lb-ft into equivalent value in N · m.

SOLUTION

$$\text{moment} = 1 \text{ lb-ft} = (1 \text{ lb-ft})\left(\frac{4.448 \text{ N}}{1 \text{ lb}}\right)\left(\frac{0.3048 \text{ m}}{1 \text{ ft}}\right) = 1.356 \text{ N} \cdot \text{m}$$

The two conversion factors (4.448 N/1 lb) and (0.3048 m/1 ft) are each equal to unity. The value of a quantity is not changed when it is multiplied by factors of unity.

TABLE 1–2   U.S. Customary Units and SI Equivalents

Quantity	U.S. Customary Unit	SI Equivalent
Length	ft	0.3048 m
	in.	25.40 mm
	mi	1.609 km
Force	lb	4.448 N
	kip	4.448 kN
Mass	slug	14.59 kg
Area	ft²	0.092 90 m²
	in.²	645.2 mm²
	mi²	2.590 km²
Volume	ft³	0.028 32 m³
Velocity	ft/s (fps)	0.3048 m/s
	mi/h (mph)	1.609 km/h
Acceleration	ft/s²	0.3048 m/s²
Stress	lb/ft² (psf)	47.88 Pa (pascal or N/m²)
(or pressure)	lb/in.² (psi)	6.895 kPa (kN/m²)
	kip/in.² (ksi)	6.895 MPa (MN/m²)
Moment	lb-ft	1.356 N · m
(of a force)	lb-in.	0.1130 N · m
Area moment	in.⁴	0.4162 × 10⁻⁶ m⁴
of inertia		
Work	ft-lb	1.356 J (joule or N · m)
Power	ft-lb/s	1.356 W (watt or N · m/s)
	hp (1 hp = 550 ft-lb/s)	745.7 W

EXAMPLE 1–2

Convert a stress (a quantity derived as force per unit area) of 1 psf (lb/ft²) into equivalent value in Pa (pascal or N/m²).

SOLUTION

$$\text{stress} = 1 \text{ psf} = \left(1 \frac{\cancel{lb}}{\cancel{ft^2}}\right)\left(\frac{4.448 \text{ N}}{1 \cancel{lb}}\right)\left(\frac{1 \cancel{ft^2}}{0.0348^2 \text{ m}^2}\right) = 47.88 \text{ N/m}^2$$

$$= 47.88 \text{ Pa}$$

■

The U.S. customary units and the SI equivalents that are used most frequently in mechanics are listed in Table 1–2.

1–6
REMARKS ON NUMERICAL ACCURACY

The accuracy of the solution of a problem depends on two factors:

1. The accuracy of the data given.
2. The accuracy of the computations performed.

The accuracy of computations made by using an electronic calculator is always greater than the accuracy of the physical data given. Therefore, the accuracy of a solution is always limited by the accuracy of the known physical data.

A practical rule of rounding off figures in the computations involved in engineering analysis and design is to retain four significant figures for numbers beginning with the figure "1" and to retain three significant figures for numbers beginning with any figures from "2" through "9." Thus the value 182.35 is rounded off to 182.4, and the value 2934 is rounded off to 2930.

PROBLEMS*

1–1 Convert the following SI units to the SI units indicated.
  (a) 6.38 Gg to kg
  (b) 900 km to m
  (c) 3.76×10^7 g to Mg
  (d) 70 mm to m
  (e) 23 400 N to kN

1–2 The specific weight (weight per unit volume) of concrete is 150 lb/ft^3. What is its equivalent value in kN/m^3?

1–3 Use the conversion factors listed in Table 1–2 to convert the following units.
  (a) 200 lb-ft to N $\cdot$ m
  (b) 60 mph to km/h
  (c) 100 hp to kW
  (d) 9.81 m/s^2 to ft/s^2
  (e) 100 MN/m^2 to ksi (kips/in.^{2})
  (f) 10 m/s to mph

1–7
BRIEF REVIEW OF STATICS

Some fundamental definitions, principles, laws, or theorems in statics are listed below.

1. Forces are vector quantities. Two concurrent forces can be added by the *parallelogram law*.
2. As far as the external effect of a force is concerned, the force can be considered to act at any point along its line of action. This is known as the *principle of transmissibility*.
3. The *moment of a force* about a point is defined as the product of the magnitude of the force and the perpendicular distance from the point to the line of action of the force. The moment is also equal to the sum of the moments produced by the components of the force about the same point. This is known as the *theorem of moments*.
4. The action and reaction forces between interactive bodies always occur in equal and opposite pairs. This is known as *Newton's third law*.
5. A body is in static equilibrium when the resultant of all the forces acting on it is zero. For a body to be in equilibrium under the action of coplanar

---

* Answers to two-thirds of the problems are given at the end of the book.

force system, the following equilibrium equations must be satisfied:

$$\Sigma F_x = 0 \qquad \Sigma F_y = 0 \qquad \Sigma M_A = 0 \qquad (1\text{–}2)$$

where A is an arbitrary point in the plane of the forces.

6. Instead of Eqs. (1–2), either one of the following sets of equilibrium equations can be used for solving static equilibrium problems.

$$\Sigma F_x = 0 \qquad \Sigma M_A = 0 \qquad \Sigma M_B = 0 \qquad (1\text{–}3)$$

where A and B are arbitrary points, except that line AB is not along the y-direction; or

$$\Sigma M_A = 0 \qquad \Sigma M_B = 0 \qquad \Sigma M_C = 0 \qquad (1\text{–}4)$$

where A, B, and C are arbitrary points not along the same straight line.

7. A free-body diagram is a sketch of an isolated body with all the external forces acting on the body shown. To solve an equilibrium problem, a free-body diagram must be drawn first.

8. A two-force body is a body acted upon by only two forces. For a two-force body to be in equilibrium, the two forces must be equal, opposite, and collinear.

9. A three-force body is a body acted upon by three coplanar nonparallel forces. For a three-force body to be in equilibrium, the three forces must pass through a common point.

For detailed information on the items listed above, the student should refer to a statics text.

───── **EXAMPLE 1–3** ──

The bracket shown is supported by a fixed support at C. Determine the reactions at C due to the given loads.

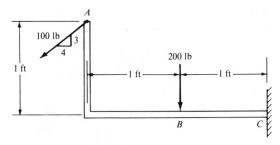

SOLUTION

The free-body diagram of the bracket is shown in the following figure. The 100-lb force is replaced by its horizontal and vertical components. At the fixed support C, three unknown reaction components, R_x, R_y, and M, are shown.

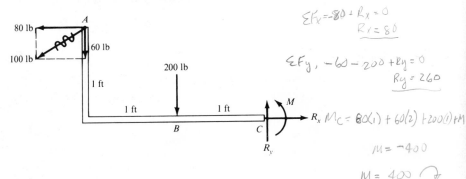

$\Sigma F_x = -80 + R_x = 0$

$R_x = 80$

$\Sigma F_y, \ -60 - 200 + R_y = 0$

$R_y = 260$

$M_C = 80(1) + 60(2) + 200(1) + M$

$M = -400$

$M = 400 \ \circlearrowright$

The equilibrium equations for the free-body diagram give

$$\xrightarrow{+}\Sigma F_x = -80 \text{ lb} + R_x = 0 \quad R_x = +80 \text{ lb} \quad R_x = 80 \text{ lb} \rightarrow$$

$$+\uparrow\Sigma F_y = -60 \text{ lb} - 200 \text{ lb} + R_y = 0 \quad R_y = +260 \text{ lb} \quad R_y = 260 \text{ lb} \uparrow$$

$$\circlearrowright\Sigma M_C = (80 \text{ lb})(1 \text{ ft}) + (60 \text{ lb})(2 \text{ ft}) + (200 \text{ lb})(1 \text{ ft}) + M = 0$$

$$M = -400 \text{ lb-ft} \quad M = 400 \text{ lb-ft} \circlearrowright$$

■

EXAMPLE 1–4

Determine the reactions at supports A and C of the wall bracket shown due to a load $P = 30$ kN.

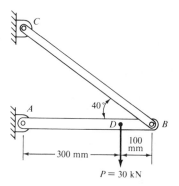

SOLUTION

Since BC is a two-force body, the two forces acting on it must be along the axial direction. If member BC is in tension, the force exerted on member AB by member BC is acting from B toward C. Member AB is subjected to three forces; therefore, it is a three-force body. The three forces must meet at the same point. In the free-body diagram of member AB shown below, the forces P and F_{BC} intersect at E; then the reaction R_A must also pass through E.

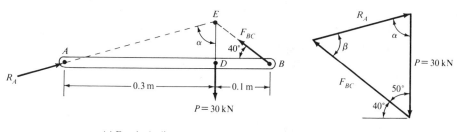

(a) Free-body diagram   (b) Force triangle

Since member AB is in equilibrium, the three forces acting on it must form a closed force triangle as shown. From triangle BDE we have

$$DE = BD \tan 40° = (0.1 \text{ m}) \tan 40° = 0.0839 \text{ m}$$

and from triangle ADE we have

$$\tan \alpha = \frac{AD}{DE} = \frac{0.3 \text{ m}}{0.0839 \text{ m}} = 3.576$$

$$\alpha = \tan^{-1} 3.576 = 74.4°$$

In the force triangle, the unknown angle β is

$$\beta = 180° - \alpha - 50° = 55.6°$$

Now we apply the law of sines to the force triangle to get

$$\frac{R_A}{\sin 50°} = \frac{F_{BC}}{\sin 74.4°} = \frac{30 \text{ kN}}{\sin 55.6°}$$

$$R_A = 27.8 \text{ kN} \nearrow^{74.4°} \qquad F_{BC} = 35.0 \text{ kN (T)}$$

The reaction at the support C is

$$R_C = F_{BC} = 35.0 \text{ kN} \; {}_{40°}\!\nwarrow$$

━━━━━━ **EXAMPLE 1–5** ━━━━━━━━━━━━━━━━━━━━━━━━━━━━━━━━━━

Determine the forces acting on each member of the frame due to the loads shown.

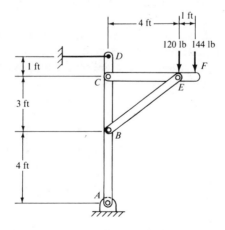

SOLUTION

To determine the reactions of the supports at D and A, consider the free-body diagram of the entire frame as shown in the following figure.

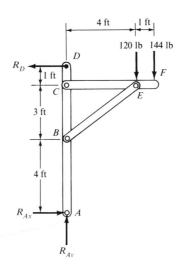

The equilibrium equations give

$$\oplus\Sigma M_A = R_D \,(8 \text{ ft}) - (120 \text{ lb})(4 \text{ ft}) - (144 \text{ lb})(5 \text{ ft}) = 0$$

$$R_D = +150 \text{ lb} \qquad R_D = 150 \text{ lb} \leftarrow$$

$$\oplus\Sigma M_D = R_{Ax} \,(8 \text{ ft}) - (120 \text{ lb})(4 \text{ ft}) - (144 \text{ lb})(5 \text{ ft}) = 0$$

$$R_{Ax} = +150 \text{ lb} \qquad R_{Ax} = 150 \text{ lb} \rightarrow$$

$$+\uparrow\Sigma F_y = R_{Ay} - 120 \text{ lb} - 144 \text{ lb} = 0$$

$$R_{Ay} = +264 \text{ lb} \qquad R_{Ay} = 264 \text{ lb} \uparrow$$

Check

$$\rightarrow\Sigma F_x = +150 - 150 = 0 \qquad\qquad \text{(checks)}$$

$$\oplus\Sigma M_E = 150(1) + 150(7) - 264(4) - 144(1) = 0 \qquad \text{(checks)}$$

Member BE is a two-force body. If we assume that BE is in compression, force F_{BE} is pushing member CEF at E, as shown in the free-body diagram of member CEF. The force at joint C exerted on the member is represented by its horizontal and vertical components, C_x and C_y.

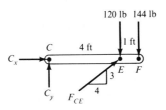

The equilibrium equations give

$$\oplus\Sigma M_C = \tfrac{3}{5}F_{BE} \,(4 \text{ ft}) - (120 \text{ lb})(4 \text{ ft}) - (144 \text{ lb})(5 \text{ ft}) = 0$$

$$F_{BE} = +500 \text{ lb} \qquad F_{BE} = 500 \text{ lb (C)}$$

$$\xrightarrow{+}\Sigma F_x = C_x + \tfrac{4}{5}F_{BE} = 0 \qquad C_x = -\tfrac{4}{5}(+500 \text{ lb}) = -400 \text{ lb}$$

$$+\uparrow\Sigma F_y = C_y - 120 \text{ lb} - 144 \text{ lb} + \tfrac{3}{5}F_{BE} = 0$$

$$C_y = 120 \text{ lb} + 144 \text{ lb} - \tfrac{3}{5}(+500 \text{ lb}) = -36 \text{ lb}$$

where the positive sign for F_{BE} means that member BE is in compression, as we assumed. The negative sign for C_x and C_y means that the assumed direction of these two components in the free-body diagram of member CEF must be reversed.

Now the forces acting on each member of the frame can be drawn as shown in the following figures. Keep in mind that the action and reaction forces between interactive bodies are equal in magnitude and opposite in directions.

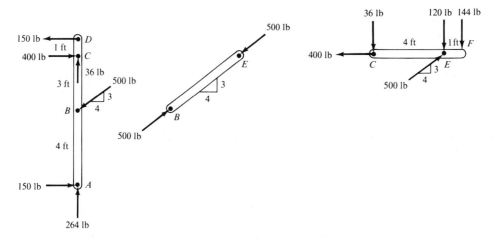

Check

The equilibrium conditions of the vertical member AD provide a very useful check to the solution of the problem. Thus

$$\xrightarrow{+}\Sigma F_x = 150 + 400 - 150 - \tfrac{4}{5}(500) = 0 \qquad \text{(checks)}$$

$$+\uparrow\Sigma F_y = 264 - \tfrac{3}{5}(500) + 36 = 0 \qquad \text{(checks)}$$

$$\circlearrowleft\Sigma M_B = 150(4) - 400(3) + 150(4) = 0 \qquad \text{(checks)}$$

■

PROBLEMS

*In Problems **1–4** to **1–7**, determine the reactions at the supports of the beam due to the loading shown.*

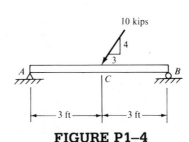

FIGURE P1–4

1-5

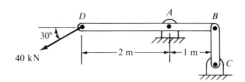

FIGURE P1–5

1-6

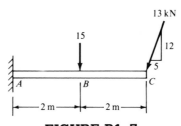

FIGURE P1–6

1-7

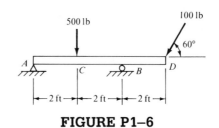

FIGURE P1–7

In Problems **1–8** *and* **1–9,** *determine the reactions at the supports of the structure due to the loading shown.*

1-8

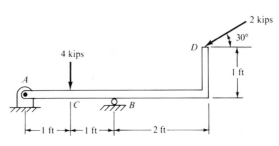

FIGURE P1–8

1-9

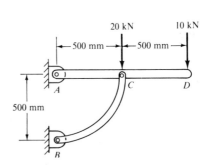

FIGURE P1–9

In Problems **1–10** *and* **1–11,** *determine the forces acting on each member of the frame due to the loading shown.*

1–10

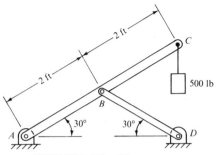

FIGURE P1–10

1–11

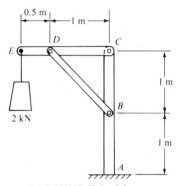

FIGURE P1–11

1–8
BRIEF DISCUSSION OF DIFFERENTIAL AND INTEGRAL CALCULUS

There are two branches of calculus: the *differential* and *integral calculus.* These are discussed briefly in the following.

Differential Calculus: The Derivative

The functional relationship

$$y = f(x)$$

can be plotted into a curve as shown in Fig. 1–1. To find the rate of change of the function y with respect to x at the point $P(x, y)$, we draw a secant line through P intersecting the curve at $Q(x + \Delta x, y + \Delta y)$. The change in the value of y is Δy, and the change in the value of x is Δx. The quotient $\Delta y/\Delta x$ is the average rate of change of y with respect to x over the interval PQ. As Δx becomes smaller and smaller, this average rate $\Delta y/\Delta x$ comes closer and closer to the exact rate at P. As Δx approaches zero, the limiting value of the quotient $\Delta y/\Delta x$ defines the instantaneous rate of change of y with respect to x at P. This limit is called the *derivative* of y with respect to x. It is denoted by the symbol dy/dx, thus:

$$\frac{dy}{dx} = \lim_{\Delta x \to 0} \frac{\Delta y}{\Delta x}$$

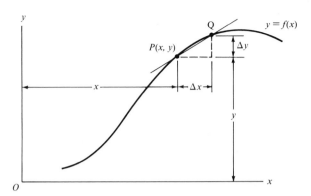

FIGURE 1–1

Geometrically, the quotient $\Delta y/\Delta x$ is the slope of the secant line PQ. As Δx approaches zero, the secant line becomes tangent to the curve at P. Hence *the value of the derivative of y with respect to x at P is equal to the slope of the tangent line drawn to the graph of y = f(x) at point P.*

The process of determining the derivative of a function $f(x)$ with respect to x is called *differentiation*. Formulas for calculating the derivatives of simple algebraic functions are as follows (where c and n are constants):

$$\frac{d}{dx}(c) = 0 \tag{1–5}$$

$$\frac{d}{dx}(x) = 1 \tag{1–6}$$

$$\frac{d}{dx}(x^n) = nx^{n-1} \tag{1–7}$$

$$\frac{d}{dx}[cf(x)] = c\frac{d}{dx}[f(x)] \tag{1–8}$$

$$\frac{d}{dx}[f(x) + g(x)] = \frac{d}{dx}[f(x)] + \frac{d}{dx}[g(x)] \tag{1–9}$$

──────── **EXAMPLE 1–6** ────────────────────────────────

Determine the angle that the tangent line drawn to the parabola $y = \frac{1}{2}x^2 + 1$ at the point (2, 3) makes with the x-axis.

SOLUTION

The general expression of the derivative of y with respect to x is

$$\frac{dy}{dx} = \frac{d}{dx}\left(\frac{1}{2}x^2 + 1\right) = \frac{d}{dx}\left(\frac{1}{2}x^2\right) + \frac{d}{dx}(1) = \frac{1}{2}\frac{d}{dx}(x^2) + 0 = \frac{1}{2}(2x) = x$$

At the point (2, 3) the value of the derivative is

$$\left[\frac{dy}{dx}\right]_{x=2} = 2$$

Since the slope of the tangent line drawn to a curve at a point is equal to the value of the derivative at the point, we see that the slope of the tangent at the point (2, 3) is

$$\text{slope} = \left[\frac{dy}{dx}\right]_{x=2} = 2$$

as indicated in the following figure.

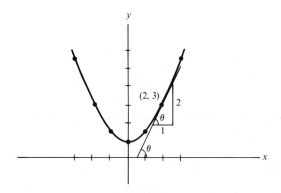

The angle θ that the tangent makes with the x-axis is

$$\theta = \tan^{-1}(2) = 63.4°$$

◼

EXAMPLE 1–7

When a body is thrown vertically upward, its vertical position is expressed as the following function of time:

$$y = y_0 + v_0 t - \tfrac{1}{2}gt^2$$

where the constants y_0, v_0, and g are

$$y_0 = \text{initial position of the body}$$

$$v_0 = \text{initial velocity of the body}$$

$$g = \text{gravitational acceleration}$$

and t is the time in seconds. (a) Find an expression for the velocity of the body as a function of time t. (b) If the initial velocity is 161 ft/s and the gravitational acceleration is $g = 32.2$ ft/s^2, determine the time when the velocity of the body is reduced to zero.

SOLUTION

(a) The velocity of a body is defined as the rate of change of its position with respect to time. That is,

$$v = \frac{dy}{dt} = \frac{d}{dt}\left(y_0 + v_0 t - \frac{1}{2}gt^2\right) = \frac{d}{dt}(y_0) + \frac{d}{dt}(v_0 t) - \frac{d}{dt}\left(\frac{1}{2}gt^2\right)$$

$$= 0 + v_0 - \frac{1}{2}g(2t) = v_0 - gt$$

which is a general expression of velocity as a function of time.

(b) To determine the time when the velocity of the body is reduced to zero, we solve t from the equation $v = 0$. Thus

$$v = v_0 - gt = 0$$

from which

$$t = \frac{v_0}{g} = \frac{161 \text{ ft/s}}{32.2 \text{ ft/s}^2} = 5 \text{ s}$$

■

Integral Calculus: The Integral

The reverse process of differentiation is known as *integration*. For example, if we ask what is the function y of x for which the derivative is $3x^2$, the answer is $y = x^3 + C$, where C is an arbitrary constant. Because

$$\frac{dy}{dx} = \frac{d}{dx}(x^3 + C) = 3x^2$$

In general, if

$$\frac{dy}{dx} = f(x)$$

we may write this derivative in the differential form as follows:

$$dy = f(x)\, dx \tag{1-10}$$

where dy is called the *differential* of y, and dx is called the differential of x. Equation (1–10) expresses the differential (or the change) of y in terms of the differential (or the change) of x. The expression $y = F(x)$ whose differential is given by Eq. (1–10) can be obtained by integration. Using the symbol $\int$ as the integral sign, we write

$$y = \int dy = \int f(x)\, dx = F(x) + C$$

where $\int f(x)\, dx$ is called the *indefinite integral* of the function $f(x)$, for which

$$\frac{d}{dx}[F(x) + C] = f(x)$$

where C is an arbitrary constant, called the *constant of integration*.

Some integration formulas for simple algebraic functions are listed below. In these formulas c and n are constants and C is the constant of integration.

$$\int dx = x \tag{1–11}$$

$$\int x^n \, dx = \frac{x^{n+1}}{n+1} + C \qquad (n \neq -1) \tag{1–12}$$

$$\int cf(x) \, dx = c \int f(x) \, dx \tag{1–13}$$

$$\int [f(x) + g(x)] \, dx = \int f(x) \, dx + \int g(x) \, dx \tag{1–14}$$

An important application of integration is to find the area under a curve. In Fig. 1–2 let $y = f(x)$ be the equation of the curve; it is desired to calculate the area bounded by the curve, the x-axis, and the vertical lines $x = x_1$ and $x = x_2$.

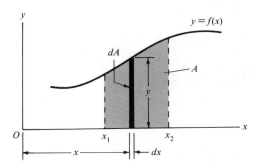

FIGURE 1–2

Consider a vertical differential area element dA with dx as the width and y as the height; we have

$$dA = y \, dx = f(x) \, dx$$

The desired area can be obtained by integrating $f(x) \, dx$ from x_1 to x_2; thus

$$A = \int dA = \int_{x_1}^{x_2} f(x) \, dx \tag{1–15}$$

The integral $\int_{x_1}^{x_2} f(x) \, dx$ is called the *definite integral* of $f(x)$ from x_1 to x_2, where x_1 and x_2 are called the *lower* and *upper limits*, respectively. The value of a definite integral is a definite number determined by the formula

$$\int_{x_1}^{x_2} f(x) = \left[F(x) \right]_{x_1}^{x_2} = F(x_2) - F(x_1) \tag{1–16}$$

where $\int f(x) \, dx = F(x)$.

The definite integral is essentially a summation process. Physically, the definite integral $\int_{x_1}^{x_2} f(x) \, dx$ can be regarded as the sum of infinite numbers of differential elements $f(x)\,dx$ between x_1 and x_2. Geometrically, when all the differential areas are summed up, it gives the value of the desired area.

EXAMPLE 1–8

Find the area between the parabolic curve $y = 9 - x^2$ and the x-axis.

SOLUTION
The curve is plotted as shown. It intersects the x-axis at $x = -3$ and $x = +3$.

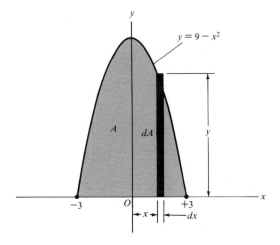

A vertical rectangular element has a differential area of

$$dA = y\,dx = (9 - x^2)\,dx$$

The definite integral of the expression $(9 - x^2)\,dx$ from -3 to $+3$ gives the desired area. Thus

$$A = \int dA = \int_{-3}^{+3} (9 - x^2)\,dx = \left[9x - \tfrac{1}{3}x^3 \right]_{-3}^{+3}$$

$$= [9(+3) - \tfrac{1}{3}(+3)^3] - [9(-3) - \tfrac{1}{3}(-3)^3] = 36$$

If the length unit had been given, the unit of area would be the length unit squared. ■

EXAMPLE 1–9

A particle is moving along a straight line with uniform acceleration a. If the initial velocity at $t = 0$ is v_0 and the initial position is s_0, express (a) the velocity v as a function of time, (b) the position s as a function of time.

SOLUTION
(a) By definition, acceleration is the rate of change of velocity with respect to time. That is,

$$\frac{dv}{dt} = a = \text{constant}$$

Expressed in differential form, the expression becomes

$$dv = a \, dt$$

Integrating the expression with the lower limits corresponding to the initial conditions $t = 0$ and $v = v_0$ and the upper limits corresponding to $t = t$ and $v = v$, we write

$$\int_{v_0}^{v} dv = \int_{0}^{t} a \, dt = a \int_{0}^{t} dt$$

which gives

$$\left[v \right]_{v_0}^{v} = a \left[t \right]_{0}^{t}$$

or

$$v - v_0 = a(t - 0) \qquad v = v_0 + at$$

(b) By definition, velocity is the rate of change of position with respect to time. That is,

$$\frac{ds}{dt} = v = v_0 + at$$

or in differential form, we write

$$ds = (v_0 + at) \, dt$$

Now we integrate both sides of the expression, the left-hand side with respect to s from $s = s_0$ to $s = s$, and the right-hand side with respect to t from $t = 0$ to $t = t$. We have

$$\int_{s_0}^{s} ds = \int_{0}^{t} (v_0 + at) \, dt$$

from which

$$s - s_0 = v_0 t + \tfrac{1}{2}at^2$$

or

$$s = s_0 + v_0 t + \tfrac{1}{2}at^2$$

PROBLEMS

In Problems **1–12** *to* **1–16,** *differentiate the given function, using Eqs. (1–5) through (1–9).*

1–12 $y = 3x^4$

1–13 $y = 7x^2 + 4x + 9$

1–14 $y = (x + 2)^2$

1–15 $s = 100 + 25t - 16.1t^2$

1–16 $w = u(u^2 - 5)$

In Problems **1–17** *to* **1–20,** *find the slope of the tangent to the curve at the point indicated. Sketch the curve and the tangent line.*

1–17 $y = x^2 + 1$ at $(2, 5)$

1–18 $y = x^2 - 4$ at $(3, 5)$

1–19 $y = (x + 1)^2$ at $(2, 9)$

1–20 $y = \dfrac{4}{x}$ at $(4, 1)$

1–21 Locate the point on the curve $y = 2x^2 - 4x$ where its tangent is along the horizontal direction.

1–22 The position of a particle moving along a straight line is given by the function $y = 6t^2 - 2t$, where y is in feet and t is in seconds. Determine the values of t at which the velocity of the particle is zero.

In Problems **1–23** *to* **1–26,** *integrate the given function of x, using Eqs. (1–11) through (1–14).*

1–23 $\displaystyle\int 3x^2 \, dx$

1–24 $\displaystyle\int (x^2 + x - 4) \, dx$

1–25 $\displaystyle\int \left(6x^2 + \dfrac{1}{x^2} \right) dx$

1–26 $\displaystyle\int \sqrt{x} \, dx$

In Problems **1–27** *to* **1–30,** *determine the value of the given definite integral.*

1–27 $\displaystyle\int_0^3 (x^2 + 1)\, dx$

1–28 $\displaystyle\int_{-2}^4 (x + 1)^2\, dx$

1–29 $\displaystyle\int_{-1}^2 \left(4x + \frac{2}{x^3}\right) dx$

1–30 $\displaystyle\int_4^9 \frac{dx}{\sqrt{x}}$

In Problems **1–31** *to* **1–34,** *find the area bounded by the curve, the x-axis, and the vertical lines indicated.*

1–31 $y = 2x + 1$ between $x = 1$ and $x = 4$

1–32 $y = x^2 + 1$ between $x = 0$ and $x = 3$

1–33 $y = x^2 - 2x + 5$ between $x = 1$ and $x = 5$

1–34 $y = x^2 + x + 1$ between $x = -3$ and $x = 4$

1–35 Sketch the curve $y = 3 + 2x - x^2$ and find the area bounded by the curve and the x-axis.

1–36 A particle is accelerating along a straight line with the acceleration expressed as function of time as $a = 3t^2$, where a is the acceleration in m/s^2 and t is the time in seconds. If at $t = 0$, the initial velocity is 2 m/s and the initial position is 10 m, determine **(a)** the velocity, **(b)** the position at the time $t = 3$ s.

CHAPTER 2

Simple Stresses

2–1
DEFINITION OF THE NORMAL AND SHEAR STRESSES

When a structural member is subjected to a load, internal resisting forces are set up within the member to balance the external forces. Consider a body subjected to a system of balanced external forces F_1, F_2, F_3, and F_4, as shown in Fig. 2–1(a). These forces tend to pull the body apart. Internal resisting forces are developed within the body that act to hold the body together.

To determine the internal resisting forces in a body, an arbitrary section m–m is passed through the body, separating the body completely into two parts. The internal forces in the section can then be determined by considering the equilibrium of either part of the body separated by the section.

Since the body in Fig. 2–1(a) is in equilibrium, each part of the body must also be in equilibrium. The free-body diagram of the left-hand side and the right-hand side of the body is shown in Fig. 2–1(b). The internal force **R** acting on the free-body is in equilibrium with the external forces. The force **R** can be resolved into two components: R_n normal to the section and R_s parallel to the section.

The force **R** is the resultant of many minute forces acting on the entire cross section, as shown in Fig. 2–2. In general, the internal resisting forces are not uniformly distributed across the section. At an area ΔA a small internal force $\Delta \mathbf{R}$ is exerted on area ΔA. This force $\Delta \mathbf{R}$ is resolved into two components: ΔR_n normal to the area and ΔR_s parallel to the area.

The average intensity of the normal force ΔR_n over the small area ΔA is equal to ΔR_n divided by ΔA. This intensity of internal normal force per unit area is called the *normal stress*. Denoting the normal stress by σ (the Greek lowercase letter sigma), we have

$$\sigma_{\text{avg}} = \frac{\Delta R_n}{\Delta A}$$

As ΔA approaches zero, we obtain the normal stress at a point. Thus

$$\sigma = \lim_{\Delta A \to 0} \frac{\Delta R_n}{\Delta A} = \frac{dR_n}{dA} \tag{2–1}$$

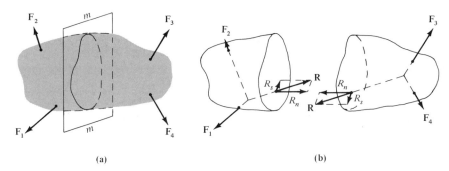

(a)   (b)

FIGURE 2–1

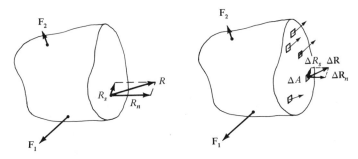

FIGURE 2–2

The intensity of internal shear force per unit area is called the *shear stress*. Denoting the shear stress by τ (the Greek lowercase letter tau), the sheer stress at a point is

$$\tau = \lim_{\Delta A \to 0} \frac{\Delta R_s}{\Delta A} = \frac{dR_s}{dA} \tag{2-2}$$

Note that in general, the normal and shear stresses may vary from point to point across the section.

There are two types of normal stress: tensile and compressive stresses. *Tensile stress* is produced by a force pulling *away from* an area. *Compressive stress* is produced by a force pushing *toward* an area. Tensile stresses are usually considered positive, and compressive stresses are considered negative.

Stress is one of the most important concepts in the study of strength of materials. Whenever a body is subjected to external loads, stress is induced inside the body. Whether the material will fail and to what extent it will deform depend on the amount of stress induced in the body.

In the U.S. customary system, the units generally used for stress are pounds per square inch (psi) or kips (kilopound or 1000 lb) per square inch (ksi). The SI units of stress are newtons per square meter (N/m^2), also designated pascal (Pa). When prefixes are used, the following units are frequently encountered:

$$1 \text{ kPa} = 10^3 \text{ Pa}$$

$$1 \text{ MPa} = 10^6 \text{ Pa}$$

$$1 \text{ GPa} = 10^9 \text{ Pa}$$

Table 1–2 in Chapter 1 gives conversion factors for psi and kPa, and for ksi and MPa.

2–2
NORMAL STRESS DUE TO AXIAL LOAD

Consider a rod of uniform cross section subjected to a pair of equal and opposite forces P acting along the axis of the rod, as shown in Fig. 2–3(a). The forces applied along the axial direction of the member are called *axial forces*. A member subjected to axial forces is called an *axially loaded member*.

Since the entire rod in Fig. 2–3(a) is in equilibrium, any portion of the rod separated by imaginary transverse cutting planes must also be in equilibrium. Figure 2–3(b) shows the free-body diagrams of the two parts of the rod separated by the plane *m–m* perpendicular to the axis of the rod. From either one of the two free-body diagrams, the equilibrium condition requires that the internal force in the section be equal to the external force P. If the axial force P is applied through the centroid of the cross-sectional area, then the internal force is uniformly distributed over the cross section. Thus, by definition, the normal tensile stress σ in the section is

$$\sigma = \frac{P}{A} \tag{2–3}$$

where P is the internal tensile force at the section and A is the cross-sectional area of the rod. Figure 2–3(c) shows the stress diagram of the section.

A similar analysis may be applied to a compression member if the length of the member is relatively short compared to the lateral dimensions of the member. When long and slender members are subjected to compressive axial loads, they tend to buckle. The problem of buckling is discussed in detail in Chapter 12.

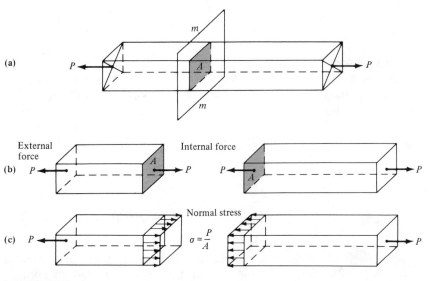

FIGURE 2–3

Variation of internal axial force along the length of a member can be depicted by an *axial force diagram* whose ordinate at any section of a member is equal to the value of the internal axial force at that section. In plotting an axial force diagram, we usually treat tensile force as positive and compressive force as negative. Example 2–2 illustrates the computation and construction of an axial force diagram.

─────── **EXAMPLE 2–1** ───────

A rod of uniform cross-sectional area $A = 2$ in.^{2} is hanging from a ceiling, as shown in Fig. (a). A weight W of 2000 lb is attached to a cable of cross-sectional area $A' = 0.25$ in.^{2}, which is firmly fastened to the lower end of the rod through the centroid of the cross section. Assuming that both the rod and the cable have negligible weight, calculate the normal stresses in the rod and in the cable.

$$\sigma = \frac{P}{A}$$

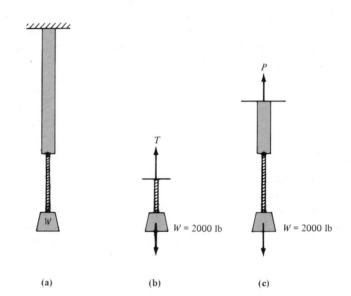

(a)   (b)   (c)

SOLUTION
From the free-body diagram shown in Fig. (b), we see that the tension T, which represents the internal resisting force in any cross section of the cable, must be equal to W, because

$$+\uparrow \Sigma F_y = T - W = 0 \qquad T = W = 2000 \text{ lb}$$

Hence the normal stress in the cable is

$$\sigma_{\text{cable}} = \frac{T}{A'} = \frac{2000 \text{ lb}}{0.25 \text{ in.}^2} = 8000 \text{ psi (T)}$$

Similarly, from the free-body diagram shown in Fig. (c), the internal force P in any cross section of the rod must also be equal to W and the normal stress in

the rod is

$$\sigma_{\text{rod}} = \frac{P}{A} = \frac{2000 \text{ lb}}{2 \text{ in.}^2} = 1000 \text{ psi (T)}$$

■

EXAMPLE 2–2

A steel bar 10 mm by 20 mm in cross section is subjected to axial loads shown in Fig. (a). Determine the normal stresses in segments AB, BC, and CD.

(a)
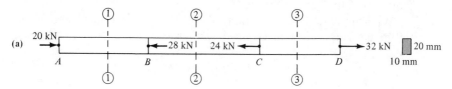

(b)
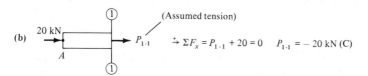

(c)
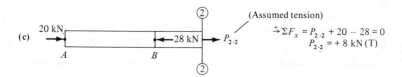

(d)
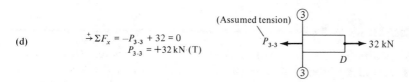

(e)
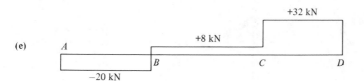

(f)

SOLUTION

In Fig. (a), the algebraic sum of the given axial forces is equal to zero; thus the bar is in equilibrium.

To find the internal force at a section of the bar, a plane is passed through that section, separating the bar into two parts. Then the internal axial force at the section can be determined by considering the equilibrium of either part of the bar separated by the plane. Figures (b), (c), and (d) show the free-body diagrams and their respective equilibrium equations. In each free-body diagram, the internal axial force is assumed to cause tension in the bar and is thus shown pulling away from the section. A positive value would indicate that the section is indeed in tension, and a negative value would indicate that the section is actually in compression.

The axial force at any section between A and B is constant and is equal to 20 kN in compression. The axial force at any section between B and C is 8 kN in tension and any section between C and D is 32 kN in tension, as plotted in the axial force diagram in Fig. (e). Figure (f) shows the axial force to which each segment of the bar is subjected.

With the axial force variations along the bar determined, the normal stresses in the bar can now be determined. The cross-sectional area of the bar is

$$A = (0.010 \text{ m})(0.020 \text{ m}) = 0.0002 \text{ m}^2$$

The normal stresses in the three segments are

$$\sigma_{AB} = \frac{P_{AB}}{A} = \frac{-20 \text{ kN}}{0.0002 \text{ m}^2} = -100\,000 \text{ kN/m}^2 \qquad \sigma_{AB} = 100 \text{ MPa (C)}$$

$$\sigma_{BC} = \frac{P_{BC}}{A} = \frac{+8 \text{ kN}}{0.0002 \text{ m}^2} = +40\,000 \text{ kN/m}^2 \qquad \sigma_{BC} = 40 \text{ MPa (T)}$$

$$\sigma_{CD} = \frac{P_{CD}}{A} = \frac{+32 \text{ kN}}{0.0002 \text{ m}^2} = +160\,000 \text{ kN/m}^2 \qquad \sigma_{CD} = 160 \text{ MPa (T)}$$

■

EXAMPLE 2–3

A 10-kip weight is supported by a rod and cables, as shown. Neglecting the weight of the rod, determine the normal stresses in the rod and cables.

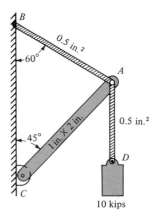

SOLUTION

Rod AC is a two-force member, since it is subjected to only two forces at the ends of the member. Hence the forces exerted on rod AC must be along the axial direction of the member. The tension in cable AD is equal to the 10-kip weight. To determine the axial forces in cable AB and rod AC, consider the equilibrium of joint A. The free-body diagram of joint A and the corresponding force triangle are as shown.

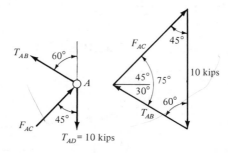

By the law of sines,

$$\frac{T_{AB}}{\sin 45°} = \frac{F_{AC}}{\sin 60°} = \frac{10}{\sin 75°}$$

from which

$$T_{AB} = 7.32 \text{ kips} \qquad F_{AC} = 8.97 \text{ kips}$$

The normal stresses in the rod and in the cables are

$$\sigma_{AB} = \frac{T_{AB}}{A_{AB}} = \frac{+7.32 \text{ kips}}{0.5 \text{ in.}^2} = +14.6 \text{ ksi} \qquad \sigma_{AB} = 14.6 \text{ ksi (T)}$$

$$\sigma_{AD} = \frac{T_{AD}}{A_{AD}} = \frac{+10 \text{ kips}}{0.5 \text{ in.}^2} = +20.0 \text{ ksi} \qquad \sigma_{AD} = 20 \text{ ksi (T)}$$

$$\sigma_{AC} = \frac{F_{AC}}{A_{AC}} = \frac{-8.97 \text{ kips}}{1 \times 2 \text{ in.}^2} = -4.49 \text{ ksi} \qquad \sigma_{AC} = 4.49 \text{ ksi (C)}$$

PROBLEMS

In Problems 2–1 to 2–3, plot the axial force diagram and determine the normal stresses in sections 1–1, 2–2, and 3–3.

2–1

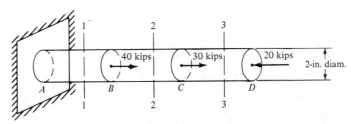

FIGURE P2–1

2-2

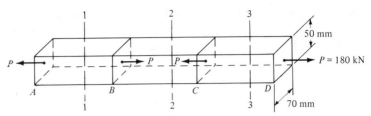

FIGURE P2-2

2-3

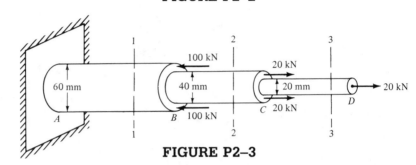

FIGURE P2-3

2-4 A short column is composed of two standard steel pipes, as shown in Fig. P2-4. The load $P = 20$ kips. Determine the compressive stress in each pipe. Neglect the weight of the pipes.

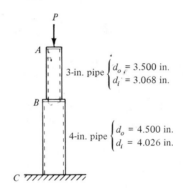

3-in. pipe $\begin{cases} d_o = 3.500 \text{ in.} \\ d_i = 3.068 \text{ in.} \end{cases}$

4-in. pipe $\begin{cases} d_o = 4.500 \text{ in.} \\ d_i = 4.026 \text{ in.} \end{cases}$

FIGURE P2-4

2-5 In Fig. P2-5, determine the normal stresses in strut AB and rod BC.

$\cos \theta = \frac{x}{h} = \frac{A}{h}$

$\sin \theta = \frac{y}{h} = \frac{o}{h}$

$\tan \theta = \frac{A}{o} \quad \frac{x}{y}$

318 (T)   318MPA

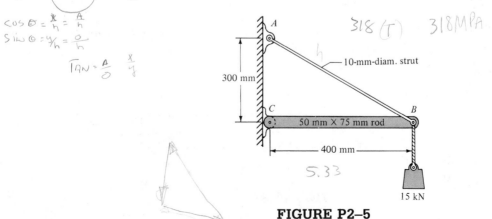

10-mm-diam. strut

300 mm

50 mm X 75 mm rod

400 mm

5.33

15 kN

FIGURE P2-5

2–6 In Fig. P2–6, weight W is supported by rod AB and cable BC. Determine the normal stresses in cable BC and rod AB.

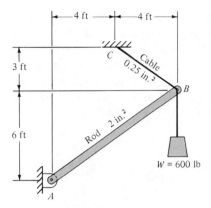

FIGURE P2–6

2–7 The maximum compression that the hydraulic compression testing machine shown in Fig. P2–7 can exert is 600 kN. Each of the two posts, A and B, has a diameter $d = 80$ mm. Determine the tensile stress in the posts.

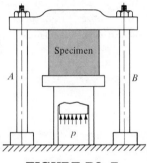

FIGURE P2–7

2–8 The force applied to the brake pedal of a car is transmitted by lever AD and connecting rod BC, as shown in Fig. P2–8. If $P = 20$ lb, $a = 10$ in., $b = 2$ in., and $d = \frac{1}{4}$ in., determine the normal stress in rod BC.

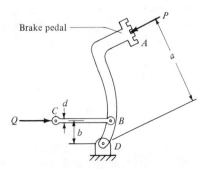

FIGURE P2–8

2–3
SHEAR STRESS DUE TO DIRECT SHEAR FORCE

Shear stress has been defined in Section 2–2 as the intensity of a force acting parallel to a section. Shear stress differs from normal stress in that the direction of shear stress is parallel to the plane rather than perpendicular to it.

Consider a block with a projection shown in Fig. 2–4(a). A horizontal force P applied to the projection tends to shear the piece off from the block along the shear plane $abcd$. The body resists the shear action of the force P by developing a resisting shear stress in the shear plane.

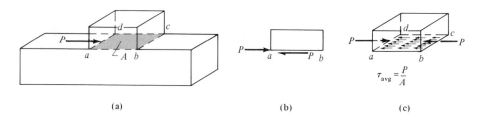

(a)   (b)   (c)

FIGURE 2–4

The resultant of the shear stress in the shear area must be equal to the applied force P, as shown in Fig. 2–4(b). The shear stress usually has a nonuniform distribution over the shear area. In engineering practice, however, the shear stress is often assumed to be uniformly distributed over the shear area. Thus the average shear stress τ_{avg} is

$$\tau_{avg} = \frac{P}{A} \qquad (2\text{–}4)$$

where P is the internal resisting shear force in the shear plane and A is the area of the shear plane $abcd$.

Many structural elements and machine parts are subjected to shear loads. A few examples will now be discussed.

The Lap Joint

The lap joint connects overlapped tension members by rivets or bolts [Fig. 2–5(a)]. From Fig. 2–5(b), we see that section m–m of the rivet is subjected to shear stress. The shear stress, in general, is not uniformly distributed in the section. The average value used in engineering practice is

$$\tau_{avg} = \frac{P}{A} \quad \frac{20}{3.5} \qquad (2\text{–}5)$$

where A is the cross-sectional area of the rivet. Since the shear stress occurs only in section m–m of the rivet, the rivet is said to be in *single shear*.

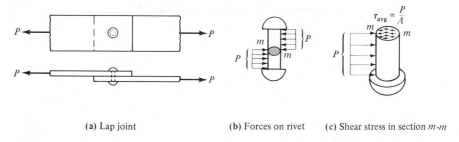

(a) Lap joint　　　　　(b) Forces on rivet　　(c) Shear stress in section m-m

FIGURE 2–5

The Butt Joint

The butt joint connects nonoverlapping tension members by rivets or bolts [Fig. 2–6(a)]. From Fig. 2–6(b), we see that rivet A is subjected to shear stresses at sections m–m and n–n. Assume that the shear force is equally shared by the two sections. Then the average shear stress [Fig. 2–6(c)] is

$$\tau_{avg} = \frac{P}{2A} \tag{2–6}$$

where A is the cross-sectional area of the rivet. Since the shear stresses occur in two sections of the rivet, the rivet is said to be in *double shear*. Rivet B is similarly loaded and is subjected to the same shear stress.

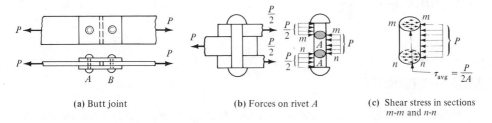

(a) Butt joint　　　　(b) Forces on rivet A　　(c) Shear stress in sections m-m and n-n

FIGURE 2–6

The Shaft Key

The shaft key connects a gear (or pulley) to a shaft [Fig. 2–7(a)]. The key has width b, height h, and length L. The shaft has radius r. To transmit the couple M, the key is subjected to shear forces P, as shown in Fig. 2–7(b). These forces are assumed to be concentrated on the rim of the shaft. The moment of P about the center of the shaft must be equal to M. Thus

$$Pr = M$$

or

$$P = \frac{M}{r} \tag{2–7}$$

Assume that the shear stress is uniformly distributed [Fig. 2–7(c)]. Then the average shear stress in section m–m of the key is

$$\tau_{\text{avg}} = \frac{P}{A} = \frac{M/r}{bL} = \frac{M}{rbL} \qquad (2\text{–}8)$$

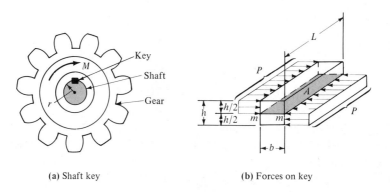

(a) Shaft key   (b) Forces on key

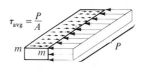

(c) Shear stress in section m-m

FIGURE 2–7

2–4
BEARING STRESS

When one body presses against another, bearing stress occurs between the two bodies. For example, Fig. 2–8(a) shows that the bottom of the block is pressed

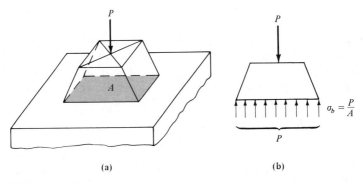

(a)   (b)

FIGURE 2–8

against the top of the pier by a compressive force P. Assume that the bearing stress in the contact area A [the shaded area in Fig. 2–8(a)] is uniformly distributed. Then the bearing stress [Fig. 2–8(b)] is

$$\sigma_b = \frac{P}{A} \tag{2-9}$$

Bearing stress occurs between the key and the gear and between the key and the shaft, as shown in Fig. 2–7(b). The compressive force P is assumed to be uniformly distributed over an area $(h/2)L$. The bearing stress is, therefore,

$$\sigma_b = \frac{P}{A} = \frac{M/r}{(h/2)L} = \frac{2M}{rhL} \tag{2-10}$$

In the lap joint or butt joint shown in Figs. 2–5 and 2–6, bearing stresses occur between the rivet and the plates. The stress is distributed over a cylindrical surface as shown in Fig. 2–9(a) and (b). The maximum bearing stress occurs at the midpoint of the cylindrical contact surface. This maximum bearing stress was found to be approximately equal to the value obtained by dividing the force transmitted by the projected area of the rivet onto the plate [the rectangular area

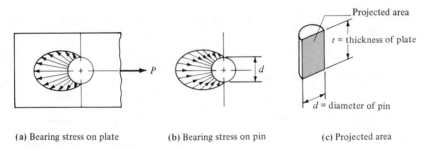

(a) Bearing stress on plate   (b) Bearing stress on pin   (c) Projected area

FIGURE 2–9

with thickness t of the plate and diameter d of the rivet as its two sides, shown as the shaded area in Fig. 2–9(c)]. Therefore, in engineering practice, the bearing stress between the rivet and the plate is computed by

$$\sigma_b = \frac{P}{\text{projected area}} = \frac{P}{td} \tag{2-11}$$

where

P = force transmitted
t = thickness of the plate
d = diameter of the pin

─── **EXAMPLE 2–4** ───

A circular blanking punch is shown in the figure. It is operated by causing shear failure in the plate. Knowing that thickness t of the steel plate is 10 mm and that the ultimate shear strength of the steel (the greatest shear stress a material can withstand before failure) is $\tau_u = 300$ MPa, determine the minimum force P required to punch a hole 50 mm in diameter.

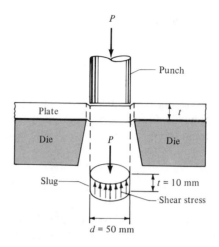

SOLUTION

The shear area that resists the punch is a cylindrical area on the side of the slug that is punched out. The shear area is

$$A = \pi dt = \pi(0.05 \text{ m})(0.01 \text{ m}) = 0.001\ 57 \text{ m}^2$$

The minimum force P is the force needed to cause shear failure over the shear area. Thus

$$P_{\min} = A\tau_u = (0.001\ 57 \text{ m}^2)(300\ 000 \text{ kN/m}^2) = 471 \text{ kN}$$

EXAMPLE 2–5

A rectangular key $b \times h \times L = \frac{3}{4}$ in. $\times \frac{1}{2}$ in. $\times$ 3 in. is used to connect a gear and a shaft of diameter $d = 3$ in., as shown. The couple transmitted by the key is 15 kip-in. Determine the shear stress in the key and the bearing stress between the key and the shaft.

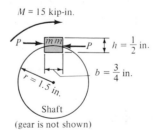

(gear is not shown)

SOLUTION

The shear force P on the key is

$$P = \frac{M}{r} = \frac{15 \text{ kip-in.}}{1.5 \text{ in.}} = 10 \text{ kips}$$

The shear area in section m–m is

$$A_s = bL = (\tfrac{3}{4} \text{ in.})(3 \text{ in.}) = 2.25 \text{ in.}^2$$

The shear stress is

$$\tau = \frac{P}{A_s} = \frac{10 \text{ kips}}{2.25 \text{ in.}^2} = 4.44 \text{ ksi}$$

The bearing area is

$$A_b = \left(\frac{h}{2}\right)L = (\tfrac{1}{4} \text{ in.})(3 \text{ in.}) = 0.75 \text{ in.}^2$$

The bearing stress is

$$\sigma_b = \frac{P}{A_b} = \frac{10 \text{ kips}}{0.75 \text{ in.}^2} = 13.3 \text{ ksi}$$

■

EXAMPLE 2–6

Two steel plates $\tfrac{1}{2}$ in. by 5 in. are fastened by means of two $\tfrac{3}{4}$-in.-diameter bolts, as shown. If the joint transmits a tensile force P of 12 kips, determine (a) the average shear stress in the bolts, (b) the bearing stress between the bolts and the plates.

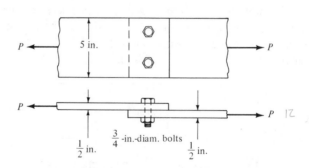

$\frac{3}{4}$-in.-diam. bolts

$\frac{1}{2}$ in.

SOLUTION
The cross-sectional area of the bolt is

$$\frac{\pi d^2}{4} = \frac{\pi(\tfrac{3}{4} \text{ in.})^2}{4} = 0.442 \text{ in.}^2$$

(a) Assume that the force P is shared equally by the two bolts. The shear force on each bolt is thus $P/2$. For the bolt in single shear, the average shear stress is

$$\tau_{avg} = \frac{P/2}{A} = \frac{12 \text{ kips}}{2(0.442 \text{ in.}^2)} = 13.6 \text{ ksi}$$

(b) The compressive force transmitted by each bolt is $P/2$. By Eq. (2–9), the bearing stress is

$$\sigma_b = \frac{P/2}{td} = \frac{12 \text{ kips}}{2(\tfrac{1}{2} \text{ in.})(\tfrac{3}{4} \text{ in.})} = 16 \text{ ksi}$$

■

PROBLEMS

2–9 Figure P2–9 shows a schematic diagram of apparatus for determining the ultimate shear strength (failure shear stress) of wood. The test specimen is 4 in. in height, 2 in. in width, and 2 in. in depth. If the load required to shear the specimen into two pieces is 8000 lb, determine the ultimate shear strength of the specimen.

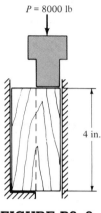

$P = 8000$ lb

4 in.

FIGURE P2–9

2–10 The lap joint shown in Fig. P2–10 is connected by four 20-mm-diameter rivets. Determine **(a)** the shear stress in the rivets, **(b)** the bearing stress between the rivets and the plates. Assume that the load $P = 120$ kN is carried equally by the four rivets.

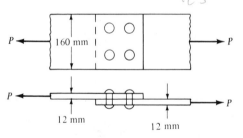

P 160 mm P

P P

12 mm

12 mm

FIGURE P2–10

2–11 The clevis shown in Fig. P2–11 is connected by a pin $\frac{3}{4}$ in. in diameter. Determine the shear stress in the pin and the bearing stress between the pin and the plates if $P = 10$ kips and $t = \frac{1}{4}$ in.

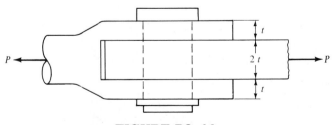

P t $2t$ t P

FIGURE P2–11

2–12 The pulley shown in Fig. P2–12 is connected to a shaft 80 mm in diameter by a 20-mm square key that is 100 mm long. If the belt tensions are $T_1 = 40$ kN and $T_2 = 120$ kN, determine **(a)** the shear stress in the key, **(b)** the bearing stress between the key and the shaft.

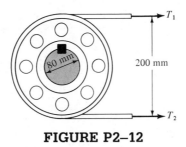

FIGURE P2–12

2–13 A force $F = 600$ lb is applied to a crank and is transmitted to a shaft through a steel key, as shown in Fig. P2–13. The key is $\frac{1}{2}$ in. square and $2\frac{1}{2}$ in. long. Determine **(a)** the shear stress in the key, **(b)** the bearing stress between the key and the shaft.

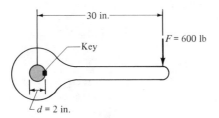

FIGURE P2–13

2–14 The geometry of a punch hole is shown in Fig. P2–14. Determine the minimum force that must be exerted on a punch to shear this hole in a steel plate 4 mm thick. The plate has an ultimate shear strength (failure shear stress) of 300 MPa.

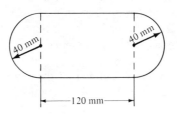

FIGURE P2–14

2–15 In the collar bearing shown in Fig. P2–15, the average bearing stress between the collar and the support is known to be 4000 psi. If $d = 2$ in., $D = 4$ in., and $t = \frac{1}{2}$ in., determine **(a)** the load P applied on the column, **(b)** the average shear stress on the area between the collar and the column.

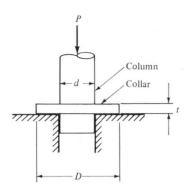

FIGURE P2–15

2–16 A wood joint is shown in Fig. P2–16. The dimensions are $a = 100$ mm, $b = 150$ mm, $c = 40$ mm, and $d = 90$ mm. Determine the shear stress and the bearing stress in the joint if $P = 50$ kN.

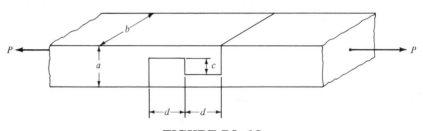

FIGURE P2–16

2–5
ALLOWABLE STRESS AND FACTOR OF SAFETY

To provide a margin of safety in design, members are usually designed for a limited stress level called the *allowable stress* or the *working stress,* a value much smaller than the ultimate strength (stress) of the material. The *ultimate strength* of a material is the greatest stress the material can withstand before rupture. The ultimate strength is determined by rupture testing of a specimen made of the given material. The maximum load that can be applied to the specimen before rupture divided by the appropriate loading area gives the value of the ultimate strength of the material.

The *factor of safety* is defined as the ratio of ultimate strength to allowable stress as expressed in the following equation:

$$\text{factor of safety} = \frac{\text{ultimate strength}}{\text{allowable stress}} \qquad (2\text{–}12)$$

With this definition, a factor of safety of 3 would mean that the member could withstand a maximum load equal to three times the load for which the member is designed before failure would occur.

There are numerous reasons for using a factor of safety in structural design. A few reasons are cited below.

1. It is difficult to estimate the exact load to which a structure is subjected, and unexpectedly large loads might occur.
2. Materials of structural members are not entirely homogeneous or of uniform quality.
3. The assumptions used in the analysis and design are often subjected to appreciable error.
4. Manufacturing processes, such as uneven cooling in different portions of the metal, often leave some residual stress within a structure member.
5. The conditions of a material may appreciably deviate from the initial state of the material due to corrosion, creep (continuous deformation due to a long-term sustained load), and fatigue (weakening of strength in a material due to repeated and alternating load applications).

2–6
DESIGN OF AXIALLY LOADED MEMBERS AND SHEAR PINS

Design as used here simply means the determination of the required cross-sectional area of a member.

For an axially loaded member, the minimum cross-sectional area required can be determined by solving Eq. (2–3) for A:

$$A_{req} = \frac{P}{\sigma_{allow}} \qquad (2\text{–}13)$$

where P is the largest internal axial force in the member and σ_{allow} is the allowable normal stress of the material.

For tension members, the area A_{req} computed from Eq. (2–13) is the required net cross-sectional area. For short compression blocks, Eq. (2–13) can also be used. For slender compression members, however, do not use Eq. (2–13); because buckling of the member may occur. Refer to Chapter 12 for a discussion of the buckling of compression members.

For pins subjected to shear stress, the minimum area required can be determined by solving Eq. (2–4) for A:

$$A_{req} = \frac{P}{\tau_{allow}} \qquad (2\text{–}14)$$

where P is the shear force in the pin due to the most severe loading condition, and τ_{allow} is the allowable shear stress of the material of the pin.

EXAMPLE 2–7

The member *AC* is supported by a round structural steel tie rod *BD* and a pin at *A*, as shown. Neglect the weight of member *AC* and assume that the ultimate strength of structural steel rod is 490 MPa in tension and that the ultimate shear strength of the pin is 315 MPa. Using a factor of safety of 3.5 for both tension and shear, determine the minimum required diameters of the tie rod and the pin.

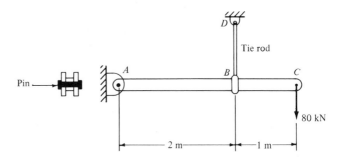

SOLUTION

To determine the tension in the rod and the reaction at A, consider the equilibrium of bar AC.

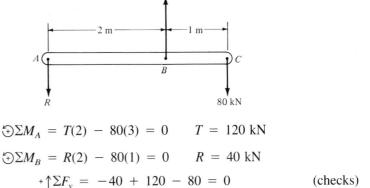

$$\circlearrowleft \Sigma M_A = T(2) - 80(3) = 0 \qquad T = 120 \text{ kN}$$

$$\circlearrowleft \Sigma M_B = R(2) - 80(1) = 0 \qquad R = 40 \text{ kN}$$

$$+\uparrow \Sigma F_y = -40 + 120 - 80 = 0 \qquad\qquad \text{(checks)}$$

The allowable stress in tension is

$$\sigma_{\text{allow}} = \frac{\text{ultimate strength in tension}}{\text{factor of safety}} = \frac{490 \text{ MPa}}{3.5} = 140 \text{ MPa}$$

The tensile stress in the rod must be less than the allowable tensile stress. Thus

$$\sigma = \frac{T}{A} = \frac{120 \text{ kN}}{\frac{1}{4}\pi(d_{\text{rod}})^2} = \frac{152.8 \text{ kN}}{(d_{\text{rod}})^2} \le \sigma_{\text{allow}} = 140\,000 \text{ kN/m}^2$$

from which

$$d_{\text{rod}} \ge \sqrt{\frac{152.8 \text{ kN}}{140\,000 \text{ kN/m}^2}} = 0.0330 \text{ m}$$

The required minimum diameter of the rod is thus 33.0 mm.

The allowable shear stress is

$$\tau_{\text{allow}} = \frac{\text{ultimate shear strength}}{\text{factor of safety}} = \frac{315 \text{ MPa}}{3.5} = 90 \text{ MPa}$$

The pin is in double shear. The shear stress in the pin must be less than the allowable shear stress. Thus

$$\tau = \frac{R}{2A} = \frac{40 \text{ kN}}{2 \times \frac{1}{4}\pi(d_{\text{pin}})^2} = \frac{25.5 \text{ kN}}{(d_{\text{pin}})^2} \leq \tau_{\text{allow}} = 90\,000 \text{ kN/m}^2$$

from which

$$d_{\text{pin}} \geq \sqrt{\frac{25.5 \text{ kN}}{90\,000 \text{ kN/m}^2}} = 0.0168 \text{ m}$$

The minimum required diameter of the pin is thus 16.8 mm. ∎

EXAMPLE 2–8

A clevis is connected by a pin, as shown in Fig. (a). If $P = 18$ kips, $\tau_{\text{allow}} = 15$ ksi, and $(\sigma_b)_{\text{allow}} = 48$ ksi, determine the required diameter d of the pin and the required thickness t.

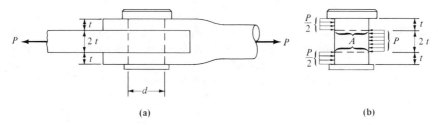

(a)   (b)

SOLUTION

The pin is in double shear, as shown in Fig. (b). Thus the shear stress in the pin is

$$\tau = \frac{P}{2A} = \frac{18 \text{ kips}}{2(\frac{1}{4}\pi d^2)} = \frac{11.46 \text{ kips}}{d^2} \leq \tau_{\text{allow}} = 15 \text{ ksi}$$

from which the required minimum diameter of the pin is

$$d \geq \sqrt{\frac{11.46 \text{ kips}}{15 \text{ kip/in.}^2}} = 0.874 \text{ in.}$$

A pin of diameter $\frac{7}{8}$ in. ($= 0.875$ in.) may be selected.
The bearing stress between the pin and the plate is

$$\sigma_b = \frac{P}{d(2t)} = \frac{18 \text{ kips}}{(\frac{7}{8} \text{ in.})(2t)} = \frac{10.3 \text{ kips/in.}}{t} \leq (\sigma_b)_{\text{allow}} = 48 \text{ ksi}$$

from which we have

$$t \geq \frac{10.3 \text{ kips/in.}}{48 \text{ kips/in.}^2} = 0.214 \text{ in.}$$

Thus the thickness $t = \frac{1}{4}$ in. may be selected. ∎

PROBLEMS

2–17 In Fig. P2–17, determine the required cross-sectional area in square millimeters of members *BD*, *BE*, and *CE* of the truss subjected to the forces shown. The allowable stresses are 140 MPa in tension and 70 MPa in compression.

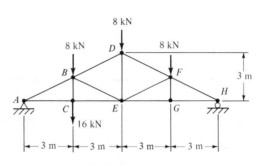

FIGURE P2–17

2–18 The bell crank mechanism shown in Fig. P2–18 is subjected to a vertical force of 10 kips applied at *C*. The force is resisted by a horizontal force *P* at *A* and a reaction at *B*. If the mechanism is in equilibrium and the allowable shear stress is 15 ksi, determine the required diameter of the pin *B*.

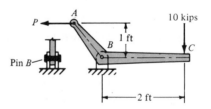

FIGURE P2–18

2–19 The structure shown in Fig. P2–19 is supported by bolts at *A* and *B*. The bolts are in double shear. If the allowable shear stress in the bolts is 120 MPa, determine the required diameters of the bolts at *A* and *B*.

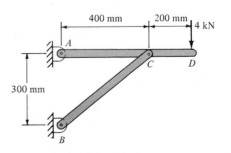

FIGURE P2–19

2-20 In Fig. P2–20, a tie rod $\frac{1}{4}$ in. diameter is used to hold a wall in place. The tensile stress in the rod caused by P is 20 ksi. Determine the required diameter d of the washer to keep the bearing stress between the wall and the washer from exceeding 300 psi.

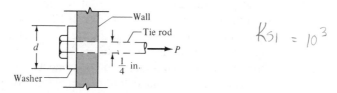

$ksi = 10^3$

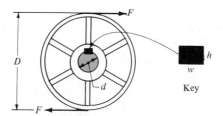

Washer

FIGURE P2-20

2-21 The control gate shown in Fig. P2–21 is operated by a wheel and shaft connected by a flat key. The allowable stresses in the key are 8000 psi in shear and 20 000 psi in bearing. If $d = 2\frac{1}{4}$ in., $D = 30$ in., $w = \frac{1}{2}$ in., $h = \frac{3}{8}$ in., and $F = 450$ lb, determine the required length of the key.

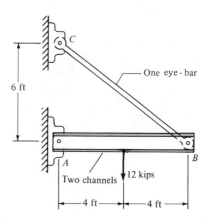

FIGURE P2-21

2-22 The wall bracket shown in Fig. P2–22 carries a load of 12 kips. The allowable tensile stress in the eye bar is 20 ksi, and the allowable shear stress in the pins is 12 ksi. Each pin at A, B, and C is in double shear. Determine **(a)** the required cross-sectional area of the eye bar, **(b)** the required diameters of the pin at A.

FIGURE P2-22

2-23 In the collar bearing of Fig. P2–15 (on p. 40), if the load P is 50 kips, the thickness of the collar is $\frac{1}{2}$ in. The allowable compressive stress in the column is 20 ksi, the allowable shear stress in the collar is 15 ksi, and the allowable bearing stress between the collar and the support is 5 ksi. Determine the required diameters d and D.

*2–7
STRESSES ON INCLINED PLANES

For an axially loaded member, it has been shown that normal stress occurs on a plane perpendicular to the axis of the member. However, on an inclined plane such as plane m–m shown in Fig. 2–10(a), both normal and shear stresses exist. The equilibrium condition of the free-body diagram in Fig. 2–10(b) requires that the internal resisting force R in section m–m be equal to the applied force P. The force can be resolved into two components: the normal component R_n perpendicular to the inclined plane and the tangential component R_s parallel to the inclined plane. The normal component R_n produces normal stress and the tangential component R_s produces shear stress.

Let the angle between the inclined plane and the cross section be θ. Then

$$R_n = R \cos \theta = P \cos \theta$$

$$R_s = R \sin \theta = P \sin \theta$$

Let the dimensions of the cross section A be b and h. Then the dimensions of the area A' of the inclined plane are b and $h/\cos \theta$.

Assume that the normal stress (σ_θ) is uniformly distributed over the inclined plane. Then σ_θ is, by definition,

$$\sigma_\theta = \frac{R_n}{A'} = \frac{P \cos \theta}{b(h/\cos \theta)} = \frac{P}{A} \cos^2 \theta$$

Since P/A is the normal stress σ over the cross section, we write

$$\sigma_\theta = \sigma \cos^2 \theta \qquad\qquad (2\text{–}15)$$

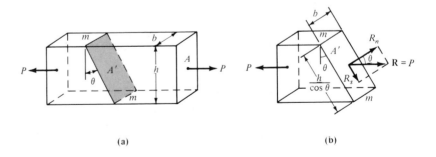

(a)   (b)

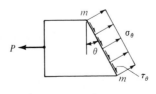

(c)

FIGURE 2–10

The average shear stress τ_θ over the inclined plane is

$$\tau_\theta = \frac{R_s}{A'} = \frac{P \sin \theta}{b(h/\cos \theta)} = \frac{P}{A} \sin \theta \cos \theta$$

Using the trigonometric identity $\frac{1}{2} \sin 2\theta = \sin \theta \cos \theta$, we write

$$\tau_\theta = \tfrac{1}{2}\sigma \sin 2\theta \qquad (2\text{–}16)$$

From Eq. (2–16), we see that the maximum shear stress is

$$\tau_{max} = \tfrac{1}{2}\sigma \qquad (2\text{–}17)$$

which occurs when $\sin 2\theta = 1$, or $\theta = 45°$, that is, on the 45° inclined plane.

The normal and shear stresses on any inclined plane in an axially loaded member can be computed by using Eqs. (2–15) and (2–16). Or these stresses can be determined simply by definition, as demonstrated in the following two examples.

EXAMPLE 2–9

Determine the normal and shear stresses on the inclined plane m–m of the axially loaded member shown in Fig. (a).

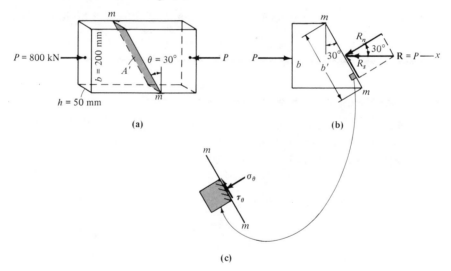

(a)

(b)

(c)

SOLUTION

In the free-body diagram of the lower part of the member shown in Fig. (b), $\Sigma F_x = 0$ requires that

$$R = P = 800 \text{ kN}$$

which acts horizontally to the left and compresses the section. The force R is resolved into two components, R_n and R_s, as shown. They are

$$R_n = R \cos 30° = (800 \text{ kN}) \cos 30° = 693 \text{ kN}$$

$$R_s = R \sin 30° = (800 \text{ kN}) \sin 30° = 400 \text{ kN}$$

The width b' along the incline is

$$b' = \frac{b}{\cos 30°} = \frac{200 \text{ mm}}{\cos 30°} = 231 \text{ mm} = 0.231 \text{ m}$$

Thus the area A' of the inclined plane is

$$A' = b'h = (0.231 \text{ m})(0.050 \text{ m}) = 0.011\ 55 \text{ m}^2$$

By definition the normal and shear stresses on the inclined plane are

$$\sigma_\theta = \frac{R_n}{A'} = \frac{693 \text{ kN}}{0.011\ 55 \text{ m}^2} = 60\ 000 \text{ kN/m}^2 = 60 \text{ MPa (C)}$$

$$\tau_\theta = \frac{R_s}{A'} = \frac{400 \text{ kN}}{0.011\ 55 \text{ m}^2} = 34\ 600 \text{ kN/m}^2 = 34.6 \text{ MPa}$$

These stresses can also be determined from Eqs. (2–15) and (2–16). For $\theta = 30°$, we have

$$\sigma = \frac{P}{A} = \frac{-800 \text{ kN}}{0.200 \times 0.050 \text{ m}^2} = -80\ 000 \text{ kN/m}^2$$

$$\sigma_\theta = \sigma \cos^2 \theta = (-80 \text{ MPa}) \cos^2 30° = -60 \text{ MPa (C)}$$

$$\tau_\theta = \tfrac{1}{2}\sigma \sin 2\theta = \tfrac{1}{2}(80 \text{ MPa}) \sin (2 \times 30°) = 34.6 \text{ MPa}$$

$\blacksquare$

EXAMPLE 2–10

Two bolts, one on each side, are used to connect the flanges of the members shown in Fig. (a). The diameter of each bolt is $\tfrac{1}{2}$ in. and the load P is 6 kips. Determine the normal and shear stresses in each bolt at the connecting plane.

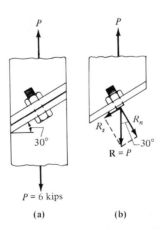

(a) (b)

SOLUTION

A free-body diagram of the upper part of the member with the bolts cut through the plane of the connection is shown in Fig. (b). The two bolts are subjected to normal and shear forces as shown. Assume that each bolt is subjected to one-half of the applied force.

The cross-sectional area of the bolt is

$$A = \tfrac{1}{4}\pi d^2 = \tfrac{1}{4}\pi(\tfrac{1}{2} \text{ in.})^2 = 0.1963 \text{ in.}^2$$

By definition, the normal and shear stresses are

$$\sigma = \frac{\tfrac{1}{2}R_n}{A} = \frac{\tfrac{1}{2}P \cos 30°}{A} = \frac{\tfrac{1}{2}(6 \text{ kips}) \cos 30°}{0.1963 \text{ in.}^2} = 13.2 \text{ ksi (T)}$$

$$\tau = \frac{\tfrac{1}{2}R_s}{A} = \frac{\tfrac{1}{2}P \sin 30°}{A} = \frac{\tfrac{1}{2}(6 \text{ kips}) \sin 30°}{0.1963 \text{ in.}^2} = 7.64 \text{ ksi}$$

■

PROBLEMS

2-24 In Fig. P2–24, determine the normal and shear stresses on the inclined plane *m–m* of a steel plate subjected to axial load *P*.

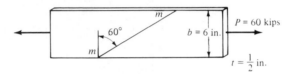

FIGURE P2–24

2-25 When subjected to an axial compressive load, bricks fail in shear approximately on a 45° inclined plane. Hence the shear strength of bricks is less than one-half of their compressive strength [refer to Eq. (2–17)]. A brick of the dimensions shown in Fig. P2–25 is tested in compression. If its shear strength is 800 psi, determine the minimum load *P* that will cause the brick to break.

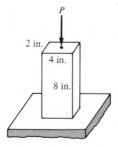

FIGURE P2–25

2-26 A short concrete post having a square section 100 mm by 100 mm is subjected to an axial load *P*. If the shear stress on an inclined plane at 30° from the cross section is 2500 kPa, determine the value of the load *P*.

2–27 A flat plate $\frac{1}{2}$ in. thick is subjected to an axial force P of 40 kips as shown in Fig. P2–27. Determine the normal and shear stresses on the section m–m.

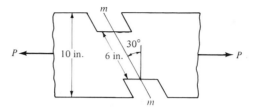

FIGURE P2–27

2–28 Dowels made of hard wood are used to connect the frame shown in Fig. P2–28. If the diameter of the dowel is 10 mm and the allowable shear stress of the dowels is 8 MPa, determine the maximum load P that can be applied. Neglect frictional effect.

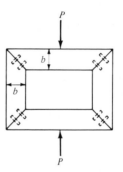

FIGURE P2–28

CHAPTER 3

Relationship Between Stress and Strain

3–1
INTRODUCTION

Structural materials deform under the action of forces. The amount of deformation depends on the values of the stresses and the mechanical properties of the material.

There are three types of deformation. An increase in length is called *elongation*. A decrease in length is called *compression*. An angular distortion is called *shear deformation*.

This chapter is devoted to the investigation of mechanical properties of materials that govern the deformations of stressed bodies. The nature of the deformations in a stressed body and the relationship between deformations and stresses are also studied in this chapter.

3–2
DEFINITION OF LINEAR STRAIN

Centrally applied axial forces on a member tend to elongate or compress the member. Figure 3–1 shows a bar elongated by a tensile force. The original dimensions of the undeformed member are shown by dashed lines. The original length L of the member is elongated to a length $L + \delta$ after the tensile load P is applied. The total deformation (change of length) is thus equal to δ (the Greek lowercase letter delta).

Linear strain is defined as the unit deformation, or the deformation per unit of original length. Thus if the linear strain is denoted by ϵ (the Greek lowercase letter epsilon), the average linear strain over the length L is

$$\epsilon_{avg} = \frac{\delta}{L} \qquad (3\text{–}1)$$

Strain is a dimensionless quantity, but it is customary to refer to strain in such units as in./in., ft/ft, or m/m.

51

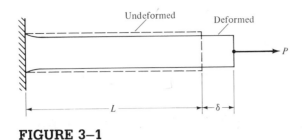

FIGURE 3–1

3–3
TENSION TEST

Information regarding the mechanical properties of materials is usually obtained through laboratory testing. Tension tests are the most commonly used tests for metals. Figure 3–2 shows a universal testing machine used for this purpose. The universal testing machine is also capable of performing compression tests, shear tests, and bending tests.

In a *tension test*, a round test specimen, made to ASTM (American Standard of Testing and Materials) specification, is clamped to the machine between the upper head and the lower head, as shown in Figs. 3–2 and 3–3. The lower head is stationary during the test, while the upper head is pushed upward by the hydraulic pressure developed in the hydraulic loading system. When the upper head rises, the specimen is stretched at a slow rate controlled by the load valve. The tensile force P acting on the specimen at any time during the test is indicated by

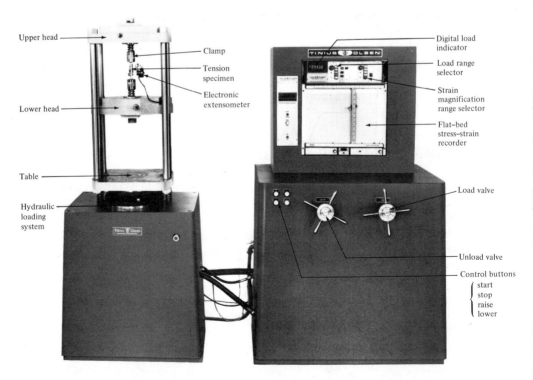

FIGURE 3–2 Universal testing machine. (Courtesy of the Tinius Olsen Testing Machine Company.)

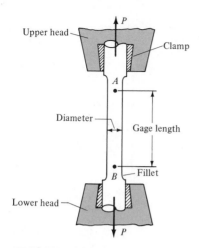

FIGURE 3–3 Tension test.

the digital load indicator. While the tensile force P is recorded, the corresponding change in length between two punched marks A and B (Fig. 3–3) on the specimen is measured. The original distance between the marks is called the *gage length*. Commonly used gage lengths are 2 in. and 8 in. To measure elongation, a mechanical or electronic extensometer capable of measuring deformation as small as 0.0001 in. is mounted on the specimen at the marks A and B. The values of load P versus elongation δ at proper intervals are recorded. The corresponding values of stress versus strain can be calculated by

$$\sigma = \frac{P}{A} \quad \text{STRESS}$$

$$\epsilon = \frac{\delta}{L} \quad \text{LINEAR STRAIN}$$

where A is the original cross-sectional area and L is the gage length.

3–4
STRESS–STRAIN DIAGRAM

In the tension test, the corresponding values of stress versus strain can be calculated from the applied load and the corresponding deformation as a specimen is stretched continuously until failure occurs. These values can be plotted to produce a stress–strain diagram. A modern universal testing machine, such as the one shown in Figure 3–2, can be set up to plot the stress–strain diagram automatically. The diagram establishes a relationship between stress and strain, and for most practical purposes, the relationship is independent of the cross-sectional area of the specimen and the gage length used.

It is customary to plot the diagram by plotting stress as the ordinate and strain as the abscissa. A typical stress–strain diagram obtained by testing a mild steel specimen is shown in Fig. 3–4. Mild steel is a steel with low carbon content and is widely used in construction.

The stress–strain diagram in Fig. 3–4 consists of the following four stages.

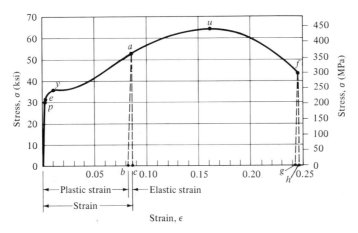

FIGURE 3–4

Elastic Stage

The plot is a straight line up to point p, the *proportional limit,* because up to this point the stress is proportional to the strain. The stress at the proportional limit is denoted by σ_p.

Beyond the proportional limit, the stress is no longer proportional to the strain, but up to point e the deformation is still elastic. This means that the specimen will return to its initial size and shape upon removal of the load. Point e is called the *elastic limit,* and the corresponding stress is denoted by σ_e. In the σ–ϵ diagram, points p and e are very close, so for all practical purposes, the elastic limit is not strictly distinguished from the proportional limit.

Beyond the elastic limit only part of the deformation can be recovered after the load is removed; the remaining part of the deformation becomes permanent deformation, or "set." The deformation that can be recovered is called *elastic deformation,* and the permanent deformation is called *plastic deformation.*

Yield Stage

Beyond point p, the slope of the curve gradually decreases. At point y the curve becomes horizontal and remains horizontal for a small increase in ϵ. At this point, y, the specimen continues to elongate without any significant increase in load. The material is said to have yielded, and the point is called the *yield point.* The corresponding stress is called the *yield strength (stress)* of the material and is denoted by σ_y.

When the stress in a material reaches σ_y, there will be an appreciable amount of plastic deformation. In many machine parts the plastic deformation may affect the useful function of the parts. Therefore, the yield stress σ_y is an important index of the strength of the material.

Strain-Hardening Stage

The ability of the material to resist deformation is regained after the yield stage is passed. Because of plastic deformation, the material "strain hardens"; thus the stresses required to yield the specimen become larger. The stress–strain diagram reaches the highest point, u; the corresponding stress at this point is denoted by

σ_u and is called the *ultimate strength,* that is, the maximum stress that a material can resist.

Stage of Localized Deformation

While the specimen is being elongated, its lateral dimension contracts. The lateral contraction is so small for low stress levels that it cannot be observed [Fig. 3–5(a)]. Beyond point u, the lateral contraction becomes more pronounced. A drastic decrease in diameter occurs in a localized area, as shown in Fig. 3–5(b). This phenomenon is called *necking.* Only ductile metals exhibit this characteristic. After initial necking occurs, the cross section at the necked-down section quickly decreases, and the tensile force required to produce further stretch of the specimen also decreases. The tensile stress, which is computed based on the original cross-sectional area, decreases accordingly. When the σ–ε curve drops to point f, the specimen suddenly breaks into two parts. Point f on the curve is called the *point of rupture.*

Because necking and rupture follow after the ultimate strength σ_u is reached, σ_u is an important index of the strength of the material.

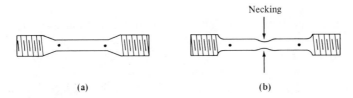

Necking

(a)　　　　　　　　　(b)

FIGURE 3–5

3–5
HOOKE'S LAW

In a stress–strain diagram, the part Op from the origin to the proportional limit is essentially a straight line (Fig. 3–6). Thus, in this region, the stress is proportional to the strain. This material behavior is known as *Hooke's law,* in honor of the English scientist Robert Hooke, who first announced this property in 1676.

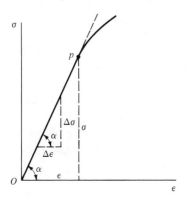

FIGURE 3–6

Hooke's law may be expressed by the equation

$$\frac{\sigma}{\epsilon} = E \tag{3-2a}$$

or

$$\sigma = E\epsilon \tag{3-2b}$$

where E is the constant of proportionality between stress and strain and is called the *modulus of elasticity*. Since ϵ is a dimensionless number, E must have the same units as those of stress. In U.S. customary units, E is usually expressed in psi or ksi. In SI units, E is expressed in GPa.

From Eq. (3–2a) and Fig. 3–6, it is seen that

$$E = \frac{\sigma}{\epsilon} = \frac{\Delta\sigma}{\Delta\epsilon} = \tan \alpha$$

Geometrically, E can be interpreted as the slope of the straight line Op in the stress–strain diagram.

The modulus of elasticity is a definite property of a given material. Physically, the stiffness of a material is represented by its modulus of elasticity. Typical moduli of elasticity are 30×10^3 ksi (or 210 GPa) for steel and 10×10^3 ksi (or 70 GPa) for aluminum. The data indicate that an aluminum bar would stretch three times more than a steel bar of the same length when subjected to the same stress.

Table A–7 of the Appendix Tables gives the ultimate strengths, the yield strengths, the moduli of elasticity, and other mechanical properties for common engineering materials.

3–6
FURTHER REMARKS ON STRESS–STRAIN DIAGRAMS

When a mild steel specimen is elongated to the plastic range at point a (Fig. 3–4) and then the load is gradually released, the unloading stress–strain curve ab has a slope that is essentially the same as that of line Op. When the load is completely removed, permanent deformation in the specimen occurs. Thus the strain at point a consists of two parts: part bc is the strain that is recovered upon unloading; it is thus the elastic strain. Part Ob is the permanent strain in the material that cannot be recovered upon unloading; it is thus the plastic strain.

After the specimen fractures at point f (Fig. 3–4), the elastic strain gh is recovered, but the plastic strain Og remains. When the specimen is removed from the testing machine and the two broken pieces are put together, the distance between the gage length marks becomes L'. The percent elongation is defined as the ratio of change of gage length to the original gage length L, expressed as a percentage. Thus

$$\text{percent elongation} = \frac{L' - L}{L} \times 100\% \tag{3-3}$$

The percent elongation is a good index of the ductility of a material. A material with a high percent of elongation indicates that the material has a high degree of plastic deformation and thus is more ductile than a material with a low percent of elongation. Mild steel, being very ductile, has a 20 to 30 percent elongation. Materials are generally classified into two categories according to their percent of elongation. Those with a percent of elongation greater than 5 percent, such as steel, copper, and aluminum, are called *ductile* materials; those with a percent of elongation of less than 5 percent, such as cast iron, concrete, glass, and ceramics, are called *brittle* materials.

After the specimen breaks, the ratio of the reduction in cross-sectional area at the fractured section to the original cross-sectional area, expressed as a percentage, is called the percent reduction in area. Thus

$$\text{percent reduction in area} = \frac{A - A'}{A} \times 100\% \tag{3–4}$$

where A is the original cross-sectional area and A' is the area at the fractured section. The percent reduction in area is also an index of the ductility of a material.

The stress–strain diagram shown in Fig. 3–4 is typical of ductile materials. Less ductile or brittle materials exhibit different stress–strain characteristics. Figure 3–7(a) and (b) shows the stress–strain diagrams of several common materials.

The stress–strain diagram for cast iron in Fig. 3–7(b) does not consist of an initial straight line for small values of ϵ. Nevertheless, up to point p the curve can, without appreciable error, be considered a straight line, and the slope of the line Op is the approximate modulus of elasticity of cast iron.

Note that the yielding phenomenon is absent in less ductile and brittle materials. For materials that do not possess a well-defined yield point, the *offset method* is used to determine the location of the yield point along the curve. In the offset method the yield point is established by drawing a line parallel to the straight-line portion of the stress–strain diagram, starting from a point on the abscissa with a given offset from the origin. The intersection of the line and the stress–strain diagram is the yield point. The most commonly used offset is 0.2 percent (i.e., $\epsilon = 0.002$). The stress at point y in Fig. 3–8 obtained by this method is called the yield strength of the material at 0.2 percent offset.

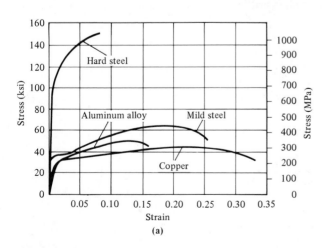

(a)

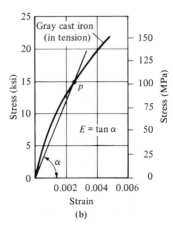

(b)

FIGURE 3–7

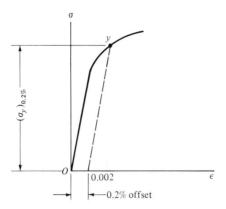

FIGURE 3–8

3–7
COMPRESSION TEST

In a compression test, the specimen used is generally a round bar with a height $1\frac{1}{2}$ to 3 times of the specimen diameter so that the specimen will not buckle under the compressive load.

The stress–strain diagram for compression test of mile steel is shown in Fig. 3–9. It is seen that the modulus of elasticity E and yield stress σ_y in compression are the same as those determined in the tension test. After the specimen has yielded, it becomes flatter; its diameter increases continuously. Thus the compressive load also increases continuously and the ultimate strength in compression cannot be obtained. Compression tests of ductile materials are usually not performed.

The compression characteristics of brittle materials differ greatly from the test results for ductile materials. Brittle materials rupture under small compressive deformations along a section at an angle of 45° from the axis of the specimen. This is evidence that the failure is caused by excessive shear stress (see Section 2–7). Figure 3–10 shows the stress–strain diagram of cast iron in a compression test. The ultimate strength of the cast iron in compression is four to five times higher than that in tension. Other brittle materials, such as concrete, brick, and

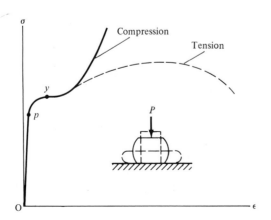

FIGURE 3–9

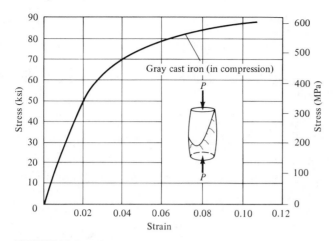

FIGURE 3–10

ceramics, also display much higher resistance to compression than to tension. For these materials, a compression test is more important than a tension test.

EXAMPLE 3–1

The stress–strain diagram shown is the result of a tension test on a steel specimen. The specimen has an original diameter of 0.502 in. and a gage length of 2 in. between two punch marks. After rupture, the diameter at the rupture reduces to 0.405 in., and the length between the two punch marks stretches to 2.55 in. Determine (a) the stress at the proportional limit, (b) the modulus of elasticity, (c) the ultimate strength, (d) the yield stress at 0.2 percent offset, (e) the percent elongation, and (f) the percent reduction in area.

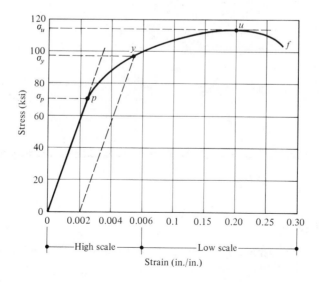

SOLUTION

(a) The stress at the proportional limit p is, from the diagram,

$$\sigma_p = 71 \text{ ksi}$$

The corresponding strain is

$$\epsilon_p = 0.0024 \text{ in./in.}$$

(b) The modulus of elasticity is the slope of the straight line from the origin to the proportional limit. Thus

$$E = \frac{\sigma_p}{\epsilon_p} = \frac{71 \text{ ksi}}{0.0024} = 29.6 \times 10^3 \text{ ksi}$$

(c) The ultimate strength is the stress at the highest point u of the stress–strain diagram. This value is

$$\sigma_u = 114 \text{ ksi}$$

(d) Since there is no well-defined yield point in the stress–strain diagram, the yield point is determined by the 0.2 percent offset method. From a point on the abscissa where the strain is equal to 0.002, draw a line parallel to the straight line. The point of intersection y of this line with the stress–strain diagram is the yield point corresponding to 0.2 percent offset. The yield stress at this point is

$$\sigma_y = 98 \text{ ksi}$$

(e) The percent elongation is, by definition,

$$\text{percent elongation} = \frac{L' - L}{L} \times 100\%$$

$$= \frac{2.55 - 2.00}{2.00} \times 100\% = 27.5\%$$

Thus the material is very ductile.
    (f) The percent reduction in area is, by definition,

$$\text{percent reduction in area} = \frac{A - A'}{A} \times 100\%$$

$$= \frac{\frac{1}{4}\pi(0.502)^2 - \frac{1}{4}\pi(0.405)^2}{\frac{1}{4}\pi(0.502)^2} \times 100\% = 34.9\%$$ ∎

PROBLEMS

3–1 Define deformation and strain. State the similarities and differences between the two definitions.

3–2 What is the difference between elastic deformation and plastic deformation?

3–3 What occurs when a mild steel bar is stretched to its yield point?

3–4 What is the necking phenomenon in a tension test?

3–5 What is the meaning and significance of the ultimate strength of a material?

3–6 Two bars, one made of aluminum and one made of copper, have the same length. The moduli of elasticity are 10×10^3 ksi for aluminum and 17×10^3 ksi for copper. Which bar stretches more if both bars are subjected to the same stress within their proportional limits?

3–7 How is the yield point at 0.2 percent offset determined?

3–8 Why is the tension test more important than the compression test for ductile materials?

3–9 Why is the compression test more important than the tension test for brittle materials?

3–10 A stress–strain diagram for a tension test of an alloy specimen is shown in Fig. P3–10. The following data are recorded:

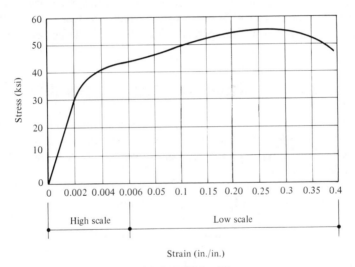

FIGURE P3–10

$$\text{initial diameter} = 0.502 \text{ in.}$$
$$\text{gage length} = 2.00 \text{ in.}$$
$$\text{diameter at the fractured section} = 0.412 \text{ in.}$$
$$\text{final length after fracture} = 2.78 \text{ in.}$$

Determine **(a)** the stress at the proportional limit, **(b)** the modulus of elasticity, **(c)** the yield stress at 0.2 percent offset, **(d)** the ultimate strength, **(e)** the percent elongation, **(f)** the percent reduction in area.

3–11 The following data were obtained from a tension test of a steel specimen. The specimen had an original diameter of 0.505 in. and a gage length of 2.00 in. After

the specimen ruptured, the gage length became 2.31 in. and the diameter at the
section of rupture decreased to 0.450 in.

Total Tensile Load (lb)	Total Elongation in 2-in. Gage Length (in.)	Total Tensile Load (lb)	Total Elongation in 2-in. Gage Length (in.)
200	0.0000	14 400	0.0118
1 000	0.0003	15 200	0.0167
2 000	0.0006	16 000	0.0212
4 000	0.0012	16 800	0.0263
6 000	0.0019	17 600	0.0327
8 000	0.0026	18 400	0.0380
10 000	0.0033	19 200	0.0440
12 000	0.0039	20 000	0.0507
13 400	0.0045	20 800	0.0580
13 600	0.0054	21 600	0.0660
13 800	0.0063	22 400	0.0780
14 000	0.0090	25 400	Specimen broke

Plot the stress–strain diagram. Determine **(a)** the stress at the proportional limit,
(b) the modulus of elasticity, **(c)** the yield stress at 0.2 percent offset, **(d)** the
ultimate strength, **(e)** the percent elongation, **(f)** the percent reduction in area.

3–8
DEFORMATION OF AXIALLY LOADED MEMBER

An axially loaded member elongates under a tensile load and contracts under a
compressive load. If the normal stress in an axially loaded member is within its
proportional limit, then the axial deformation may be completely recovered when
the load is removed.

Consider a bar of constant cross-sectional area A subjected to an axial tensile
force P, as shown in Fig. 3–11. By definition,

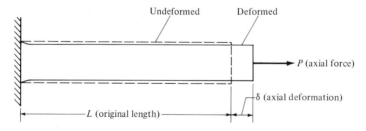

FIGURE 3–11

$$\epsilon = \frac{\delta}{L}$$

or

$$\delta = \epsilon L \qquad\qquad\text{(a)}$$

If the stress in the member is within the proportional limit, Hooke's law applies.

Thus

$$\epsilon = \frac{\sigma}{E} = \frac{P}{AE} \qquad \text{(b)}$$

Substituting Eq. (b) in Eq. (a) gives

$$\delta = \frac{PL}{AE} \qquad \text{(3–5)}$$

The axial deformation δ calculated from Eq. (3–5) is called the *elastic deformation.* This equation may also be used to determine deformation in a member subjected to axial compression. Then δ denotes the total axial contraction. For most structural materials, the moduli of elasticity for tension and for compression are the same.

The stress, strain, and deformation caused by tensile forces are usually considered as positive; those caused by compressive force are considered as negative.

───── **EXAMPLE 3–2** ─────

Determine the total deformation of the steel bar between sections A and D. The bar has a cross section $\frac{1}{2}$ in. by 1 in. and is subjected to the axial loads shown. The modulus of elasticity of steel is $E = 30 \times 10^3$ ksi.

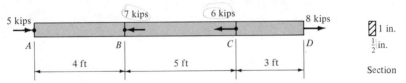

SOLUTION

Using the method of sections as illustrated in Example 2–2, it can be determined that each of the segments AB, BC, and CD is subjected to the axial force shown in the following figure.

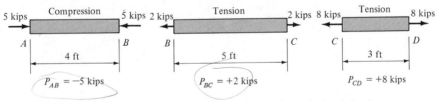

The cross-sectional area multiplied by the modulus of elasticity is

$$AE = (\tfrac{1}{2} \times 1 \text{ in.}^2)(30\ 000 \text{ kips/in.}^2) = 15\ 000 \text{ kips}$$

Thus the axial deformation of each segment is

$$\delta_{AB} = \frac{P_{AB}L_{AB}}{AE} = \frac{(-5 \text{ kips})(4 \text{ ft})}{15\ 000 \text{ kips}} = -0.001\ 33 \text{ ft}$$

$$\delta_{BC} = \frac{P_{BC}L_{BC}}{AE} = \frac{(+2 \text{ kips})(5 \text{ ft})}{15\ 000 \text{ kips}} = +0.000\ 66 \text{ ft}$$

$$\delta_{CD} = \frac{P_{CD}L_{CD}}{AE} = \frac{(+8 \text{ kips})(3 \text{ ft})}{15\ 000 \text{ kips}} = +0.001\ 60 \text{ ft}$$

The total deformation of the bar is the algebraic sum of the axial deformations of the three segments. Thus

$$\delta_{AD} = \delta_{AB} + \delta_{BC} + \delta_{CD}$$

$$= -0.001\ 33 \text{ ft} + 0.000\ 66 \text{ ft} + 0.001\ 60 \text{ ft}$$

$$= (0.000\ 93 \text{ ft})\left(\frac{12 \text{ in.}}{1 \text{ ft}}\right) = 0.0112 \text{ in. (elongation)}$$

■

EXAMPLE 3–3

A circular steel bar 20 in. long is subjected to an axial tensile load of 4 kips. Determine the required diameter of the bar if the allowable tensile stress is 20 ksi and the total elongation is limited to 0.0055 in. The modulus of elasticity of steel is $E = 30 \times 10^3$ ksi.

SOLUTION
The tensile stress in the bar is

$$\sigma = \frac{P}{A} = \frac{4 \text{ kips}}{\frac{1}{4}\pi d^2} = \frac{5.093 \text{ kips}}{d^2} \leq \sigma_{\text{allow}} = 20 \text{ ksi}$$

from which

$$d \geq \sqrt{\frac{5.093 \text{ kips}}{20 \text{ kips/in.}^2}} = 0.505 \text{ in.}$$

The total elongation of the bar is

$$\delta = \frac{PL}{AE} = \frac{(4 \text{ kips})(20 \text{ in.})}{(\frac{1}{4}\pi d^2)(30 \times 10^3 \text{ kips/in.}^2)} = \frac{0.003\ 395 \text{ in.}^3}{d^2}$$

$$\leq \text{ allowable elongation} = 0.0055 \text{ in.}$$

from which

$$d \geq \sqrt{\frac{0.003\ 395 \text{ in.}^3}{0.0055 \text{ in.}}} = 0.786 \text{ in.}$$

To satisfy both conditions for stress and deformation, the diameter of the bar must be no less than 0.786 in.

■

EXAMPLE 3–4

A beam of negligible weight is supported in a horizontal position by two rods, one of steel and one of aluminum, as shown. Determine the distance x that locates the point of application of load P so that the two rods are deformed by the same amount and the beam remains in the horizontal position.

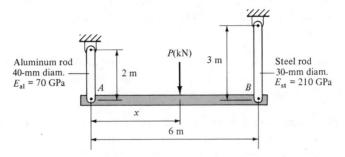

SOLUTION

The condition to be satisfied is

$$\delta_{al} = \delta_{st}$$

which requires that

$$\frac{F_{al}L_{al}}{A_{al}E_{al}} = \frac{F_{st}L_{st}}{A_{st}E_{st}}$$

$$\frac{F_{al}(2 \text{ m})}{\frac{1}{4}\pi(0.040 \text{ m})^2(70 \times 10^6 \text{ kN/m}^2)} = \frac{F_{st}(3 \text{ m})}{\frac{1}{4}\pi(0.030 \text{ m})^2(210 \times 10^6 \text{ kN/m}^2)}$$

This reduces to

$$F_{al} = 0.889F_{st} \qquad (a)$$

Now consider the equilibrium of the beam:

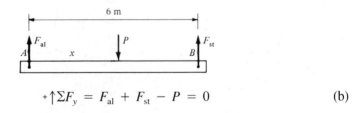

$$+\uparrow\Sigma F_y = F_{al} + F_{st} - P = 0 \qquad (b)$$

When we substitute Eq. (a) in Eq. (b), we have

$$(1 + 0.889)F_{st} = P \qquad F_{st} = 0.529P \qquad (c)$$

$$\circlearrowleft\Sigma M_A = F_{st}(6 \text{ m}) - P(x) = 0 \qquad x = \frac{(6 \text{ m})F_{st}}{P} \qquad (d)$$

Next, we substitute Eq. (c) in Eq. (d):

$$x = \frac{(6 \text{ m})(0.529P)}{P} = 3.17 \text{ m}$$

PROBLEMS

3–12 A 10-ft steel bar is subjected to a tensile stress of 20 ksi. Determine **(a)** the linear strain and **(b)** the total deformation of the bar. The modulus of elasticity of steel is 30 × 10³ ksi.

3–13 An aluminum rod of 20-mm diameter is elongated 3.5 mm along its longitudinal direction by a load of 25 kN. If the modulus of elasticity of aluminum is $E = 70$ GPa, determine the original length of the bar.

3–14 A 20-ft wrought-iron bar of $\frac{1}{2}$-in. diameter is subjected to a tensile force of 3 kips. Determine the stress, strain, and the elongated length of the bar. The modulus of elasticity of wrought iron is $E = 29 \times 10^3$ ksi.

3–15 A metal wire is 10 m long and 2 mm in diameter. It is elongated 6.06 mm by a tensile force of 400 N. Determine the modulus of elasticity of the material and indicate a possible material of which the wire is made.

3–16 A steel tape used in surveying is designed to be exactly 100 ft long when fully supported on a horizontal frictionless plane and subjected to a tensile force of 10 lb. Determine the stretched length of the tape if it is subjected to an axial tensile force of 20 lb when supported in the same way. The tape is $\frac{1}{32}$ in. thick and $\frac{3}{8}$ in. wide.

3–17 Determine the total elongation of strut AB in Problem 2–5 (on p. 30) if the material of the strut is steel. $E = 210$ GPa.

3–18 Determine the total elongation of cable BC in Problem 2–6 (on p. 31) if the cable is made of steel. $E = 30 \times 10^6$ psi.

3–19 An aluminum bar 30 mm in diameter is suspended as shown in Fig. P3–19. Determine the total displacement of the lower end C after the load is applied. The modulus of elasticity of aluminum is $E = 70$ GPa.

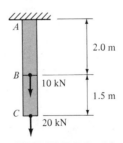

FIGURE P3–19

3–20 The brass bar shown in Fig. P3–20, which has a uniform cross-sectional area of 2 in.², is subjected to the forces shown. Determine the total deformation of the bar. The modulus of elasticity of brass is $E = 17 \times 10^3$ ksi.

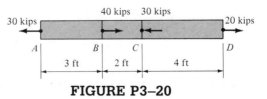

FIGURE P3–20

3–21 In Fig. P3–21, determine the total elongation of the 10-mm-diameter steel eye bar *BC* due to the load $P = 8$ kN. $E = 210$ GPa.

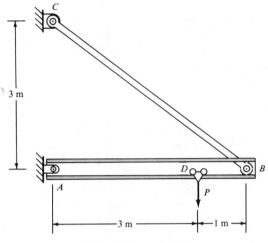

FIGURE P3–21

3–22 Determine the total deformation between points *A* and *D* of a stepped steel bar subjected to the axial forces shown in Fig. P3–22. The modulus of elasticity of steel is $E = 210$ GPa.

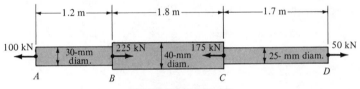

FIGURE P3–22

3–23 The two wires shown in Fig. P3–23 support a heavy bar weighing 900 lb. The wires *AC* and *BD* are identical, having the same $\frac{3}{8}$-in. diameter, the same original 5-ft length, and the same modulus of elasticity $E = 30 \times 10^6$ psi. Determine the deformation of each wire.

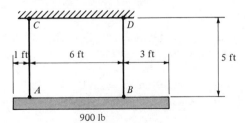

FIGURE P3–23

3–24 Determine the diameter of wire *BD* in Problem 3–23 so that the deformations of the two wires are equal. Other data remain unchanged.

3–25 A steel rod used in a control mechanism must transmit a tensile force of 10 kN without exceeding an allowable stress of 150 MPa or stretching more than 1 mm per 1 meter of length. $E = 210$ GPa. Determine the required diameter of the bar.

3–9
POISSON'S RATIO

When a bar is subjected to an axial tensile load, it is elongated in the direction of the applied load; at the same time its transverse dimension decreases, as shown in Fig. 3–12(a). Similarly, if an axial compressive load is applied to the bar, the bar contracts along the axial direction while its transverse dimension increases, as shown in Fig. 3–12(b).

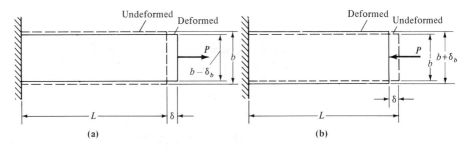

FIGURE 3–12

Experimental results show that the absolute value of the ratio of the transverse strain ϵ_t to the axial strain ϵ_a is a constant for a given material subjected to axial stresses within the elastic range. Thus

$$\mu = \left| \frac{\text{transverse strain}}{\text{axial strain}} \right| = \left| \frac{\epsilon_t}{\epsilon_a} \right| \tag{3–6}$$

where

$$\epsilon_a = \text{axial strain} \quad = \frac{\delta}{L}$$

$$\epsilon_t = \text{transverse strain} = \frac{\delta_b}{b}$$

This relationship was established in the early nineteenth century by the French mathematician Poisson. The constant μ (the Greek lowercase letter mu) is called *Poisson's ratio*. Poisson's ratio is a distinct material constant. For most structural materials, Poisson's ratio ranges from 0.25 to 0.35.

Because ϵ_a and ϵ_t are of opposite sign, Eq. (3–6) can also be written as

$$\epsilon_t = -\mu\epsilon_a \tag{3–7}$$

———— **EXAMPLE 3–5** ————

A steel rod 4 in. in diameter is subjected to an axial tensile force of 200 kips. Given $E = 30 \times 10^3$ ksi and $\mu = 0.29$, determine the change in diameter of the rod after the load is applied.

SOLUTION

The tensile stress in the rod is

$$\sigma = \frac{P}{A} = \frac{200 \text{ kips}}{\frac{1}{4}\pi(4 \text{ in.})^2} = +15.9 \text{ ksi (T)}$$

This stress is within the proportional limit of steel, which is approximately 30 ksi. Hence Hooke's law applies, and the axial strain is

$$\epsilon_a = \frac{\sigma}{E} = \frac{+15.9 \text{ ksi}}{30 \times 10^3 \text{ ksi}} = +0.000\ 53 \text{ (elongation)}$$

From Eq. (3–7) the transverse strain is

$$\epsilon_t = -\mu\epsilon_a = -0.29(+0.000\ 53) = -0.000\ 154 \text{ (contraction)}$$

By definition,

$$\epsilon_t = \frac{\delta_D}{D}$$

Thus

$$\delta_D = D\epsilon_t = (4 \text{ in.})(-0.000\ 154) = -0.000\ 62 \text{ in.}$$

Hence the diameter contracts by 0.000 62 in. ■

3–10
SHEAR DEFORMATION AND SHEAR STRAIN

A shear force causes shape distortion of a body. Figure 3–13 shows that a square element (shown by dashed lines) is distorted into a rhombus after the shear force F_s is applied.

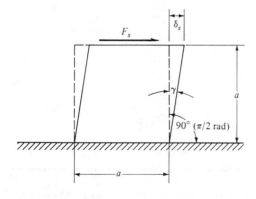

FIGURE 3–13

The total deformation δ_s occurs over a length a. The deformation per unit length, δ_s/a, is equivalent to tan γ. Since the angle γ is very small, tan γ is equal to γ in radians. The shear strain is thus the angle γ (the Greek lowercase letter gamma) in radians. The shear strain is the distortion of a right angle, that is, the amount of change in radians of the angle between two lines originally at a 90° ($\pi/2$ radians) angle before the shear force is applied.

3–11
HOOKE'S LAW FOR SHEAR STRESS AND SHEAR STRAIN

When the shear stress is within the elastic limit of the material, it is found experimentally that, for most materials, the shear stress is proportional to shear strain. This is known as Hooke's law for shear stress and shear strain. Mathematical expression of this law is

$$\tau = G\gamma \qquad (3\text{–}8)$$

where G is a constant of proportionality called the *shear modulus of elasticity* or the *modulus of rigidity*.

Like E and μ, G is a constant for a given material. Since γ is measured in radians (which are dimensionless), G is measured in the same units as those for stresses.

It can be proved that the three elastic constants E, μ, and G are related by the equation

$$G = \frac{E}{2(1 + \mu)} \qquad (3\text{–}9a)$$

or

$$\mu = \frac{E}{2G} - 1 \qquad (3\text{–}9b)$$

For example, if the moduli of elasticity and of rigidity of steel have been determined experimentally to be

$$E = 30 \times 10^6 \text{ psi} \qquad \text{and} \qquad G = 11.6 \times 10^6 \text{ psi}$$

Poisson's ratio for steel must be

$$\mu = \frac{E}{2G} - 1 = \frac{30 \times 10^6 \text{ ksi}}{2(11.6 \times 10^6 \text{ ksi})} - 1 = 0.29$$

──────── **EXAMPLE 3–6** ────────────────────────────────

An aluminum alloy rod 10 mm in diameter is subjected to an axial pull of 6 kN. Given $E = 70$ GPa and $G = 26.3$ GPa, determine the axial and transverse strain in the rod.

SOLUTION

The axial tensile stress is

$$\sigma = \frac{P}{A} = \frac{6 \text{ kN}}{\frac{1}{4}\pi(0.01 \text{ m})^2} = 76\ 400 \text{ kN/m}^2 = 76.4 \text{ MPa}$$

This stress is within the proportional limit of aluminum, which is approximately 200 MPa. Thus Hooke's law applies, and the axial strain is

$$\epsilon_a = \frac{\sigma}{E} = \frac{+76.4 \text{ MPa}}{70 \times 10^3 \text{ MPa}} = +0.001\ 09 \text{ m/m (elongation)}$$

From Eq. (3–9b), Poisson's ratio is

$$\mu = \frac{E}{2G} - 1 = \frac{70 \text{ GPa}}{2(26.3 \text{ GPa})} - 1 = 0.33$$

From Eq. (3–7), the transverse strain is

$$\epsilon_t = -\mu\epsilon_a = -(0.33)(+0.001\ 09) = -0.000\ 36 \text{ m/m (contraction)} \blacksquare$$

PROBLEMS

3–26 A common mistake is to consider Poisson's ratio to be the ratio of the lateral deformation to the axial deformation. What is wrong?

3–27 Usually, only two of the three elastic constants of a material, E, G, and μ, need to be determined experimentally. Why?

3–28 Consider a carefully conducted tensile test of a copper specimen 10 mm in diameter and 50 mm in gage length. When a load of 10 kN is applied, the elastic deformation in the gage length is 0.0544 mm and the diameter is decreased by 0.0039 mm. Calculate the three elastic constants, E, μ, and G.

3–29 A steel tensile specimen 0.505 in. in diameter and 2 in. in gage length has stress and strain at the proportional limit equal to 42.0 ksi and 0.0014 in./in., respectively. The shear modulus is $G = 11.6 \times 10^3$ ksi. Determine the change in diameter of the specimen at the proportional limit.

3–30 An aluminum plate is subjected to an axial tensile force $P = 10$ kN. The plate has the following dimensions: length $L = 100$ mm, width $b = 20$ mm, and thickness $t = 5$ mm. $E = 70$ GPa and $G = 26.3$ GPa. Determine the deformations in the length L, in the width b, and in the thickness t.

*3–12
STRESS CONCENTRATIONS

When a member of uniform cross section is subjected to axial tension or compression, the normal stress is uniformly distributed in the cross section. However, an abrupt change in geometry of a member, such as that caused by a notch or a hole,

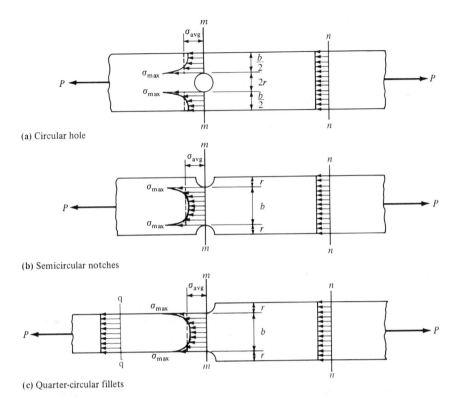

(a) Circular hole

(b) Semicircular notches

(c) Quarter-circular fillets

FIGURE 3–14

results in a nonuniform stress distribution, as shown in Fig. 3–14. The maximum stress in section *m–m* occurs at the edge of the hole [Fig. 3–14(a)], at the edge of the notch [Fig. 3–14(b)], or at the edge of the fillet [Fig. 3–14(c)]. The maximum stress may be several times greater than the average stress over the net cross-sectional area at section *m–m*. The stress distribution is uniform over sections at a considerable distance away from the region of abrupt geometric change, such as sections *n–n* and *q–q*. The abrupt increase in stress at localized regions is called *stress concentration*.

The ratio of the maximum stress to the average stress over the net cross-sectional area at *m–m* is called the *stress concentration factor*. Thus

$$K = \frac{\sigma_{max}}{\sigma_{avg}} \tag{3–10}$$

where K is the stress concentration factor and σ_{avg} is the average stress over the net cross-sectional area. If K is known, the maximum stress is

$$\sigma_{max} = K\sigma_{avg} = K\frac{P}{A_{net}} = K\frac{P}{bt} \tag{3–11}$$

where b is the net width at section *m–m* and t is the thickness of the plate.

Theoretical analysis as well as experimental results shows that the value K is a function of the ratio r/b in each of the three cases shown in Fig. 3–14. The

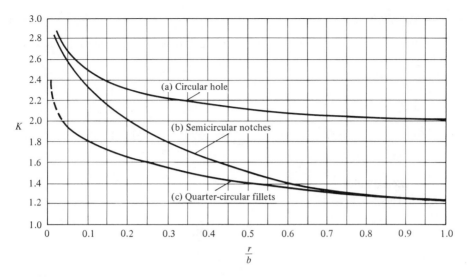

FIGURE 3–15

variation of K with respect to the ratio r/b is plotted for the three cases in Fig. 3–15.

From Fig. 3–15, we see that a smaller ratio of r/b gives a larger value of K, and accordingly, a higher stress concentration. Therefore, the radius of holes, notches, and fillets should be reasonably large to avoid high stress concentrations but not too large so that it will leave a large enough net cross-sectional area.

In some ductile metals, such as mild structural steel, the material at the point of maximum stress will yield when the yield point is reached. Additional load causes additional points to yield while the maximum stress remains at σ_y, thereby distributing the load more evenly over the net cross section, as shown in Fig. 3–16. When every point in the critical section reaches the yield point, the stress distribution in the section is essentially uniform and equal to σ_y at every point. Therefore, the effect of stress concentrations due to static load is not an important design factor for ductile materials.

The lack of a yield point in brittle materials, however, causes a continuous increase in the maximum stress until σ_u is finally reached. The material will then begin to crack at the point of maximum stress. Therefore, for brittle materials, stress concentration is an important factor, and it should not be overlooked in structural design.

Stress concentrations are of particular importance in the design of machine parts subjected to cyclic stress variations or repetitive reversals of stress. Under these conditions, progressive cracks are likely to start gradually from the points of stress concentration for both ductile and brittle materials.

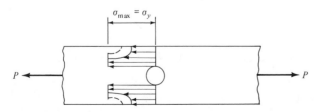

FIGURE 3–16

EXAMPLE 3–7

Find the maximum stress in the $\frac{1}{2}$-in.-thick plate with semicircular notches shown.

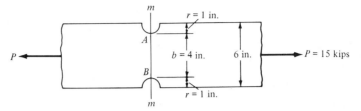

SOLUTION

The r/b ratio is

$$\frac{r}{b} = \frac{1 \text{ in.}}{4 \text{ in.}} = 0.25$$

From Fig. 3–15, for semicircular notches and $r/b = 0.25$,

$$K = 1.9$$

The average stress in section m–m is

$$\sigma_{avg} = \frac{P}{bt} = \frac{15 \text{ kips}}{4 \times \frac{1}{2} \text{ in.}^2} = 7.5 \text{ ksi}$$

The maximum stress due to stress concentration occurring at points A and B is

$$\sigma_{max} = K\sigma_{avg} = 1.9(7.5 \text{ ksi}) = 14.3 \text{ ksi}$$

EXAMPLE 3–8

Determine the safe load P that can be applied to the 10-mm-thick plate shown without causing the tensile stress at any point to exceed 100 MPa.

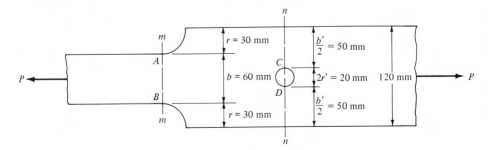

SOLUTION

First, consider stress concentration at the fillets, section m–m:

$$\frac{r}{b} = \frac{30 \text{ mm}}{60 \text{ mm}} = 0.5$$

From Fig. 3–15, $K = 1.42$. The maximum stress at A and B is

$$\sigma_{max} = K\sigma_{avg} = K\frac{P}{bt} = 1.42\frac{P}{0.060 \times 0.010 \text{ m}^2} \leq 100\,000 \text{ kN/m}^2$$

from which

$$P \leq \frac{(100\,000 \text{ kN/m}^2)(0.060 \times 0.010 \text{ m}^2)}{1.42} = 42.3 \text{ kN}$$

Next, consider stress concentration at the circular hole, section n–n:

$$\frac{r'}{b'} = \frac{10 \text{ mm}}{100 \text{ mm}} = 0.10$$

From Fig. 3–15, $K = 2.5$. The maximum stress at C and D is

$$\sigma_{max} = K\sigma_{avg} = K\frac{P}{b't'} = 2.5\frac{P}{0.10 \times 0.010 \text{ m}^2} \leq 100\,000 \text{ kN/m}^2$$

from which

$$P \leq \frac{(100\,000)(0.10 \times 0.010 \text{ m}^2)}{2.5} = 40.0 \text{ kN}$$

Thus the load P must be no more than 40.0 kN. ∎

PROBLEMS

3–31 What is stress concentration? What is the stress concentration factor?

3–32 Why is the effect of stress concentration more important for brittle materials than for ductile materials when designing a member subjected to static load?

3–33 What is the significance of stress concentration in the situation where there are cyclic stress variations?

3–34 Determine the maximum stress in the $\frac{1}{2}$-in.-thick plate shown in Fig. P3–34 if **(a)** $r = \frac{1}{4}$ in., **(b)** $r = \frac{1}{2}$ in., **(c)** $r = 1$ in.

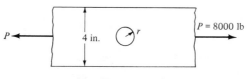

$P \longleftarrow$ 4 in. r $P = 8000$ lb $\longrightarrow$

FIGURE P3–34

3–35   Determine the maximum stress in the plate shown in Fig. P3–35 and indicate the points where the maximum stress occurs.

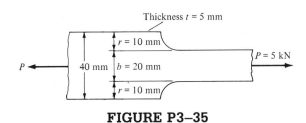

FIGURE P3–35

3–36   Determine the maximum permissible static load P that may be applied to the plate with semicircular notches shown in Fig. P3–36 if the tensile stress must not exceed 15 ksi.

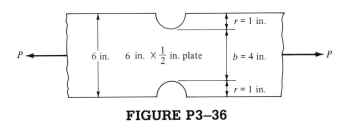

FIGURE P3–36

3–37   Determine the static axial load P that may be applied to the 10-mm-thick plate shown in Fig. P3–37 without causing the maximum stress in the plate to exceed 160 MPa.

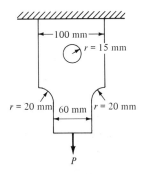

FIGURE P3–37

3–38   The two links shown in Fig. P3–38 are of the same material and have the same thickness. Determine the ratio of the allowable load P of the straight link to the allowable load P' of the link with an enlarged section. Assume that the links are subjected to cyclic stress variation; thus the stress concentration is an important factor to be considered.

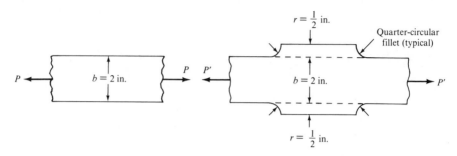

FIGURE P3–38

3–39   A plate 5 mm thick has two semicircular notches, as shown in Fig. P3–39. The material of the plate is mild steel with a yield stress σ_y of 250 MPa. Determine the maximum stress at section m–m if **(a)** $P = 40$ kN, **(b)** $P = 70$ kN. In each case sketch the stress distribution in the section. (**HINT:** Mild steel yields and stress remains at constant value when the yield stress is reached.)

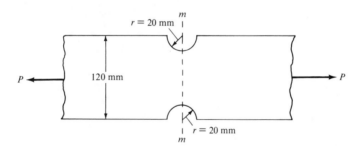

FIGURE P3–39

3–40   In Problem 3–39, determine the load P applied to the plate if **(a)** only one point in section m–m reaches the yield point, **(b)** all points in section m–m reach the yield point.

CHAPTER 4

Centroids and Moments of Inertia of Areas

4–1
INTRODUCTION

Both centroid and moment of inertia are geometric properties of an area. The geometric center of an area is its centroid, the point at which the area can be balanced if it is supported at the point. The moment of inertia of an area depends on the distribution of the area with respect to an axis; it is a quantity that appears frequently in mathematical expressions in mechanics. For example, the moment of inertia of a cross-sectional area is encountered in the analysis of the strength and the deformations of shafts and beams, and in the stability of columns. This chapter is concerned with the determination of centroids and moments of inertia of simple and composite areas.

4–2
CENTROIDS OF SIMPLE AREAS

Consider an area A as shown in Fig. 4–1, in which ΔA denotes the an area element. The sum of all the area elements over the entire area gives the area A. Thus

$$A = \Sigma \Delta A$$

If we increase the number of elements while decreasing the size of each element, we obtain, at the limit when ΔA approaches zero, the following integral:

$$A = \int_A dA \qquad (4–1)$$

The first moment of the element ΔA with respect to the y-axis is $x \, \Delta A$. The sum of the first moment of each element over the entire area, $\Sigma x \, \Delta A$, represents the first moment of the area A with respect to the y-axis. This sum must be equal to

79

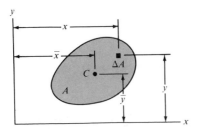

FIGURE 4–1

$\bar{x}A$, where $\bar{x}$ defines the location of the centroid of the area. Thus we have

$$\bar{x}A = \Sigma x\, \Delta A \qquad (4\text{–}2)$$

Similarly, consider the first moment with respect to the x-axis; we have

$$\bar{y}A = \Sigma y\, \Delta A \qquad (4\text{–}3)$$

In the limit, when ΔA approaches zero, the sum over an infinite number of infinitesimal quantities pertaining to the area element is, by definition, the integration over the entire area. That is,

$$\bar{x}A = \int_A x\, dA \qquad (4\text{–}4)$$

$$\bar{y}A = \int_A y\, dA \qquad (4\text{–}5)$$

These equations define the coordinates $\bar{x}$ and $\bar{y}$ of the centroid of the area A.

The area in Fig. 4-2(a) is symmetrical about axis aa', since for every point P of the area there is a point P' that is the mirror image of the point P with respect to the aa' axis. If aa' is chosen to be the y-axis, then the coordinate $\bar{x}$ of the centroid is found to be zero, because to every product $x\Delta A$ in Eq. (4–2) there is a corresponding product of equal magnitude but opposite in sign. Hence the sum $x\Delta A$ is zero over the entire area. Therefore, *the centroid of an area must be located on its axis of symmetry*. It follows that if an area possesses two axes of symmetry, the centroid of the area is located at the point of intersection of the two axes of symmetry [Fig. 4–2(b)].

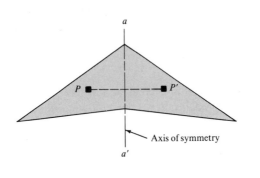

(a) Area with one axis of symmetry

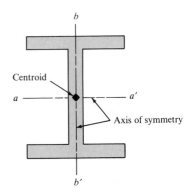

(b) Area with two axes of symmetry

FIGURE 4–2

━━━━━ **EXAMPLE 4–1** ━━━━━

Determine the centroid of a triangle by integration.

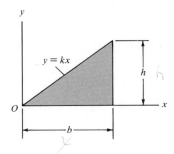

SOLUTION
As shown in the figure above, the equation of the straight line is $y = kx$. The value of k is determined by substituting $x = b$ and $y = h$ in the equation. We have

$$h = kb$$

or

$$k = \frac{h}{b}$$

Then the equation of the straight line becomes

$$y = \frac{h}{b} x$$

Now let us consider the vertical differential element dA shown in the following figure. We have

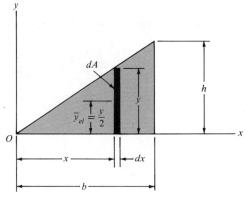

$$dA = y \, dx = \frac{h}{b} x \, dx$$

Integrating this expression from 0 to b gives the area of the triangle:

$$A = \int_A dA = \int_0^b \frac{h}{b} x \, dx = \frac{h}{b} \left[\frac{x^2}{2} \right]_0^b = \frac{1}{2} bh$$

The first moment of the differential area with respect to the y-axis is

$$x \, dA = \frac{h}{b} x^2 \, dx$$

Integrating this expression gives the first moment of the entire area with respect to the y-axis. Thus

$$\int_A x \, dA = \int_0^b \frac{h}{b} x^2 \, dx = \frac{h}{b} \left[\frac{x^3}{3} \right]_0^b = \frac{b^2 h}{3}$$

From Eq. (4–4),

$$\bar{x} A = \int_A x \, dA$$

or

$$\bar{x} \left(\frac{bh}{2} \right) = \frac{b^2 h}{3}$$

from which

$$\bar{x} = \tfrac{2}{3} b$$

The first moment of the differential element with respect to the x-axis is

$$\bar{y}_{\text{el}} \, dA = \frac{1}{2} y \, dA = \frac{1}{2} \left(\frac{h}{b} x \right) \left(\frac{h}{b} x \, dx \right) = \frac{h^2}{2b^2} x^2 \, dx$$

The first moment of the entire area with respect to the x-axis is

$$\int_A \bar{y}_{\text{el}} \, dA = \int_0^b \frac{h^2}{2b^2} x^2 \, dx = \frac{h^2}{2b^2} \left[\frac{x^3}{3} \right]_0^b = \frac{bh^2}{6}$$

From Eq. (4–5),

$$\bar{y} A = \int_A \bar{y}_{\text{el}} \, dA$$

or

$$\bar{y} \left(\frac{bh}{2} \right) = \frac{bh^2}{6}$$

from which

$$\bar{y} = \tfrac{1}{3} h$$

■

Determine by integration the area and the centroid of the parabolic spandrel shown.

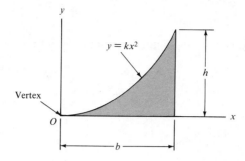

SOLUTION

The equation of the parabola is $y = kx^2$. The value of k is determined by substituting $x = b$ and $y = h$ in the equation. Thus

$$h = kb^2$$

or

$$k = \frac{h}{b^2}$$

The parabolic equation becomes

$$y = \frac{h}{b^2} x^2$$

Consider the vertical differential element dA shown in the following figure:

$$dA = y\, dx = \frac{h}{b^2} x^2\, dx$$

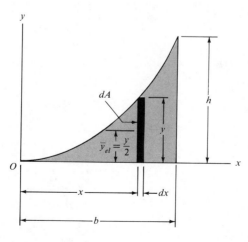

The area of the parabolic spandrel is

$$A = \int_A dA = \int_0^b \frac{h}{b^2} x^2\, dx = \frac{h}{b^2} \left[\frac{x^3}{3} \right]_0^b = \frac{bh}{3}$$

The first moment of the differential element with respect to the y-axis is

$$x \, dA = \frac{h}{b^2} x^3 \, dx$$

The first moment of the entire area with respect to the y-axis is

$$\int_A x \, dA = \int_0^b \frac{h}{b^2} x^3 \, dx = \frac{h}{b^2} \left[\frac{x^4}{4} \right]_0^b = \frac{b^2 h}{4}$$

From Eq. (4–4),

$$\bar{x} A = \int_A x \, dA$$

or

$$\bar{x} \left(\frac{bh}{3} \right) = \frac{b^2 h}{4}$$

from which

$$\bar{x} = \tfrac{3}{4} b$$

The first moment of the differential element with respect to the x-axis is

$$\bar{y}_{el} \, dA = \frac{1}{2} y \, dA = \frac{1}{2} \left(\frac{h}{b^2} x^2 \right) \left(\frac{h}{b^2} x^2 \, dx \right) = \frac{h^2}{2b^4} x^4 \, dx$$

The first moment of the entire area with respect to the x-axis is

$$\int_A \bar{y}_{el} \, dA = \int_0^b \frac{h}{2b^4} x^4 \, dx = \frac{h^2}{2b^4} \left[\frac{x^5}{5} \right]_0^b = \frac{bh^2}{10}$$

From Eq. (4–5),

$$\bar{y} A = \int_A y_{el} \, dA$$

or

$$\bar{y} \left(\frac{bh}{3} \right) = \frac{bh^2}{10}$$

from which

$$\bar{y} = \tfrac{3}{10} h$$

∎

Centroids of several simple areas, such as the triangle, the semicircle, the semiparabolic area, the parabolic spandrel, and so on, are listed in Table 4–1.

TABLE 4–1   Centroids of Simple Areas

Shape	Centroid	Area
Triangle		$\dfrac{1}{2}\,bh$
Semicircle		$\dfrac{1}{2}\,\pi R^2$
Quarter-circle		$\dfrac{1}{4}\,\pi R^2$
Sector		αR^2
Semiparabolic area		$\dfrac{2}{3}\,bh$
Parabolic spandrel		$\dfrac{1}{3}\,bh$

4–3
CENTROIDS OF COMPOSITE AREAS

In many cases, a composite area may be divided into simple areas listed in Table 4–1. The area in Fig. 4–3 is divided into a rectangle, two triangles, and a semicircle, as shown. These areas are denoted by A_1, A_2, A_3, and A_4. The coordinates

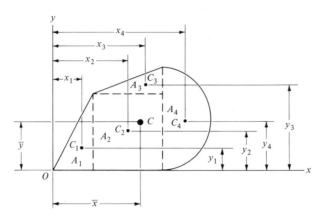

FIGURE 4–3

of the centroid of each area are denoted by x's and y's, with the subscripts corresponding to the area they represent. From Eqs. (4–2) and (4–3) the coordinates of the centroid of the composite area $\bar{x}$ and $\bar{y}$ can be determined by the following equations:

$$\bar{x} = \frac{\Sigma Ax}{A} = \frac{A_1x_1 + A_2x_2 + A_3x_3 + A_4x_4}{A_1 + A_2 + A_3 + A_4} \tag{4–6}$$

$$\bar{y} = \frac{\Sigma Ay}{A} = \frac{A_1y_1 + A_2y_2 + A_3y_3 + A_4y_4}{A_1 + A_2 + A_3 + A_4} \tag{4–7}$$

The x and y axes are conveniently chosen so that the entire area is in the first quadrant and the coordinates of the centroids of all the areas have positive values. If there is an axis of symmetry, the centroid is located along this axis. The areas of holes, notches, and so on, should be treated as negative values, since the areas in these parts are absent.

——— **EXAMPLE 4–3** ———————————————————————————————

Determine the location of the centroid of the area shown.

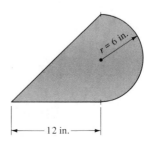

SOLUTION

The coordinate axes are chosen and the area is divided into two parts: a triangular area A_1 and a semicircular area A_2, as shown in the figure.

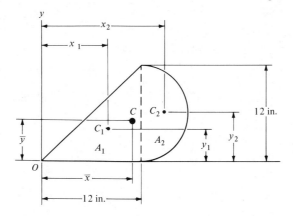

The areas are

$$A_1 = \tfrac{1}{2} (12 \text{ in.})(12 \text{ in.}) = 72 \text{ in.}^2$$

$$A_2 = \tfrac{1}{2} \pi(6 \text{ in.})^2 \qquad = 56.55 \text{ in.}^2$$

$$A = A_1 + A_2 \qquad = 128.6 \text{ in.}^2$$

The centroid of each area (using the centroid locations in Table 4–1) is located by the following coordinates:

$$x_1 = \tfrac{2}{3} (12 \text{ in.}) = 8 \text{ in.}$$

$$x_2 = 12 \text{ in.} + 0.4244(6 \text{ in.}) = 14.55 \text{ in.}$$

$$y_1 = \tfrac{1}{3} (12 \text{ in.}) = 4 \text{ in.}$$

$$y_2 = 6 \text{ in.}$$

The centroid of the entire area is then

$$\bar{x} = \frac{A_1 x_1 + A_2 x_2}{A} = \frac{(72 \text{ in.}^2)(8 \text{ in.}) + (56.55 \text{ in.}^2)(14.55 \text{ in.})}{128.6 \text{ in.}^2}$$

from which

$$\bar{x} = 10.88 \text{ in.}$$

$$\bar{y} = \frac{A_2 y_2 + A_2 y_2}{A} = \frac{(72 \text{ in.}^2)(4 \text{ in.}) + (56.55 \text{ in.}^2)(6 \text{ in.})}{128.6 \text{ in.}^2}$$

from which

$$\bar{y} = 4.88 \text{ in.}$$

The computations above can be tabulated as shown in the following table.

Part	Shape	A (in.²)	x (in.)	Ax (in.³)	y (in.)	Ay (in.³)
1	(triangle, 12 in. × 12 in.)	$\frac{1}{2}(12)(12) = 72$	$\frac{2}{3}(12) = 8$	576	$\frac{1}{3}(12) = 4$	288
2	(semicircle, 6 in.)	$\frac{1}{2}\pi(6)^2 = 56.55$	$12 + 0.4244(6)$ $= 14.55$	823	6	339
Σ		128.6		1399		627

$$\bar{x} = \frac{\Sigma Ay}{\Sigma A} = \frac{1399 \text{ in.}^3}{128.6 \text{ in.}^2} = 10.88 \text{ in.}$$

$$\bar{y} = \frac{\Sigma Ay}{\Sigma A} = \frac{627 \text{ in.}^3}{128.6 \text{ in.}^2} = 4.88 \text{ in.}$$

EXAMPLE 4–4

Determine the location of the centroid of the composite area shown.

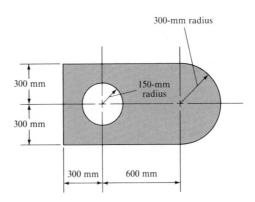

300-mm radius

150-mm radius

300 mm

300 mm

300 mm   600 mm

SOLUTION

The area is obtained by adding a rectangle A_1 and a semicircle A_2 and subtracting a circle A_3. The coordinate axes are chosen with the origin at the lower left corner of the plate. By inspection it is clear that the line $y = 0.3$ m is an axis of symmetry. The centroid must be located along this line of symmetry. The value of the area A_3 is negative, and it is subtracted from the other areas.

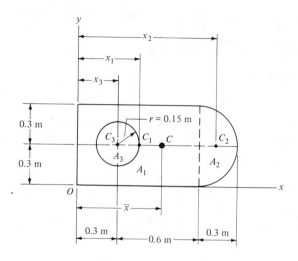

$$A_1 = (0.9 \text{ m})(0.6 \text{ m}) = 0.54 \text{ m}^2$$

$$A_2 = \tfrac{1}{2} \pi (0.3 \text{ m})^2 = 0.1414 \text{ m}^2$$

$$A_3 = -\pi (0.15 \text{ m})^2 = -0.0707 \text{ m}^2$$

$$A = A_1 + A_2 + A_3 = 0.6107 \text{ m}^2$$

The centroid of each area A_1, A_2, and A_3 is located by the following x distances:

$$x_1 = \tfrac{1}{2} (0.3 \text{ m} + 0.6 \text{ m}) = 0.45 \text{ m}$$

$$x_2 = 0.9 \text{ m} + 0.4244(0.3 \text{ m}) = 1.027 \text{ m}$$

$$x_3 = 0.3 \text{ m}$$

The centroid of the entire area is located by

$$\bar{x} = \frac{A_1 x_1 + A_2 x_2 + A_3 x_3}{A}$$

$$= \frac{(0.54 \text{ m}^2)(0.45 \text{ m}) + (0.1414 \text{ m}^2)(1.027 \text{ m}) + (-0.0707 \text{ m}^2)(0.3 \text{ m})}{0.6107 \text{ m}^2}$$

from which

$$\bar{x} = 0.601 \text{ m} = 601 \text{ mm}$$

By symmetry

$$\bar{y} = 0.3 \text{ m} = 300 \text{ mm}$$

The tabular form of the solution is as follows:

Part	Shape	A (m²)	x (m)	Ax (m³)
1	0.6 m, 0.9 m	$0.9 \times 0.6 =$ 0.540	0.45	0.2430
2	0.3 m	$\frac{1}{2}\pi(0.3)^2 =$ 0.1414	$0.9 + 0.4244(0.3) = 1.027$	0.1452
3	0.15 m	$-\pi(0.15)^2 = -0.0707$	0.3	−0.0212
Σ		0.6107		0.3670

$$\bar{x} = \frac{\Sigma Ax}{\Sigma A} = \frac{0.3670 \text{ m}^3}{0.6107 \text{ m}^2} = 0.601 \text{ m} = 601 \text{ mm}$$

By symmetry,

$$\bar{y} = 0.3 \text{ m} = 300 \text{ mm}$$

PROBLEMS

In Problems **4–1** *to* **4–8**, *locate the centroid of the plane areas shown.*

4–1

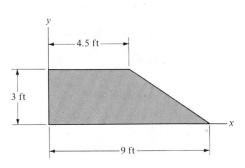

FIGURE P4–1

4–2

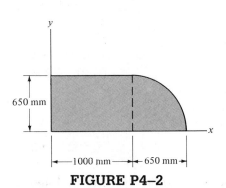

FIGURE P4–2

4–3

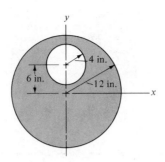

FIGURE P4–3

4–4

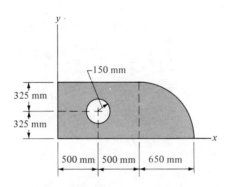

FIGURE P4–4

4–5

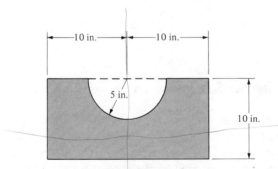

FIGURE P4–5

4–6

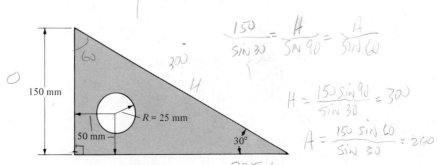

$$\frac{150}{\sin 30} = \frac{H}{\sin 90} = \frac{A}{\sin 60}$$

$$H = \frac{150 \sin 90}{\sin 30} = 300$$

$$A = \frac{150 \sin 60}{\sin 30} = 260$$

FIGURE P4–6   335.4

H

259.8

4–7

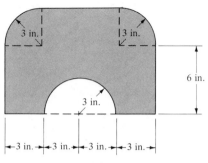

FIGURE P4–7

4–8

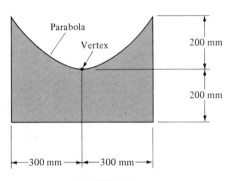

FIGURE P4–8

4–4
DEFINITION OF AREA MOMENTS OF INERTIA

The moment of inertia of an area is computed with respect to an axis. When the axis lies in the plane of the area, the moment of inertia of the area is called the *rectangular moment of inertia* (or simply the moment of inertia).When the axis is perpendicular to the plane of the area, it is called the *polar moment of inertia*.

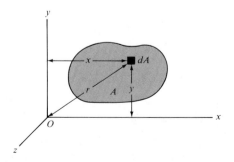

FIGURE 4–4

For the differential area element dA in the area shown in Fig. 4–4, the differential moment of inertia of the element with respect to an axis is equal to the second moment of the area with respect to the axis. That is, it is the product of the area of the element and the square of its perpendicular distance to the axis. Thus the differential moments of inertia, denoted by dI_x and dI_y, of the differential element dA with respect to the x and y axes, respectively, are

$$dI_x = y^2 \, dA$$

$$dI_y = x^2 \, dA$$

The differential polar moment of inertia of the element, denoted by dJ, with respect to the z-axis is

$$dJ = dI_z = r^2 \, dA$$

It should be noted that the moment of inertia is always positive regardless of the location of the axis. The moment of inertia of an area has units of length raised to the fourth power, in.^{4} or m^4.

For the entire area in Fig. 4–4, the moment of inertia of the area with respect to an axis is obtained by integrating the differential moment of inertia with respect to the axis. Thus

$$I_x = \int_A dI_x = \int_A y^2 \, dA \qquad\qquad (4\text{--}8)$$

$$I_y = \int_A dI_y = \int_A x^2 \, dA \qquad\qquad (4\text{--}9)$$

$$J = \int_A dJ = \int_A r^2 \, dA \qquad\qquad (4\text{--}10)$$

Since $r^2 = x^2 + y^2$, we have

$$J = \int_A r^2 \, dA = \int_A (x^2 + y^2) \, dA = \int_A x^2 \, dA + \int_A y^2 \, dA$$

Therefore,

$$J = I_x + I_y \qquad\qquad (4\text{--}11)$$

———— **EXAMPLE 4–5** ————————————————————————————

Determine the moment of inertia of a rectangular area of width b and height h with respect to its horizontal centroidal axis.

SOLUTION

A rectangle of width b and height h is shown in the following figure. The $\bar{x}$-axis is the horizontal centroidal axis. A horizontal differential strip dA is chosen as shown.

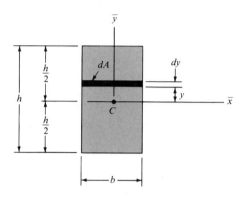

Since all points of the strip are at the same distance from the $\bar{x}$-axis, the moment of inertia of dA with respect to the x-axis is

$$d\bar{I}_x = y^2 \, dA = y^2(b \, dy)$$

Integrating this expression from $y = -h/2$ to $y = +h/2$ gives the moment of inertia of the rectangle with respect to the centroidal $\bar{x}$-axis. Thus

$$\bar{I}_x = \int_A d\bar{I}_x = \int_{-h/2}^{+h/2} by^2 \, dy = b\left[\frac{y^3}{3}\right]_{-h/2}^{+h/2} = \frac{b}{3}\left[\left(+\frac{h}{2}\right)^3 - \left(-\frac{h}{2}\right)^3\right]$$

from which

$$\bar{I}_x = \frac{bh^3}{12}$$

■

EXAMPLE 4–6

Determine (a) the polar moment of inertia, (b) the rectangular moment of inertia of a circular area with respect to the centroidal axes.

SOLUTION

(a) *Centroidal polar moment of inertia:* A circle of radius R is shown in the following figure. The centroidal axes $\bar{x}$ and $\bar{y}$ are drawn through the centroid O. A differential annular area element dA is chosen as shown. The element has a differential thickness $d\rho$ and its length is $2\pi\rho$. Hence the differential area is

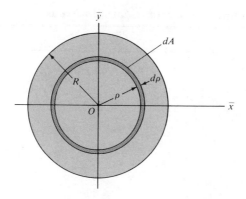

$$dA = 2\pi\rho \, d\rho$$

Since all points in the element are at the same distance from the centroidal $\bar{z}$-axis through O, the centroidal polar moment of inertia of dA with respect to the $\bar{z}$-axis is

$$d\bar{J} = \rho^2 \, dA$$

Substituting $dA = 2\pi\rho \, d\rho$, we have

$$d\bar{J} = 2\pi\rho^3 \, d\rho$$

Integrating $d\bar{J}$ from $\rho = 0$ to $\rho = R$ gives the centroidal polar moment of inertia of the circular area. Thus

$$\bar{J} = \int_A d\bar{J} = \int_0^R 2\pi\rho^3 \, d\rho = 2\pi\left[\frac{\rho^4}{4}\right]_0^R$$

$$\bar{J} = \frac{\pi R^4}{2}$$

(b) *Centroidal rectangular moment of inertia:* Due to symmetry we have $\bar{I}_x = \bar{I}_y$. From Eq. (4–11) we write

$$\bar{J} = \bar{I}_x + \bar{I}_y = 2\bar{I}_x$$

Thus

$$\bar{I}_x = \bar{I}_y = \frac{1}{2}\bar{J} = \frac{\pi R^4}{4}$$

Formulas for moments of inertia of simple areas, such as those derived in Examples 4–5 and 4–6, are listed in Table 4–2.

TABLE 4–2   Moments of Inertia of Simple Areas

Shape	Figure	$\bar{I}_x$	$\bar{I}_y$	$\bar{J}$
Rectangle		$\dfrac{bh^3}{12}$	$\dfrac{hb^3}{12}$	$\dfrac{bh}{12}(h^2 + b^2)$
Triangle		$\dfrac{bh^3}{36}$		
Circle		$\dfrac{\pi d^4}{64}$ $\dfrac{\pi R^4}{4}$	$\dfrac{\pi d^4}{64}$ $\dfrac{\pi R^4}{4}$	$\dfrac{\pi d^4}{32}$ $\dfrac{\pi R^4}{2}$
Circular ring		$\dfrac{\pi(d_0^4 - d_i^4)}{64}$ $\dfrac{\pi(R_0^4 - R_i^4)}{4}$	$\dfrac{\pi(d_0^4 - d_i^4)}{64}$ $\dfrac{\pi(R_0^4 - R_i^4)}{4}$	$\dfrac{\pi(d_0^4 - d_i^4)}{32}$ $\dfrac{\pi(R_0^4 - R_i^4)}{2}$
Semicircle		$\left(\dfrac{\pi}{8} - \dfrac{8}{9\pi}\right)R^4$ $= 0.1098R^4$	$\dfrac{\pi R^4}{8}$	$\left(\dfrac{\pi}{4} - \dfrac{8}{9\pi}\right)R^4$ $= 0.5025R^4$
Quartercircle		$\left(\dfrac{\pi}{16} - \dfrac{4}{9\pi}\right)R^4$ $= 0.0549R^4$	$\left(\dfrac{\pi}{16} - \dfrac{4}{9\pi}\right)R^4$ $= 0.0549R^4$	$\left(\dfrac{\pi}{8} - \dfrac{8}{9\pi}\right)R^4$ $= 0.1098R^4$

4–5
PARALLEL-AXIS THEOREM

The moment of inertia of an area with respect to a noncentroidal axis may be expressed in terms of the moment of inertia with respect to a parallel centroidal axis. In Fig. 4–5 the $\bar{x}$-axis passes through the centroid C of the area. By definition, the moment of inertia about a parallel noncentroidal x-axis is

$$I_x = \int (y + d)^2 \, dA = \int (y^2 + 2yd + d^2) \, dA$$

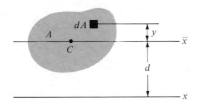

FIGURE 4–5

or

$$I_x = \int y^2 \, dA + 2d \int y \, dA + d^2 \int dA$$

It is seen that the first integral represents moment of inertia $\bar{I}_x$ about the centroidal axis $\bar{x}$. The second integral is zero, since $\int y \, dA = A\bar{y}$, but $\bar{y}$ is zero with respect to the centroidal $\bar{x}$-axis. The third integral is simply Ad^2. Thus the expression for I_x becomes

$$I_x = \bar{I}_x + Ad^2 \tag{4–12}$$

Equation (4–12) is the *parallel-axis theorem*. It states that *the moment of inertia of an area with respect to a noncentroidal axis is equal to the moment of inertia of the area with respect to the parallel centroidal axis plus the product of the area and the square of the distance between the two axes*. Two points should be noted. First, the two axes must be parallel for the theorem to apply. Second, one of the axes must pass through the centroid of the area.

4–6
RADIUS OF GYRATION

The *radius of gyration* of an area with respect to an axis is that distance from the axis to a point at which the area could be concentrated to produce the moment of inertia of the area with respect to the axis. Let the radius of gyration with respect to the x-axis be denoted by r_x. Then, by definition,

$$I_x = Ar_x^2 \tag{4–13a}$$

or

$$r_x = \sqrt{\frac{I_x}{A}} \tag{4–13b}$$

The radius of gyration is a useful parameter in the design of columns, which is discussed in Chapter 12.

──────── **EXAMPLE 4–7** ────────────────────────────────────

In the following figure, determine (a) the moments of inertia $\bar{I}_x$, $\bar{I}_y$, and $\bar{J}$ of the rectangular area about its centroidal axes, (b) the moment of inertia I_x of the area about the x-axis, (c) the radii of gyration of the area with respect to the $\bar{x}$ and x axes.

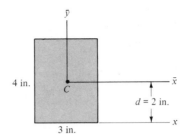

SOLUTION

(a) Using formulas from Table 4–2, we get

$$\bar{I}_x = \frac{bh^3}{12} = \frac{(3 \text{ in.})(4 \text{ in.})^3}{12} = 16 \text{ in.}^4$$

$$\bar{I}_y = \frac{hb^3}{12} = \frac{(4 \text{ in.})(3 \text{ in.})^3}{12} = 9 \text{ in.}^4$$

By Eq. (4–11),

$$\bar{J} = \bar{I}_x + \bar{I}_y = 16 \text{ in.}^4 + 9 \text{ in.}^4 = 25 \text{ in.}^4$$

(b) Use the parallel-axis theorem to transfer from the $\bar{x}$-axis to the x-axis:

$$I_x = \bar{I}_x + Ad^2 = 16 \text{ in.}^4 + (3 \times 4 \text{ in.}^2)(2 \text{ in.})^2 = 64 \text{ in.}^4$$

(c) By definition, the radii of gyration are

$$\bar{r}_x = \sqrt{\frac{\bar{I}_x}{A}} = \sqrt{\frac{16 \text{ in.}^4}{12 \text{ in.}^2}} = 1.15 \text{ in.}$$

$$r_x = \sqrt{\frac{I_x}{A}} = \sqrt{\frac{64 \text{ in.}^4}{12 \text{ in.}^2}} = 2.31 \text{ in.}$$

Note that the radius of gyration r_x with respect to the x-axis is not the same as the distance from the centroid of the area to the x-axis. ∎

EXAMPLE 4–8

In the following figure, determine (a) the moment of inertia, (b) the radius of gyration of the semicircular area with respect to the x-axis.

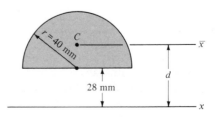

SOLUTION

(a) From Table 4–2, the moment of inertia of a semicircular area about its centroidal axis $\bar{x}$ is

$$\bar{I}_x = 0.1098r^4 = 0.1098(0.040 \text{ m})^4 = 2.81 \times 10^{-7} \text{ m}^4$$

The distance d between the $\bar{x}$-axis and the x-axis is

$$d = 0.028 \text{ m} + \frac{4(0.040 \text{ m})}{3\pi} = 0.0450 \text{ m}$$

Using the parallel-axis theorem, we get the moment of inertia of the area about the x-axis:

$$I_x = \bar{I}_x + Ad^2 = 2.81 \times 10^{-7} \text{ m}^4 + [\tfrac{1}{2}\pi(0.040 \text{ m})^2](0.045 \text{ m})^2$$

$$I_x = 5.37 \times 10^{-6} \text{ m}^4$$

(b) By definition, the radius of gyration with respect to the x-axis is

$$r_x = \sqrt{\frac{I_x}{A}} = \sqrt{\frac{5.37 \times 10^{-6} \text{ m}^4}{\tfrac{1}{2}\pi(0.040 \text{ m})^2}} = 46.2 \times 10^{-3} \text{ m}$$

or

$$r_x = 46.2 \text{ mm} \qquad\blacksquare$$

PROBLEMS

 4–9 In Fig. P4–9, verify that the radii of gyration $\bar{r}_x$ and $\bar{r}_y$ of the rectangle with respect to the centroidal axes are $\bar{r}_x = h/\sqrt{12} = 0.289h$ and $\bar{r}_y = b/\sqrt{12} = 0.289b$.

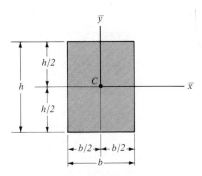

FIGURE P4–9

4–10 Verify that the radius of gyration for a circle of diameter d with respect to a centroidal axis is $\bar{r} = d/4$.

4–11   Determine the moments of inertia I_x and the radius of gyration r_x of the circular area shown in Fig. P4–11.

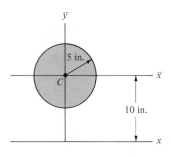

FIGURE P4–11

4–12   If the radius of gyration of the shaded area (Fig. P4–12) with respect to the y-axis is $r_y = 12.4$ in., determine its centroidal moment of inertia $\bar{I}_y$ and the radius of gyration $\bar{r}_y$ of the area.

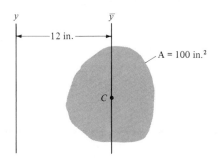

FIGURE P4–12

4–13   If the moment of inertia I_x of the rectangular area shown in Fig. P4–13 is 7320 in.^{4}, determine $I_{x'}$ of the area. (**HINT:** Do not apply the parallel-axis theorem directly to x and x' axes, because for the theorem to apply, one of the axes has to be a centroidal axis.)

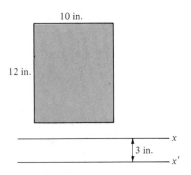

FIGURE P4–13

4–14   Determine the centroidal polar moment of inertia of a rectangle 100 mm wide by 200 mm high.

4–15 Determine the polar moment of inertia of the 18-in.-diameter circular area shown in Fig. P4–15 with respect to an axis perpendicular to the plane of the circle and passing through a point on its circumference. (**HINT:** Determine I_x and I_y and then calculate J by the formula $J = I_x + I_y$.)

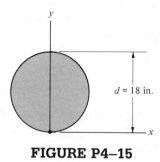

$d = 18$ in.

FIGURE P4–15

4–16 Determine the radii of gyration r_x and r_y of the semicircular area shown in Fig. P4–16.

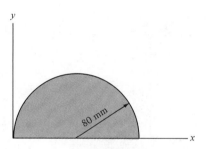

80 mm

FIGURE P4–16

4–17 Determine the moments of inertia I_x and I_y and the radii of gyration r_x and r_y of the quarter-circular area shown in Fig. P4–17.

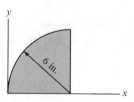

6 in.

FIGURE P4–17

4–7
MOMENTS OF INERTIA OF COMPOSITE AREAS

It is frequently necessary to determine the moments of inertia of an area composed of several simple areas. Since a moment of inertia of an area about an axis is the sum of the moments of inertia of the elements that comprise the area, it follows that *the moment of inertia of a composite area with respect to an axis is simply the sum of the moments of inertia of all the component parts with respect to the same axis.*

Subsequent applications to the strength and deflection of beams require determination of the moment of inertia of a composite area with respect to a centroidal axis of the area. In these cases, the centroid of a composite area must first be located.

In calculating the moment of inertia of an area that contains holes or notches, it is convenient to treat these areas and their moments of inertia as negative values, as shown in Example 4–10.

──────── **EXAMPLE 4–9** ────────────────────────────────────

The area shown is a frequently encountered composite beam section. Determine the moment of inertia and the radius of gyration of the area with respect to the horizontal centroidal axis.

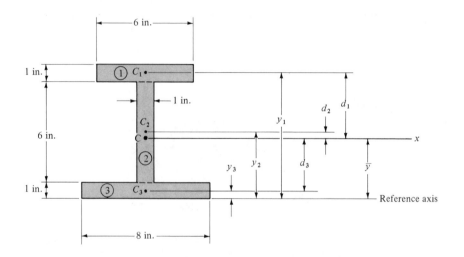

SOLUTION

The section is subdivided into three rectangular areas, ①, ②, and ③, as shown. To locate the centroid of the area, a reference axis is chosen at the bottom of the section. Thus

$$A = A_1 + A_2 + A_3 = 6 \times 1 + 1 \times 6 + 8 \times 1 = 20 \text{ in.}^2$$

$$\bar{y} = \frac{A_1 y_1 + A_2 y_2 + A_3 y_3}{A} = \frac{(6)(7.5) + (6)(4) + (8)(0.5)}{20}$$

from which

$$\bar{y} = 3.65 \text{ in.}$$

To find the moment of inertia of each area with respect to the centroidal x-axis, the parallel-axis theorem must be used to transfer the axis from the centroidal axis of each area to the centroidal axis of the entire area.

The distance d_i from the centroid of each component area to the centroidal axis of the entire area is

$$d_i = y_i - \bar{y}$$

Thus

$$d_1 = y_1 - \bar{y} = 7.5 - 3.65 = \quad 3.85 \text{ in.}$$

$$d_2 = y_2 - \bar{y} = \quad 4 - 3.65 = \quad 0.35 \text{ in.}$$

$$d_3 = y_3 - \bar{y} = 0.5 - 3.65 = -3.15 \text{ in.}$$

Hence

$$I_x^{①} = \bar{I}_x^{①} + A_1 d_1^2 = \frac{(6)(1)^3}{12} + (6)(3.85)^2 = 89.4 \text{ in.}^4$$

$$I_x^{②} = \bar{I}_x^{②} + A_2 d_2^2 = \frac{(1)(6)^3}{12} + (6)(0.35)^2 = 18.7 \text{ in.}^4$$

$$I_x^{③} = \bar{I}_x^{③} + A_3 d_3^2 = \frac{(8)(1)^3}{12} + (8)(-3.15)^2 = 80.1 \text{ in.}^4$$

The moment of inertia of the entire area with respect to its centroidal axis is the sum of the moments of inertia of the three component parts about the axis. Thus

$$I_x = I_x^{①} + I_x^{②} + I_x^{③} = 89.4 + 18.7 + 80.1 = 188 \text{ in.}^4$$

By definition, the radius of gyration of the section with respect to the x-axis is

$$r_x = \sqrt{\frac{I_x}{A}} = \sqrt{\frac{188 \text{ in.}^4}{20 \text{ in.}^2}} = 3.07 \text{ in.}$$

The computation performed above can be presented in the tabular form as follows. Before the table is set up, the moment of inertia of each component area about its own centroidal axis is determined. Thus

$$\bar{I}_x^{①} = \frac{(6)(1)^3}{12} = 0.5 \text{ in.}^4$$

$$\bar{I}_x^{②} = \frac{(1)(6)^3}{12} = 18.0 \text{ in.}^4$$

$$\bar{I}_x^{③} = \frac{(8)(1)^3}{12} = 0.7 \text{ in.}^4$$

Then the table is set up, and the computations are carried out in a systematic manner.

(1) Part	(2) Shape	(3) A (in.²)	(4) y (in.)	(5) Ay (in.³)	(6) $d = y - \bar{y}$ (in.)	(7) Ad^2 (in.⁴)	(8) $\bar{I}_x$ (in.⁴)
①	1 in.　6 in.	6	7.5	45.0	3.85	88.9	0.5
②	6 in.　1 in.	6	4.0	24.0	0.35	0.7	18.0
③	1 in.　8 in.	8	0.5	4.0	−3.15	79.4	0.7
Σ		20		73.0		169.0	19.2

Columns (1) through (5) must be set up first; then

$$\bar{y} = \frac{\Sigma Ay}{\Sigma A} = \frac{73.0 \text{ in.}^3}{20 \text{ in.}^2} = 3.65 \text{ in.}$$

The d's in column (6) are computed by $d_i = y_i - \bar{y}$ and columns (7) and (8) can subsequently be set up; then

$$I_x = \Sigma Ad^2 + \Sigma \bar{I}_x = 169.0 + 19.2 = 188 \text{ in.}^4$$

■

EXAMPLE 4–10

Determine the moment of inertia and the radius of gyration of the shaded area shown with respect to the *x*-axis.

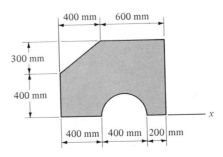

SOLUTION

The composite area shown is comprised of the positive rectangular area ①, the negative triangular area ②, and the negative semicircular area ③, as shown in the following figure.

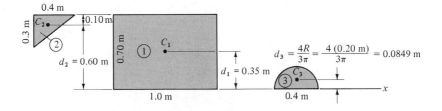

The area of each part is

$$A_1 = (1.00 \text{ m})(0.70 \text{ m}) = 0.70 \text{ m}^2$$

$$A_2 = -\tfrac{1}{2}(0.40 \text{ m})(0.30 \text{ m}) = -0.06 \text{ m}^2$$

$$A_3 = -\tfrac{1}{2}\pi R^2 = -\tfrac{1}{2}\pi (0.20 \text{ m})^2 = -0.063 \text{ m}^2$$

Using the formulas from Table 4–2, we get the moments of inertia of the areas with respect to their own centroidal axes:

$$\bar{I}_x^{①} = \frac{bh^3}{12} = \frac{(1.0 \text{ m})(0.70 \text{ m})^3}{12} = 0.0286 \text{ m}^4$$

$$\bar{I}_x^{②} = -\frac{bh^3}{36} = -\frac{(0.40 \text{ m})(0.30 \text{ m})^3}{12} = -0.0003 \text{ m}^4$$

$$\bar{I}_x^{③} = -0.1098R^4 = -0.1098(0.20 \text{ m})^4 = -0.0002 \text{ m}^4$$

The computations for I_x are tabulated as follows:

Part	Shape	A (m²)	d (m)	Ad^2 (m⁴)	$\bar{I}_x$ (m⁴)
①	Rectangle	0.70	0.35	0.0858	0.0286
②	Triangle	−0.06	0.60	−0.0216	−0.0003
③	Semicircle	−0.063	0.0849	−0.0005	−0.0002
Σ		0.577		0.0637	0.0281

$$I_x = \Sigma Ad^2 + \Sigma \bar{I}_x = 0.0637 \text{ m}^4 + 0.0281 \text{ m}^4 = 0.0918 \text{ m}^4$$

By definition, the radius of gyration is

$$r_x = \sqrt{\frac{I_x}{A}} = \sqrt{\frac{0.0918 \text{ m}^4}{0.577 \text{ m}^2}} = 0.399 \text{ m} = 399 \text{ mm}$$

PROBLEMS

4–18 Determine the moments of inertia of the area shown in Fig. P4–18 with respect to the horizontal and vertical centroidal axes.

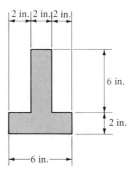

FIGURE P4–18

4–19 A double-T beam section is fabricated from three 2 in. by 10 in. full-sized timber planks, as shown in Fig. P4–19. Determine the moment of inertia I_x of the section with respect to the horizontal centroidal axis.

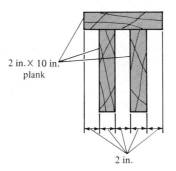

FIGURE P4–19

4–20 The timber beam section shown in Fig. P4–20 is made up of boards 50 mm thick glued together as shown. Determine the moment of inertia of the section with respect to the horizontal centroidal axis.

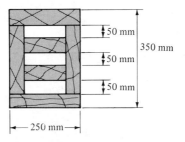

FIGURE P4–20

4–21 Determine the moment of inertia I_x and the radius of gyration r_x of the shaded area shown in Fig. P4–21.

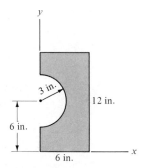

FIGURE P4–21

4–22 Determine the moment of inertia of the shaded area in Fig. P4–22 with respect to the horizontal centroidal axis.

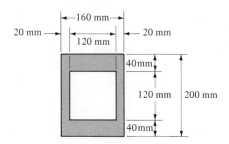

FIGURE P4–22

4–23 Find the expression for the centroidal polar moment of inertia of the area of a ring (Fig. P4–23) in terms of R_o and R_i. Use the formula for circles in Table 4–2.

FIGURE P4–23

4–24 Determine the radius of gyration of the shaded area in Fig. P4–24 with respect to the x-axis.

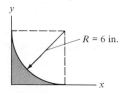

FIGURE P4–24

4–25   Determine the polar moment of inertia of the shaded area in Fig. P4–25 with respect to the z-axis through O.

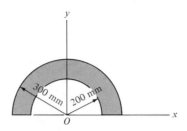

FIGURE P4–25

4–26   Determine the moment of inertia I_x of the shaded area in Fig. P4–26.

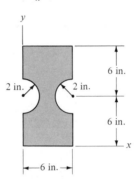

FIGURE P4–26

4–27   Determine the radius of gyration r_x of the shaded area in Fig. P4–27.

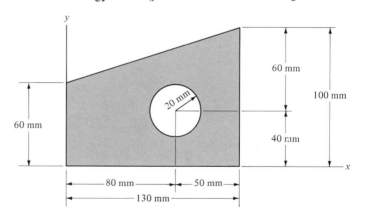

FIGURE P4–27

4–8
PROPERTIES OF STRUCTURAL STEEL SHAPES

Structural steel is rolled into a wide variety of shapes and sizes. The *AISC Manual* (the *Manual of Steel Construction* published by the American Institute of Steel Construction) provides detailed information for structural steel shapes. The properties of the cross sections of selected structural steel shapes are given in Tables A–1 to A–4 in the Appendix Tables.

(a) Wide-flange steel beam   (b) American standard steel I-beam   (c) American standard steel channel   (d) Equal-leg steel angle

FIGURE 4–6

Structural steel shapes are designated by letters that specify their shapes, followed by numbers that specify their sizes. Examples of this system follow:

1. A W21 × 83 is a wide-flange steel beam [Fig. 4–6(a)] having a nominal depth of 21 in. and weighing 83 lb/ft.
2. An S20 × 75 is an American standard steel I-beam [Fig. 4–6(b)] 20 in. deep weighing 75 lb/ft.
3. A C12 × 30 is an American standard steel channel section [Fig. 4–6(c)] 12 in. deep weighing 30 lb/ft.
4. An L6 × 6 × $\frac{1}{2}$ is an equal-leg steel angle [Fig. 4–6(d)], each leg being 6 in. long and $\frac{1}{2}$ in. thick.

Tables A–1 to A–4 list dimensions, weights, areas, locations of centroidal axes, and moments of inertia as well as radii of gyration about the centroidal axes for structural steel shapes of many different sizes.

4–9
MOMENTS OF INERTIA OF BUILT-UP STRUCTURAL STEEL SECTIONS

A built-up structural steel member may be composed of several steel shapes welded, riveted, or bolted together to form a single member. The centroidal moments of inertia of the cross section are required in the design of a built-up member. Using the information from the Appendix Tables and treating the section as a composite area, we can calculate the moments of inertia of a built-up section, as discussed in Section 4–8.

EXAMPLE 4–11

Determine the moment of inertia and the radius of gyration with respect to the horizontal centroidal axis for the structural steel section built up by a W18 × 50 and a C15 × 33.9 as shown.

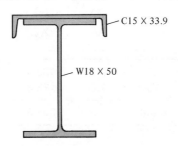

SOLUTION

The properties of W18 × 50 and C15 × 33.9 shapes from Tables A–1 and A–3 are shown in the following figure.

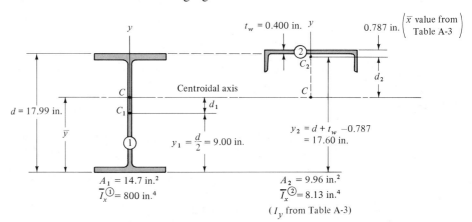

The location of the centroid of each area is indicated in the figure. The distance from the centroid C of the entire section to the bottom of the section is

$$A = A_1 + A_2 = 14.7 + 9.96 = 24.66 \text{ in.}^2$$

$$\bar{y} = \frac{A_1 y_1 + A_2 y_2}{A} = \frac{(14.7)(9.00) + (9.96)(17.60)}{24.66} = 12.47 \text{ in.}$$

Using the parallel-axis theorem, we get the moment of inertia of each area with respect to the centroidal x-axis of the entire section through C:

$$I_x^① = \bar{I}_x^① + A_1 d_1^2 = 800 + (14.7)(12.47 - 9.00)^2 = 977 \text{ in.}^4$$

$$I_x^② = \bar{I}_x^② + A_2 d_2^2 = 8.13 + (9.96)(17.60 - 12.47)^2 = 270 \text{ in.}^4$$

The moment of inertia of the entire section about its centroidal axis is then:

$$I_x = I_x^① + I_x^② = 977 + 270 = 1247 \text{ in.}^4$$

The radius of gyration r_x is, by definition,

$$r_x = \sqrt{\frac{I_x}{A}} = \sqrt{\frac{1247 \text{ in.}^4}{24.66 \text{ in.}^2}} = 7.11 \text{ in.}$$

The tabular form of the preceding computations is shown as follows:

(1)	(2)	(3)	(4)	(5)	(6)	(7)	(8)
		A	y	Ay	$d = y - \bar{y}$	Ad^2	$\bar{I}_x$
Part	Shape	(in.²)	(in.)	(in.³)	(in.)	(in.⁴)	(in.⁴)
①	W18 × 50	14.7	9.00	132.3	−3.47	177	800
②	C15 × 33.9	9.96	17.60	175.3	5.13	262	8.13
Σ		24.66		307.6		439	808

Columns (1) to (5) of the table are set up first. From which

$$\bar{y} = \frac{\Sigma A y}{\Sigma A} = \frac{307.6 \text{ in.}^3}{24.66 \text{ in.}^2} = 12.47 \text{ in.}$$

Then the last three columns are set up. We get

$$I_x = \Sigma A d^2 + \Sigma \bar{I}_x = 439 \text{ in.}^4 + 808 \text{ in.}^4 = 1247 \text{ in.}^4$$

PROBLEMS

*For the built-up structural steel section shown in Problems **4–28** to **4–33**, determine the moment of inertia and the radius of gyration of the cross-sectional area about the horizontal centroidal axis.*

4–28

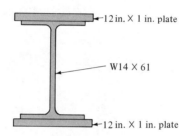

12 in. X 1 in. plate

W14 X 61

12 in. X 1 in. plate

FIGURE P4–28

4–29

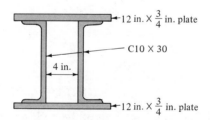

12 in. X $\frac{3}{4}$ in. plate

C10 X 30

4 in.

12 in. X $\frac{3}{4}$ in. plate

FIGURE P4–29

4–30

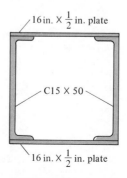

16 in. X $\frac{1}{2}$ in. plate

C15 X 50

16 in. X $\frac{1}{2}$ in. plate

FIGURE P4–30

4–31

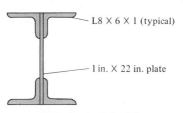

L8 × 6 × 1 (typical)

1 in. × 22 in. plate

FIGURE P4–31

4–32

C12 × 20.7

W16 × 36

FIGURE P4–32

4–33

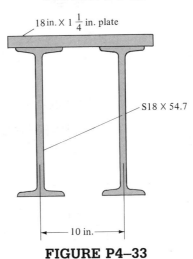

18 in. × $1\frac{1}{4}$ in. plate

S18 × 54.7

10 in.

FIGURE P4–33

4–34   Determine the radius of gyration of the built-up section shown in Fig. 4–34 with respect to the horizontal centroidal axis. The dashed line at the bottom of the section represents lacings for connecting the component parts at the open side. The areas of the lacings are not considered as effective areas in calculating the moment of inertia.

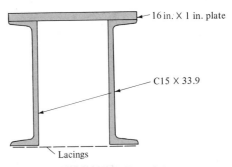

16 in. × 1 in. plate

C15 × 33.9

Lacings

FIGURE P4–34

CHAPTER 5

Stresses and Deformations in Torsional Shafts

5–1
INTRODUCTION

In this chapter the effect of a twisting moment about the longitudinal axis of a shaft is discussed. *Shafts* are members that are subjected to twisting moments, commonly called *torques*. Shafts subjected to torques are said to be under *torsional loads*.

Only solid and hollow circular shafts are considered in this chapter. Torsion in noncircular sections will not be discussed. A majority of important applications in engineering involve shafts of solid or hollow circular sections; the formulas developed in this chapter will be useful in a wide variety of problems.

Shear stress distribution in shafts is discussed first. Next, the formula for computing the torque of a rotating shaft is established, followed by computation of the angle of twist of a circular shaft and a discussion of the design of circular shafts. Finally, shaft couplings are considered.

5–2
EXTERNAL TORQUE ON SHAFTS

Consider the steering wheel and its mounting shaft shown in Fig. 5–1. The driver applies two equal and opposite forces F to the steering wheel, which transmits the torque T (equal to the moment of the couple or $T = Fd$) to the shaft at end A. The applied torque is called *external torque*. External torques can also be caused by the forces acting on the teeth of gears or the belt tensions on pulleys that are mounted on the shaft.

5–3
INTERNAL RESISTING TORQUE AND
THE TORQUE DIAGRAM

Before stress and deformation in a shaft are studied, the internal torque required to resist the external torque at sections of the shaft must be determined. The

113

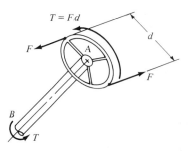

$T = Fd$

FIGURE 5-1

method of sections is used in determining the internal torque. First, the entire system is analyzed using equilibrium conditions. The value of an unknown reaction can be determined by equating the sum of moments about the axis of the shaft to zero. Next, pass an imaginary cutting plane through the section where the internal torque is to be determined. The cutting plane separates the shaft into two parts. Either part can be used as a free body to determine the internal torque at the section.

For example, Fig. 5–2(a) shows a shaft subjected to three balanced external torques. To determine the internal torque at a section between A and B, a plane m–m is passed through the section, cutting the shaft and separating it into two parts. If we consider the equilibrium of the shaft to the left of section m–m [Fig. 5–2(b)], the internal torque in section m–m is found to be 1 kip-in. If the equilibrium of the shaft to the right of the section is considered [Fig. 5–2(c)], the internal torque is also found to be 1 kip-in., but acting in an opposite direction. Since such a value represents the internal torque at the same section, the torques should have the same sign. Hence the following sign convention is adopted:

Express the internal torque as a vector according to the right-hand rule. That is, curl the fingers of the right hand in the direction of the internal torque; then the vector representation of the internal torque is indicated by the direction of the thumb. The internal torque is positive if the thumb points outward or away from the section, as shown in Fig. 5–3(a). Otherwise, the internal torque is negative, as shown in Fig. 5–3(b).

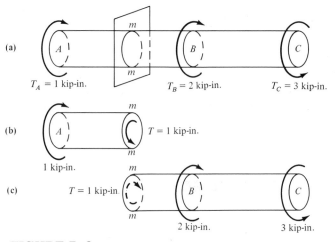

FIGURE 5-2

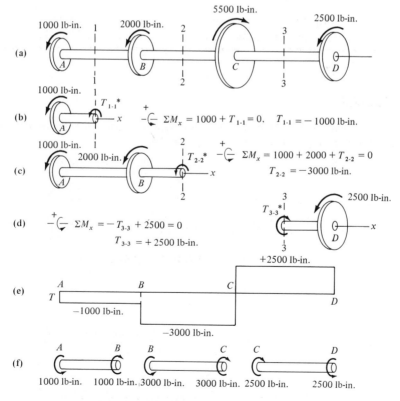

(a) Positive internal torque

(b) Negative internal torque

FIGURE 5-3

According to this sign convention, the internal torque in section m–m in both Figs. 5–2(b) and (c) is positive.

An axial force diagram has been used in Chapter 2 to show the variation of internal axial force along the length of an axially loaded member. A torque diagram can also be plotted to show the variation of the internal torque along a shaft. The following example demonstrates the computation and construction of a torque diagram.

EXAMPLE 5-1

Draw the torque diagram for the shaft acted upon by the four external torques shown in Fig. (a).

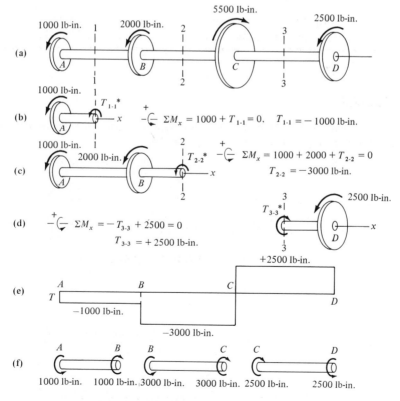

$$\Sigma M_x = 1000 + T_{1\text{-}1} = 0. \quad T_{1\text{-}1} = -1000 \text{ lb-in.}$$

$$\Sigma M_x = 1000 + 2000 + T_{2\text{-}2} = 0$$
$$T_{2\text{-}2} = -3000 \text{ lb-in.}$$

$$\Sigma M_x = -T_{3\text{-}3} + 2500 = 0$$
$$T_{3\text{-}3} = +2500 \text{ lb-in.}$$

*The internal torques, $T_{1\text{-}1}$, $T_{2\text{-}2}$, and $T_{3\text{-}3}$ are assumed to act in the positive direction.

SOLUTION

Since the algebraic sum of the given external torques about the axis of the shaft is zero, the external torques acting on the shaft are balanced.

To determine the internal torque T_{1-1} at section 1–1, consider the free-body diagram of the portion of shaft to the left of section 1–1, as shown in Fig. (b). The internal torque T_{1-1} is assumed to act in the positive direction according to the sign convention. The equilibrium equation written in Fig. (b) requires that

$$T_{1-1} = -1000 \text{ lb-in.}$$

The minus sign indicates that the assumed direction of the torque should be reversed. Since the internal torque for any section between A and B is the same, the torque diagram plotted in Fig. (e) is a horizontal line between A and B.

Similarly, from the free-body diagram and the equilibrium equation shown in Fig. (c), the internal torque in section 2–2 is

$$T_{2-2} = -3000 \text{ lb-in.}$$

and since the torque is the same for any section between B and C, the torque diagram in Fig. (e) is a horizontal line between B and C.

From Fig. (d), the torque in section 3–3 is

$$T_{3-3} = +2500 \text{ lb-in.}$$

and the torque diagram between C and D in Fig. (e) consists of a horizontal line. Figure (f) shows the torque acting on each segment of the shaft. ∎

PROBLEMS

In Problems 5–1 to 5–5, plot the torque diagram for the shaft, and show the torque acting on each segment of the shaft.

5–1

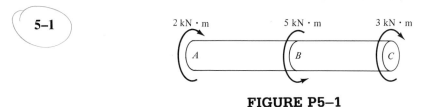

2 kN · m   5 kN · m   3 kN · m

FIGURE P5–1

5–2

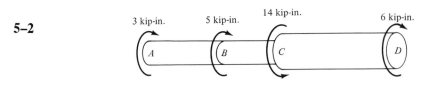

3 kip-in.   5 kip-in.   14 kip-in.   6 kip-in.

FIGURE P5–2

5–3

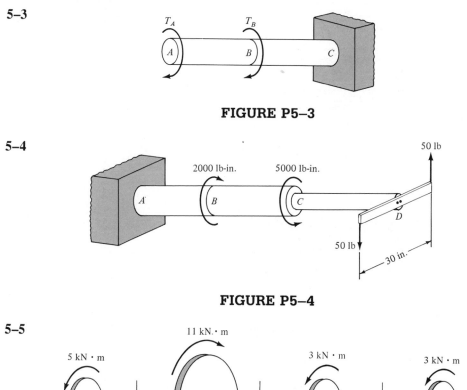

FIGURE P5–3

5–4

FIGURE P5–4

5–5

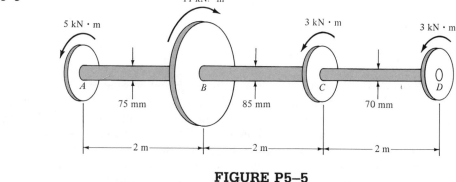

FIGURE P5–5

5–4
TORSION FORMULA

Figure 5–4(a) shows a circular member that is fixed to a support at the left-hand end and is free at the right-hand end. Figure 5–4(b) shows the deformation of the member after a torque T is applied to the free end B. We see that the longitudinal line AB on the surface of the shaft is twisted into a helix AB'. The radius OB rotated through an angle ϕ (the Greek lowercase letter phi) to a new position OB', but the radius remains a straight line.

The section at the free end rotates through the same angle ϕ while the size and shape of the section and distance to the adjacent section are unchanged. A square element bounded by the adjacent longitudinal and circumferential lines on the surface of the shaft deforms into a rhombus. This deformation is evidence that the element is subjected to shear stresses. Since the dimensions of all sides of the element are unchanged, there are no normal stresses in the element along the longitudinal and transverse directions. Thus the element is subjected only to shear stresses and is said to be in *pure shear*.

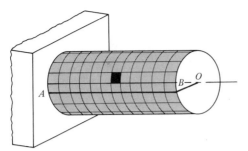

(a) Circular member before torque is applied

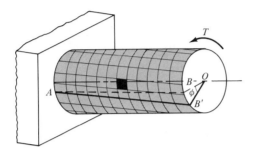

(b) Circular member after torque is applied

FIGURE 5–4

If the maximum shear stress due to the torque in the shaft is within the elastic range of the shaft material, it has been found that shear stresses vary linearly from the longitudinal centroidal axis to the outside surface of a circular shaft. The shear stresses in a cross section are in the plane of the section and act in the direction perpendicular to the radial direction. Figure 5–5 shows the variation of shear stress along a radius. The maximum shear stress occurs at points on the periphery of a section. These points, such as point C in Fig. 5–5, are located at the largest distance c from the center. We denote the maximum shear stress by τ_{max}. Then, by virtue of the linear stress variation, the shear stress τ on a differential element dA located at a distance ρ (the Greek lowercase letter rho) from the center is

$$\tau = \frac{\rho}{c}\, \tau_{max} \tag{a}$$

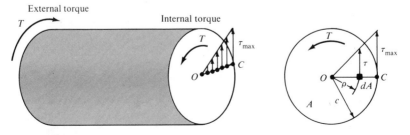

FIGURE 5–5

The shear force on the element dA is

$$dF = \tau \, dA = \frac{\rho}{c} \tau_{max} \, dA$$

The resisting torque produced by the shear force dF about the axis of the shaft is

$$dT = \rho \, dF = \rho(\tau \, dA) = \frac{\tau_{max}}{c} \rho^2 \, dA$$

The total resisting torque produced by the shear forces on all the elements over the entire section must be equal to the internal torque T in the section; thus

$$T = \int dT = \int_A \frac{\tau_{max}}{c} \rho^2 \, dA \qquad\qquad (b)$$

At any given section, c and τ_{max} are both constant, hence

$$T = \frac{\tau_{max}}{c} \int_A \rho^2 \, dA \qquad\qquad (c)$$

where $\int \rho^2 \, dA$ is, by the definition given in Chapter 4, the polar moment of inertia of the cross-sectional area, which is a constant for a given section. We denote the polar moment of inertia by J and Eq. (c) becomes

$$\tau_{max} = \frac{Tc}{J} \qquad\qquad (5\text{–}1)$$

Substitution of Eq. (5–1) in Eq. (a) gives

$$\tau = \frac{T\rho}{J} \qquad\qquad (5\text{–}2)$$

Equations (5–1) and (5–2) are two forms of the well-known torsion formula for circular shafts.

From Table 4–2, the centroidal polar moment of inertia of the cross-sectional area of a solid circular shaft in terms of the radius R, or the diameter d, is equal to

$$J = \frac{\pi R^4}{2} = \frac{\pi d^4}{32} \qquad\qquad (5\text{–}3)$$

The centroidal polar moment of inertia of the cross-sectional area of a hollow circular shaft of outside radius R_o and inside radius R_i, or outside diameter d_o and inside diameter d_i, is equal to

$$J = \frac{\pi(R_o^4 - R_i^4)}{2} = \frac{\pi(d_o^4 - d_i^4)}{32} \qquad\qquad (5\text{–}4)$$

The polar moment of inertia of a thin-walled hollow circular section of mean radius $\overline{R}$ and wall thickness t is approximately

$$J \approx \overline{R}^2(2\pi\overline{R}t) = 2\pi\overline{R}^3t \tag{5-5}$$

5-5
LONGITUDINAL SHEAR STRESS IN SHAFTS

In the preceding section it was shown that the shear stress at any point on a transverse section of the shaft can be determined by the torsion formula. It will now be shown that shear stresses also exist in longitudinal planes of a shaft.

Consider a thin-walled tubular shaft of wall thickness t subjected to torque as shown in Fig. 5–6(a). An infinitesimal element $abcd$ bounded by two transverse sections and two longitudinal axial planes is isolated from the shaft, as shown in Fig. 5–6(c). Since the shaft is in equilibrium, so is the element. The shear stress τ on side ab can be determined from the torsion formula. The equilibrium condition along the y-direction requires that the shear stress on side cd be equal to the shear stress on side ab, but acting in the opposite direction. The equilibrium condition along the x-direction requires that the shear stresses on the opposite sides ad and bc be equal and opposite; thus both shear stresses are denoted by τ'. The equilibrium equation $\Sigma M_z = 0$ of the element is

$$\circlearrowleft\Sigma M_z = \underbrace{(\tau')}_{\substack{\text{shear} \\ \text{stress}}} \underbrace{(tdx)}_{\substack{\text{area}}} (dy) - \underbrace{(\tau)}_{\substack{\text{shear} \\ \text{stress}}} \underbrace{(tdy)}_{\substack{\text{area}}} (dx) = 0$$

$$\underbrace{\text{shear}\quad\text{moment}}_{}\qquad\underbrace{\text{shear}\quad\text{moment}}_{}$$
$$\text{force}\qquad\text{arm}\qquad\qquad\text{force}\qquad\text{arm}$$
$$\underbrace{\qquad\qquad}_{\text{moment}}\qquad\qquad\underbrace{\qquad\qquad}_{\text{moment}}$$

When we divide both terms by the common factor $t\, dx\, dy$, the equation above is simplified to

$$\tau' = \tau$$

Thus we conclude that *for an infinitesimal element at a point of a stressed body, if shear stress occurs on one side of the element, there exists a shear stress*

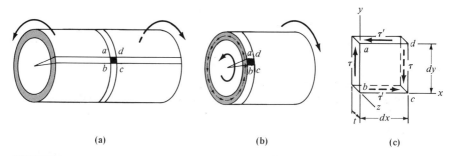

(a)   (b)   (c)

FIGURE 5–6

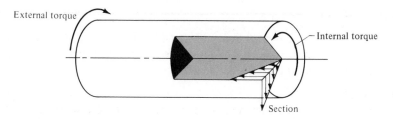

External torque

Internal torque

Section

FIGURE 5–7

of equal magnitude on the perpendicular side of the element. The shear stresses on the two perpendicular sides must point either toward the corner where the two sides meet or away from it.

The shear stress variation along a radius is shown in Fig. 5–7. Note, in particular, the directions of shear stresses. In the cross section, the direction of the shear stress coincides with the direction of the internal resisting couple.

Materials whose properties are the same in all directions are said to be *isotropic materials*. Most structural materials are isotropic. Some materials, such as wood, exhibit drastically different properties in different directions. The shear strength of wood on planes parallel to the grain is much less than that on planes perpendicular to the grain. Therefore, wood shafts fail along the direction of the grain when the shafts are overloaded in torsion.

EXAMPLE 5–2

A solid steel shaft 40 mm in diameter is subjected to the torsional loads shown. Determine the maximum shear stress in the shaft.

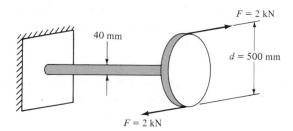

$F = 2$ kN

40 mm

$d = 500$ mm

$F = 2$ kN

SOLUTION
The torque applied to the shaft is

$$T = Fd = (2 \text{ kN})(0.5 \text{ m}) = 1 \text{ kN·m}$$

From Eq. (5–1), the maximum shear stress on the periphery of the shaft is

$$\tau_{max} = \frac{Tc}{J} = \frac{(1 \text{ kN·m})(0.040 \text{ m}/2)}{\pi(0.040 \text{ m})^4/32} = 79\,600 \text{ kN/m}^2 = 79.6 \text{ MPa}$$

──────── **EXAMPLE 5–3** ────────────────────────────────

A shaft of hollow circular section with outside diameter $d_o = 3\frac{1}{2}$ in. and inside diameter $d_i = 3\frac{1}{4}$ in. is subjected to a torque of 30 kip-in. Determine the maximum and minimum shear stresses in the shaft.

SOLUTION
From Eq. (5–4) the polar moment of inertia of the hollow section is

$$J = \frac{\pi}{32}(d_o^4 - d_i^4) = \frac{\pi}{32}(3.5^4 - 3.25^4) = 3.78 \text{ in.}^4$$

Or if we use Eq. (5–5), with mean radius $\overline{R} = \frac{1}{2}(1.75 + 1.625) = 1.69$ in., and thickness $t = \frac{1}{2}(3.5 - 3.25) = 0.125$ in., the polar moment of inertia is approximately

$$J \approx 2\pi \overline{R}^3 t = 2\pi(1.69 \text{ in.})^3(0.125 \text{ in.}) = 3.79 \text{ in.}^4$$

From Eq. (5–1), the maximum shear stress on the periphery of the shaft is

$$\tau_{max} = \frac{T(d_o/2)}{J} = \frac{(30 \text{ kip-in.})(3.5 \text{ in.}/2)}{3.78 \text{ in.}^4} = 13.9 \text{ ksi}$$

From Eq. (5–2), the minimum shear stress on the inner circumference is

$$\tau_{min} = \frac{T(d_i/2)}{J} = \frac{(30 \text{ kip-in.})(3.25 \text{ in.}/2)}{3.78 \text{ in.}^4} = 12.9 \text{ ksi}$$

■

──────── **EXAMPLE 5–4** ────────────────────────────────

If the shaft in Example 5–3 is replaced by a solid shaft subjected to the same torque $T = 30$ kip-in. and having the same strength, determine the required diameter of the solid shaft.

SOLUTION
To have the same strength, the solid shaft must have a maximum shear stress equal to 13.9 ksi when subjected to a torque of 30 kip-in., as in Example 5–3. Thus

$$\tau_{max} = \frac{Tc}{J} = \frac{(30 \text{ kip-in.})(d/2)}{\pi d^4/32} = \frac{152.8 \text{ kip-in.}}{d^3} = 13.9 \text{ kips/in.}^2$$

from which

$$d = \sqrt[3]{\frac{152.8 \text{ kip-in.}}{13.9 \text{ kips/in.}^2}} = 2.22 \text{ in.}$$

The cross-sectional area of the solid shaft is

$$A_{solid} = \frac{\pi(2.22)^2}{4} = 3.87 \text{ in.}^2$$

The cross-sectional area of the hollow shaft in Example 5–3 is

$$A_{\text{hollow}} = \frac{\pi}{4}(3.5^2 - 3.25^2) = 1.33 \text{ in.}^2$$

The weight of a shaft is equal to the product of its volume and its specific weight. For the same length and the same material, the ratio of the weights of the shafts is

$$\frac{W_{\text{solid}}}{W_{\text{hollow}}} = \frac{\gamma L\, A_{\text{solid}}}{\gamma L\, A_{\text{hollow}}} = \frac{A_{\text{solid}}}{A_{\text{hollow}}} = \frac{3.87}{1.33} = 2.91$$

The weight of the solid shaft is almost three times the weight of the hollow shaft of the same strength. The weight saved by using a hollow shaft is obvious. ■

PROBLEMS

5–6   For each shaft segment shown in Fig. P5–6(a) and (b), sketch the shear stress distribution on the section along the horizontal diameter.

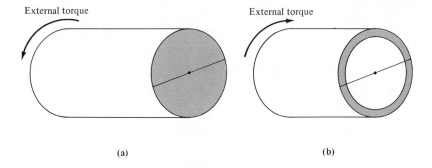

External torque   External torque

(a)   (b)

FIGURE P5–6

5–7   In the segment of shaft shown in Fig. P5–7, sketch the shear stress distribution on the cross section and the longitudinal axial planes $OACO'$ and $OBDO'$ along the radii OA and OB. The external torque T is applied as shown.

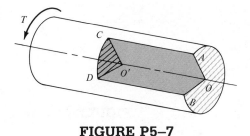

FIGURE P5–7

5–8 In Fig. P5–8, determine the shear stresses at points A and B in the cross section of the shaft due to the torque shown. Sketch the shear stress distribution along OA.

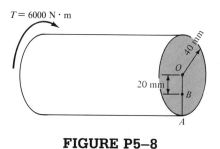

FIGURE P5–8

5–9 A hollow steel shaft has outside radius 6 in. and inside radius 4 in. If the maximum shear stress due to a torsional load is 9000 psi, find the minimum shear stress in the section.

5–10 A tubular steel shaft is subjected to the torsional loads shown in Fig. P5–10. Determine the maximum and minimum shear stresses in the shaft.

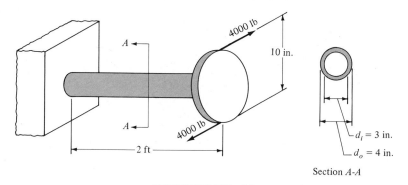

FIGURE P5–10

5–11 Determine the maximum shear stress in the steel shaft shown in Fig. P5–11.

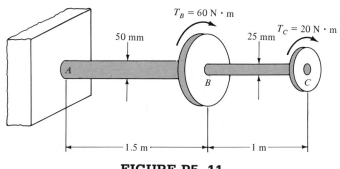

FIGURE P5–11

5–12   Determine the maximum shear stress in the shaft of Problem 5–5 (on p. 117).

5–13   A 6-in.-diameter oak shaft with the grain parallel to the longitudinal axis is used in a water mill. If the allowable shear stress is 140 psi parallel to the grain and 300 psi perpendicular to the grain, determine the maximum allowable torque to which the shaft can be subjected.

5–14   Show that the torque transmitting capacity of a solid shaft is reduced by about 25 percent if an axial hole is bored to remove half of the shaft material.

5–15   Determine the minimum diameter of a shaft subjected to a torque of 1500 N·m without exceeding an allowable shear stress of 50 MPa.

5–6
POWER TRANSMISSION BY SHAFTS

Rotating shafts are widely used for transmitting power. The drive shaft of an automobile is an example of a power transmission shaft; it transmits power from the transmission to the differential.

Power is defined as the work done per unit time. Work done by a force acting on a body is equal to the force multiplied by the displacement of the body in the direction of the force. Work done by a torque acting on a rotating shaft is equal to the torque T multiplied by the angular displacement θ of the shaft. Thus

$$\text{work done} = T\theta$$

By definition, the power P transmitted by the rotating shaft is

$$P = \frac{\text{work done}}{\text{time}} = T\frac{\theta}{t} = T\omega$$

When we solve for T, we get

$$T(\text{lb-in.}) = \frac{P(\text{lb-in.}/\text{s})}{\omega(\text{rad}/\text{s})} \tag{a}$$

which gives the torque T in lb-in. to which the shaft is subjected when transmitting power P (lb-in./s) at angular velocity ω(rad/s).

In U.S. customary units, the angular velocity of a shaft is expressed as N rpm (revolutions per minute). Then

$$\omega = \left(N\frac{\text{rev}}{\text{min}}\right)\left(\frac{2\pi\ \text{rad}}{1\ \text{rev}}\right)\left(\frac{1\ \text{min}}{60\ \text{s}}\right) = \left[\frac{\pi}{30}N(\text{rpm})\right]\text{rad}/\text{s} \tag{b}$$

The unit commonly used for power is hp (horsepower), which is equivalent to

$$1\ \text{hp} = 550\ \text{lb-ft}/\text{s} = 6600\ \text{lb-in.}/\text{s}$$

Thus

$$P = [6600P(\text{hp})] \text{ lb-in./s} \tag{c}$$

Substituting Eqs. (b) and (c) in Eq. (a) gives

$$T(\text{lb-in.}) = \frac{6600P(\text{hp})}{(\pi/30)N(\text{rpm})}$$

or

$$T(\text{lb-in.}) = \frac{63\ 000P(\text{hp})}{N(\text{rpm})} \tag{5-6}$$

Equation (5–6) gives the torque T in lb-in. to which a shaft is subjected when transmitting P horsepower rotating at N rpm.

In SI units, the speed of a shaft is also expressed in rpm.* The power is in units of kilowatts (kW) or horsepower.

Basically, in SI units Eq. (a) becomes

$$T(\text{N} \cdot \text{m}) = \frac{P(\text{N} \cdot \text{m/s})}{\omega(\text{rad/s})} \tag{d}$$

The conversion factors of kW and hp into $\text{N} \cdot \text{m/s}$ are

$$1 \text{ kW} = 1000 \text{ N} \cdot \text{m/s} \tag{e}$$

$$1 \text{ hp} = 745.7 \text{ N} \cdot \text{m/s} \tag{f}$$

Substituting the conversion factors in (b) and (e) in Eq. (d) gives

$$T(\text{N} \cdot \text{m}) = \frac{9550P(\text{kW})}{N(\text{rpm})} \tag{5-7a}$$

Substituting the conversion factors in (b) and (f) in Eq. (d) gives

$$T(\text{N} \cdot \text{m}) = \frac{7120P(\text{hp})}{N(\text{rpm})} \tag{5-7b}$$

──────── **EXAMPLE 5–5** ──

A 2-in.-diameter solid steel line shaft is used for power transmission purposes in a manufacturing plant. A motor inputs 100 hp to a pulley at A, which is transmitted by the shaft to pulleys at B, C, and D. The output horsepowers from pulleys located at B, C, and D are 45 hp, 25 hp, and 30 hp, respectively. (a) Plot the torque diagram of the shaft. (b) Determine the maximum shear stress in the shaft.

─────────────

* In SI units, the speed of shaft may also be expressed in hertz (Hz), which is cycles per second.

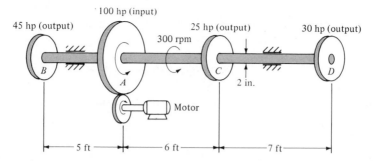

SOLUTION

(a) The torque exerted on each gear is first determined by Eq. (5–6). Thus

$$T_A = \frac{63\ 000(100)}{300} = 21\ 000 \text{ lb-in.}$$

$$T_B = \frac{63\ 000(45)}{300} = 9450 \text{ lb-in.}$$

$$T_C = \frac{63\ 000(25)}{300} = 5250 \text{ lb-in.}$$

$$T_D = \frac{63\ 000(30)}{300} = 6300 \text{ lb-in.}$$

These torques act on the pulley as shown in Fig. (a). Figures (b), (c), and (d) show the determination of the internal torque along the shaft by the method of sections. Figure (e) shows the torque diagram of the shaft.

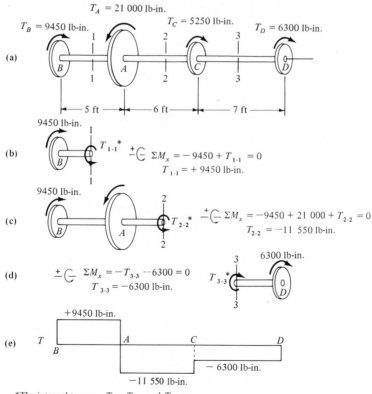

*The internal torques, $T_{1\text{-}1}$, $T_{2\text{-}2}$, and $T_{3\text{-}3}$ are assumed to act in the positive direction.

(b) Since the shaft is of uniform cross section, the maximum shear stress occurs on the periphery of segment AC, where the absolute value of the internal torque is a maximum. Thus

$$\tau_{max} = \frac{Tc}{J} = \frac{(11\ 550\ \text{lb-in.})(1\ \text{in.})}{\frac{1}{2}\pi(1\ \text{in.})^4} = 7350\ \text{psi}$$

∎

PROBLEMS

5–16 Find the horsepower that a solid steel shaft 1-in. in diameter can safely transmit without exceeding a maximum allowable shear stress of 8000 psi while rotating at 1000 rpm.

5–17 A 4-in.-diameter solid steel shaft is transmitting 200 hp at 100 rpm. Determine the maximum shear stress in the shaft and the reduction in the maximum shear stress that would occur if the speed of the shaft were increased to 300 rpm.

5–18 Determine the maximum horsepower that a hollow shaft with outside diameter 50 mm and inside diameter 35 mm can transmit at 250 rpm without exceeding an allowable shear stress of 50 MPa.

5–19 The hydraulic turbine shown in Fig. P5–19 generates 30 000 kW of electric power when rotating at 250 rpm. Determine the maximum shear stress in the tubular generator shaft with outside and inside diameters indicated as shown.

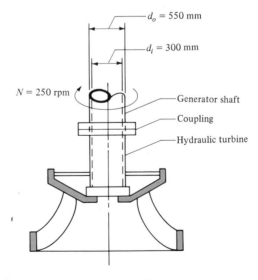

FIGURE P5–19

5–20 A motor inputs 150 hp to gear A and drives a line shaft, as shown in Fig. P5–20. The solid steel shaft has a uniform cross section of 50 mm diameter. The shaft rotates at 500 rpm and delivers 80 hp to gear B and 70 hp to gear C. Determine the maximum shear stress in the shaft.

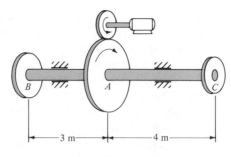

FIGURE P5–20

5–21 The steel line shaft shown in Fig. P5–21 transmits the input power of 35 kW at pulley C to pulleys A and B. Pulley A output 15 kW and pulley B output 20 kW. Determine the maximum shear stress in the shafts.

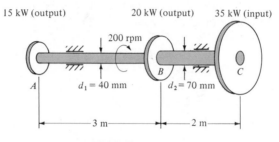

FIGURE P5–21

5–22 The solid steel shafts with the diameters indicated in Fig. P5–22 are driven by a 50-hp motor. The output horsepower at A, C, and D is 10 hp, 20 hp, and 20 hp, respectively. The shaft rotates at a constant speed of 200 rpm. **(a)** Plot the torque diagram along the shaft. **(b)** Determine the maximum shear stress in the shafts.

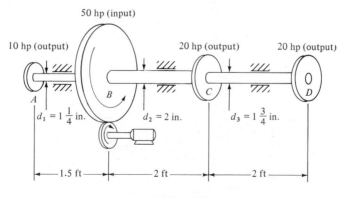

FIGURE P5–22

5-7
ANGLE OF TWIST OF CIRCULAR SHAFTS

When a shaft is subjected to torque, two sections in the shaft rotate through an angular displacement relative to each other. This relative angular displacement between two sections in the shaft is called the *angle of twist*. Figure 5–8 shows a shaft of length L that is fixed at the left-hand end and is subjected to a torque T at the free end. The longitudinal line AB on the surface of the shaft is twisted by the torque T into a helix AB'. The radius OB at the free end rotates through an angle ϕ (the Greek lowercase letter phi) to OB'. The angle ϕ is called the angle of twist of the shaft over the length L.

Attention is now directed to a differential length dL isolated from the shaft, as shown in Fig. 5–9. The longitudinal line PQ assumes a new position PQ' after the torque is applied. At the same time, the radius OQ rotates through a differential angle $d\phi$ to a new position OQ'. The angle $\angle QPQ'$ in radians represents the angular distortion between two lines that were perpendicular before twisting. This angle is identified as the shear strain according to the definition in Section 3–10. This shear strain occurs on the periphery of the shaft, where the shear stress is a maximum, and accordingly, it is a maximum shear strain, and it is denoted by γ_{max}. Both γ_{max} and $d\phi$ are measured in radians. If γ_{max} is small, which is the case in the elastic range, we have

$$\text{arc } QQ' = \gamma_{max} \, dL = c \, d\phi$$

from which

$$d\phi = \frac{\gamma_{max} \, dL}{c} \qquad \text{(a)}$$

Within the elastic range, Hooke's law applies. Then, according to Eq. (3–8),

$$\gamma_{max} = \frac{\tau_{max}}{G} = \frac{Tc}{JG} \qquad \text{(b)}$$

Substitute Eq. (b) in Eq. (a):

$$d\phi = \frac{T}{JG} \, dL$$

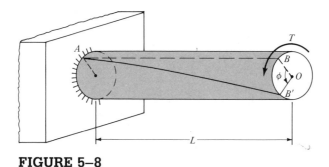

FIGURE 5–8

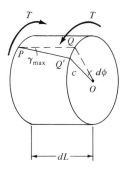

FIGURE 5–9

which gives the angle of twist of the shaft over a small length dL. The total angle of twist of the shaft of uniform cross section subjected to a constant torque T over a length L is

$$\phi = \int d\phi = \int \frac{T}{JG}\, dL = \frac{T}{JG}\int dL$$

But $\int dL = L$; thus

$$\phi = \frac{TL}{JG}\qquad (5\text{–}8)$$

which gives the angle of twist ϕ (in radians) over a length L of a shaft subjected to torque T, where J is the polar moment of inertia of the shaft section and G is the shear modulus of the shaft material. Equation (5–8) is valid for both solid and hollow circular shafts.

In Eq. (5–8), consistent units must be used for quantities on the right-hand side of the equation so that the expression is dimensionless in order to yield the angle of twist ϕ in radians.

The direction of the angle of twist ϕ coincides with the direction of the internal torque T, so the following convenient sign convention for the angle of twist is used:

$$\begin{cases} +\phi & \text{produced by positive internal torque } +T \\ -\phi & \text{produced by negative internal torque } -T \end{cases}$$

where the sign of the internal torque T is determined by the sign convention discussed in Section 5–3 (i.e., the right-hand rule).

──────── **EXAMPLE 5–6** ────────────────────────────────

A steel shaft 2 m long and 50 mm in diameter is subjected to a torque of 2000 N · m. Determine the maximum shear stress in the shaft and the angle of twist of the shaft. The shear modulus of steel is $G = 84$ GPa.

SOLUTION

The polar moment of inertia of the cross-sectional area is

$$J = \frac{\pi}{32}\,(0.050 \text{ m})^4 = 6.14 \times 10^{-7}\text{ m}^4$$

The maximum shear stress is

$$\tau_{max} = \frac{Tc}{J} = \frac{(2000 \text{ N} \cdot \text{m})(0.025 \text{ m})}{6.14 \times 10^{-7} \text{ m}^4} = 81.5 \times 10^6 \text{ N/m}^2 = 81.5 \text{ MPa}$$

which is within the elastic range of steel.
The given shear modulus is

$$G = 84 \text{ GPa} = 84 \times 10^9 \text{ N/m}^2$$

From Eq. (5–8) the angle of twist is

$$\phi = \frac{TL}{JG} = \frac{(2000 \text{ N} \cdot \text{m})(2 \text{ m})}{(6.14 \times 10^{-7} \text{ m}^4)(84 \times 10^9 \text{ N/m}^2)} = 0.0776 \text{ rad} = 4.45°$$

∎

━━━━━ **EXAMPLE 5–7** ━━━━━

Determine the relative angle of twist between D and B of the shaft in Example
5–5 (on p. 126). The shear modulus of steel is $G = 12 \times 10^6$ psi.

SOLUTION
From the torque diagram in Example 5–5, the internal torques in the three
segments are $T_{BA} = +9450$ lb-in., $T_{AC} = -11\,550$ lb-in., and $T_{CD} = -6300$
lb-in.
Since the shaft has a uniform cross section and is made of the same material,
the constant value of JG is

$$JG = \frac{\pi}{2}(1 \text{ in.})^4(12 \times 10^6 \text{ lb/in.}^2) = 1.885 \times 10^7 \text{ lb-in.}^2$$

From Eq. (5–7) the angle of twist of each segment is

$$\phi_{BA} = \frac{T_{BA}L_{BA}}{JG} = \frac{(+9450 \text{ lb-in.})(5 \times 12 \text{ in.})}{1.885 \times 10^7 \text{ lb-in.}^2} = +0.0301 \text{ rad}$$

$$\phi_{AC} = \frac{T_{AC}L_{AC}}{JG} = \frac{(-11\,550 \text{ lb-in.})(6 \times 12 \text{ in.})}{1.885 \times 10^7 \text{ lb-in.}^2} = -0.0441 \text{ rad}$$

$$\phi_{CD} = \frac{T_{CD}L_{CD}}{JG} = \frac{(-6300 \text{ lb-in.})(7 \times 12 \text{ in.})}{1.885 \times 10^7 \text{ lb-in.}^2} = -0.0281 \text{ rad}$$

The relative angle of twist of the shaft between D and B is the algebraic sum
of the angle of twist of each of the three segments between D and B. Thus

$$\phi_{D/B} = \phi_{BA} + \phi_{AC} + \phi_{CD} = +0.0301 - 0.0441 - 0.0281$$

$$= -0.0421 \text{ rad} = -2.41°$$

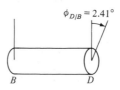

∎

PROBLEMS

In the following problems, unless otherwise specified, use the following values for the shear modulus of steel: G = 12 × 10⁶ psi in U.S. customary units, and G = 84 GPa in SI units.

5–23 Calculate the total angle of twist over the 2-ft length of the tubular shaft in Fig. P5–10 (on p. 124).

5–24 Determine the angle of twist in degrees per foot of a 2-in.-diameter steel shaft subjected to a torque of 1200 lb-ft.

5–25 Find the angular displacement in degrees of pulley C in Fig. P5–11 (on p. 124).

5–26 Find the relative angle of twist between A and D in degrees of the shaft in Fig. P5–5 (on p. 117).

5–27 Find the angle of twist in degrees of section B relative to section C of the shaft in Problem 5–20 (on p. 128).

5–28 Determine the relative angle of twist of end D relative to end A in Fig. P5–22 (on p. 129).

5–8
DESIGN OF CIRCULAR SHAFTS

A circular shaft is usually designed to satisfy the requirements for strength and stiffness. When designing a shaft for strength, the shaft material and its allowable shear stress must be selected first. Then the required size of the shaft is selected so that the maximum shear stress in the shaft is within the allowable shear stress. That is,

$$\tau_{max} = \frac{T_{max} r}{J} \leq \tau_{allow} \tag{5–9}$$

The maximum shear stress occur at the periphery of the section where the absolute value of the internal resisting torque, T_{max}, is a maximum along the shaft. If the shaft is subjected to a combined loading of torsion and bending, stresses due to the combined loading must be considered. In this chapter we are concerned with torsional loads only.

The allowable shear stress of a material is lower than the allowable normal stress. For steel, the allowable shear stress is about 50 to 60 percent of the allowable normal stress. The ASME (American Society of Mechanical Engineers) code specifies that the allowable sheer stress be 0.3 of yield shear stress or 0.18 of ultimate shear strength, whichever is smaller. For unspecified steel, the value of the allowable shear stress recommended is 8000 psi. If a shaft is subjected to shock loads or repeated stress reversal, then the allowable shear stress should be considerably lower, especially when a large stress concentration occurs on the keyway.

When designing a shaft for stiffness, the angle of twist of the shaft is limited to an allowable value. That is,

$$\phi = \frac{T_{max}L}{JG} \leq \phi_{allow} \tag{5-10}$$

where ϕ and ϕ_{allow} are both in radians. The allowable angle of twist per unit length depends on the precision requirement of the machine. For an ordinary power transmission shaft,

$$\frac{\phi_{allow}}{L} = (0.5 - 1.0) \text{ deg/m} \quad \text{or} \quad (0.15 - 0.3) \text{ deg/ft}$$

—————— **EXAMPLE 5–8** ——————————————————————————

Select the size of a solid steel shaft to transmit 150 hp at 300 rpm without exceeding an allowable shear stress of 8000 psi, or having a relative angle of twist beyond an allowable value of 0.3°/ft. The shear modulus of steel is $G = 12 \times 10^6$ psi.

SOLUTION

The torque that the shaft is subjected to when transmitting 150 hp at 300 rpm is

$$T = \frac{63\ 000\ P}{N} = \frac{63\ 000(150)}{300} = 31\ 500 \text{ lb-in.}$$

From Eq. (5–9),

$$\tau_{max} = \frac{Tc}{J} = \frac{(31\ 500 \text{ lb-in.})(d/2)}{\pi d^4/32} = \frac{160\ 400 \text{ lb-in.}}{d^3} \leq \tau_{allow} = 8000 \text{ psi}$$

from which

$$d \geq \sqrt[3]{\frac{160\ 400 \text{ lb-in.}}{8000 \text{ lb/in.}^2}} = 2.71 \text{ in.}$$

For a 1-ft length of shaft,

$$\phi_{allow} = (0.3°/\text{ft})(1 \text{ ft}) = 0.3° = (0.3°)\left(\frac{\pi \text{ rad}}{180°}\right) = 0.005\ 24 \text{ rad}$$

From Eq. (5–10),

$$\phi = \frac{TL}{JG} = \frac{(31\ 500 \text{ lb-in.})(12 \text{ in.})}{(\pi d^4/32)(12 \times 10^6 \text{ lb/in.}^2)}$$

$$= \frac{0.3209 \text{ in.}^4}{d^4} \leq \phi_{allow} = 0.005\ 24 \text{ rad}$$

from which

$$d \geqslant \sqrt[4]{\frac{0.3209 \text{ in.}^4}{0.005\ 24}} = 2.80 \text{ in.}$$

$$d_{req} = 2.80 \text{ in.}$$

which is governed by the stiffness requirement of the shaft.

■

EXAMPLE 5-9

If the shaft in Example 5–8 is rotating 16 times faster at 4800 rpm while transmitting the same horsepower and all the other conditions remain unchanged, select the size of the solid steel shaft.

SOLUTION

At higher speed, the torque is reduced to

$$T = \frac{63\ 000(150)}{4800} = 1969 \text{ lb-in.}$$

Design for strength:

$$\tau_{max} = \frac{Tc}{J} = \frac{(1969 \text{ lb-in.})(d/2)}{\pi d^4/32} = \frac{10\ 030 \text{ lb-in.}}{d^3} \leqslant \tau_{allow} = 8000 \text{ psi}$$

$$d \geqslant \sqrt[4]{\frac{10\ 030 \text{ lb-in.}}{8000 \text{ lb/in.}^2}} = 1.08 \text{ in.}$$

Design for stiffness:

$$\phi = \frac{TL}{JG} = \frac{(1969 \text{ lb-in.})(12 \text{ in.})}{(\pi d^4/32)(12 \times 10^6 \text{ lb/in.}^2)}$$

$$= \frac{0.0201 \text{ in.}^4}{d^4} \leqslant \phi_{allow} = 0.005\ 24 \text{ rad}$$

$$d \geqslant \sqrt[4]{\frac{0.0201 \text{ in.}^4}{0.005\ 24}} = 1.40 \text{ in.}$$

$$d_{req} = 1.40 \text{ in.}$$

which is one-half of the required size in the preceding example. The weights of solid shafts are proportional to their diameters squared; hence the shaft weight in this example is only one-fourth of the shaft weight in Example 5–8. There is a 75 percent saving of material. This is the reason for the tendency toward increased use of high-speed shafts in modern machinery.

■

PROBLEMS

5-29 Determine the minimum diameter of a solid steel shaft that will not twist through more than 1° per meter of length when subjected to a torque of 4 kN · m. For steel, $G = 84$ GPa.

5–30 Design a solid steel shaft to transmit 100 hp at 200 rpm. Given: τ_{allow} = 8000 psi, ϕ_{allow}/L = 0.24°/ft, and G = 12 × 10^6 psi.

5–31 Select the size of a solid steel shaft to transmit 250 hp at 450 rpm. The shear stress in the shaft must not exceed 70 MPa and the shaft must not twist more than 1° per meter of length. For steel, G = 84 GPa.

5–32 Design a hollow steel shaft to transmit 100 hp at 200 rpm. Let the ratio of inside diameter to outside diameter be d_i/d_o = 0.8. Given: τ_{allow} = 8000 psi, ϕ_{allow}/L = 0.24°/ft, and G = 12 × 10^6 psi. Compare the weight of the hollow shaft designed in this problem with the weight of the solid shaft designed in Problem 5–30.

*5–9
SHAFT COUPLINGS

It is often desirable to build a long shaft from several short shafts because a single piece of required length is not available, or because of difficulties in shipping and handling. There are many methods by which the shaft can be joined. The *flange shaft coupling* shown in Fig. 5–10 provides a practical method of shaft connection. The two pieces of shaft are connected to the coupling through keys. The two pieces of coupling, in turn, are connected by bolts. The shear stress in the key and the bearing stress between the key and the shaft have been discussed in Sections 2–3 and 2–4.

In determining the shear stress in the bolts, the following assumptions are made:

1. Friction between the two pieces of coupling is neglected and the torque is transmitted entirely by the bolts, which are subjected to shear stresses.

2. The shear stresses in the bolts are uniform, and the shear forces on each bolt on the same *bolt circle* are equal. The shear force on each bolt is denoted by F in Fig. 5–10. The resisting torque of F about the axis of the shaft is Fr. For n

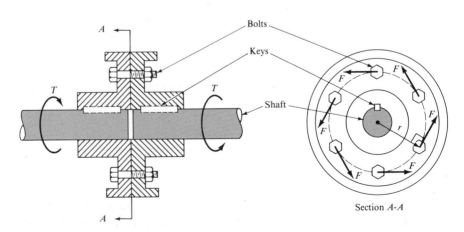

FIGURE 5–10 Flange Shaft Coupling

bolts on the same bolt circle, the total resisting torque is nFr, which must be equal to the external torque T. Thus

$$T = nFr$$

or (5–11)

$$F = \frac{T}{nr}$$

Then the uniform shear stress in the bolts is

$$\tau = \frac{F}{A} = \frac{T}{nrA} \qquad (5–12)$$

which gives the shear stress in bolts having the same cross-sectional area A and located on a single bolt circle.

3. If identical bolts are located on two or more bolt circles, their shear forces are directly proportional to the radius of the bolt circles, as shown in Fig. 5–11, we have

$$\frac{F_2}{F_1} = \frac{r_2}{r_1} \qquad (5–13)$$

or

$$F_2 = \frac{r_2}{r_1} F_1$$

The resisting torque of F_1 is $F_1 r_1$ and of F_2 is $F_2 r_2$. For n_1 bolts on bolt circle 1 and n_2 bolts on bolt circle 2, the total resisting torque about the axis of the shaft is $n_1 F_1 r_1 + n_2 F_2 r_2$, which must be equal to the external torque T. Thus

$$T = n_1 F_1 r_1 + n_2 F_2 r_2 = n_1 F_1 r_1 + n_2 \left(\frac{r_2}{r_1} F_1 \right) r_2$$

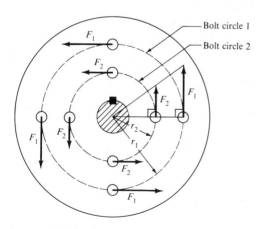

FIGURE 5–11

or

$$T = \frac{F_1}{r_1} (n_1 r_1^2 + n_2 r_2^2) \tag{5-14}$$

from which

$$F_1 = \frac{T r_1}{n_1 r_1^2 + n_2 r_2^2}$$

Therefore,

$$\tau_1 = \frac{F_1}{A} = \frac{T r_1}{A(n_1 r_1^2 + n_2 r_2^2)} \tag{5-15}$$

and τ_2 can be determined by proportion. Thus

$$\tau_2 = \frac{r_2}{r_1} \tau_1 \tag{5-16}$$

EXAMPLE 5–10

In the flange shaft coupling shown in Fig. 5–11, the bolts are $\frac{1}{4}$ in. in diameter and are located at a bolt circle of radius $r = 5$ in. The shaft is transmitting 80 hp at 400 rpm. If the allowable shear stress in the bolts is 10 000 psi, determine the number of bolts required in the coupling.

SOLUTION

The torque that the shaft is subjected to is

$$T = \frac{63\ 000\ P}{N} = \frac{63\ 000(80)}{4000} = 12\ 600\ \text{lb-in.}$$

The allowable force of each bolt is

$$F_{\text{allow}} = \tau_{\text{allow}} A = (10\ 000\ \text{lb/in.}^2)[\tfrac{1}{4}\pi(0.25\ \text{in.})^2] = 491\ \text{lb}$$

From Eq. (5–12) the allowable resisting torque is

$$T_{\text{allow}} = nFr = n(491\ \text{lb})(5\ \text{in.}) = 2455\ n\ (\text{lb-in.}) \geq T = 12\ 600\ \text{lb-in.}$$

From which

$$n \geq \frac{12\ 600\ \text{lb-in.}}{2455\ \text{lb-in.}} = 5.1$$

Therefore, six $\frac{1}{4}$-in.-diameter bolts are required in the shaft coupling.

─── **EXAMPLE 5–11** ───

Twelve 20-mm-diameter high-strength steel bolts are arranged in two concentric circles of radius r_1 = 200 mm and r_2 = 120 mm in a flange coupling shown. Determine the maximum horsepower that can be transmitted by the coupling if the shaft is rotating at 350 rpm and the allowable shear stress in the bolts is 75 MPa.

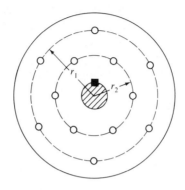

SOLUTION

The cross-sectional area of the bolts is

$$A = \tfrac{1}{4}\pi(0.020 \text{ m})^2 = 0.000\ 414 \text{ m}^2$$

The allowable shear force in bolts located on the outer bolt circle is

$$F_1 = \tau_{\text{allow}}A = (75 \times 10^6 \text{ N/m}^2)(0.000\ 314 \text{ m}^2) = 23\ 550 \text{ N}$$

From Eq. (5–14), the allowable torque is

$$T_{\text{allow}} = \frac{F_1}{r_1}(n_1 r_1^2 + n_2 r_2^2)$$

$$= \frac{23\ 550 \text{ N}}{0.2 \text{ m}}[6(0.2 \text{ m})^2 + 6(0.12 \text{ m})^2] = 38\ 430 \text{ N} \cdot \text{m}$$

From Eq. (5–7b), the maximum allowable horsepower is

$$P_{\text{allow}} = \frac{TN}{7120} = \frac{(18\ 430)(350)}{7120} = 1890 \text{ hp}$$

─────────────────────────────────── ■

PROBLEMS

5–33 A flange coupling with six $\tfrac{3}{4}$-in.-diameter bolts in an 8-in.-diameter bolt circle is subjected to a torque of 100 kip-in. Compute the shear stress in the bolts.

5–34 Determine the number of 10-mm-diameter bolts required to connect two shafts rotating at 300 rpm and transmitting 100 hp. The bolts are arranged on a bolt circle of 150 mm diameter and have an allowable shear stress of 70 MPa.

5–35 A flange coupling has twelve 20-mm-diameter high-strength bolts, with six bolts on a bolt circle of 250 mm diameter and another six bolts on a bolt circle of 150 mm diameter. If the allowable shear stress in the bolts is 90 MPa, determine the horsepower that can be transmitted by the coupling at a 300-rpm shaft speed.

5–36 A flange coupling connects two pieces of 4-in.-diameter shaft. The $\frac{3}{4}$-in. bolts are arranged in a single bolt circle 8 in. in diameter. If the allowable shear stress in the shaft is 8000 psi and the allowable shear stress in the bolts is 10 000 psi, determine the minimum number of bolts required so that the strength of the coupling is at least equal to that of the shaft.

5–37 A flange coupling with bolts arranged in three bolt circles as shown in Fig. P5–37 transmits 500 hp at 600 rpm. All bolts have 10-mm diameters. Determine the maximum shear stress in the bolts.

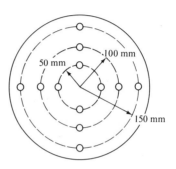

FIGURE P5–37

CHAPTER 6

Shear Force and Bending Moment in Beams

6–1
INTRODUCTION

In this chapter the effects of forces applied in the transverse direction to a member are investigated. A member subjected to loads in transverse direction is called a *beam*. The main objective of this chapter is the determination of internal forces at sections along a beam.

First, the types of beam supports and beam loadings and the calculation of external reactions are reviewed. Next, the internal forces in the beam—shear forces and bending moments—are determined. Finally, the shear force and bending moment diagrams are plotted.

The beams considered in this book are those that satisfy the following limitations:

1. The beam is straight and of uniform cross section with a vertical axis of symmetry, as shown in Fig. 6–1.
2. The beam is horizontal, although in actual situations beams may be placed in inclined or vertical positions.
3. The forces applied to the beam lie in the vertical plane that passes through the vertical axis of symmetry, as shown in Fig. 6–1.

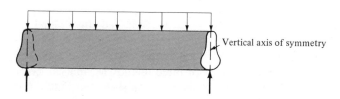
Vertical axis of symmetry

FIGURE 6–1

6-2
BEAM SUPPORTS

The three types of beam supports are roller supports, pin supports, and fixed supports.

Roller Supports

A roller (or link) support resists motion of the beam only along the direction perpendicular to the plane of the support (or along the axis of the link). Hence the reaction at a roller support acts along the known direction, as shown in Figure 6–2.

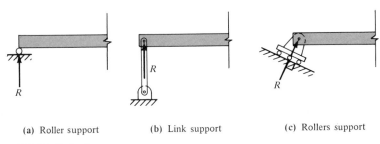

(a) Roller support   (b) Link support   (c) Rollers support

FIGURE 6–2

Pin Supports

A pin support resists motion of the beam at the support in any direction on the plane of loading. Hence the reaction at a pin support may act in any direction. The reaction can be represented by horizontal and vertical components, as shown in Figure 6–3.

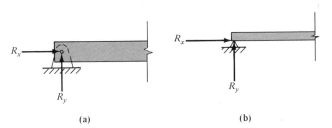

(a)   (b)

FIGURE 6–3

Fixed Supports

At a fixed support, a beam is either built-in as an integral part of a concrete column or welded to a steel column. The end of the beam at the fixed support is prevented from displacement in any direction and also from rotation. In general, the reaction at a fixed support consists of three unknowns, that is, two unknown components of force and one unknown moment, as shown in Figure 6–4.

At a roller or pin support a beam is free to rotate. Hence roller and pin supports are termed *simple supports* to differentiate them from fixed supports.

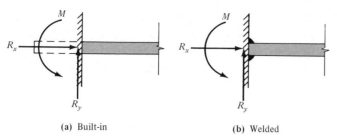

(a) Built-in   (b) Welded

FIGURE 6–4

6–3
TYPES OF LOADING

Beams are subjected to a variety of loads; among them are concentrated loads, uniform loads, linearly varying loads, and concentrated moments.

Concentrated Loads

A *concentrated load* is applied at a specific point on the beam and is considered as a discrete force acting at a point, as shown in Fig. 6–5(a). For example, a weight fastened to a beam by a cable applies a concentrated load to the beam.

Uniform Loads

When a load is distributed over a finite length along a beam, it is called a *distributed load*. If the intensity of a distributed load is a constant value, it is called a *uniform load*. The load intensity is expressed as force per unit length of the beam, such as lb/ft or N/m. For computing the reactions, the distributed load

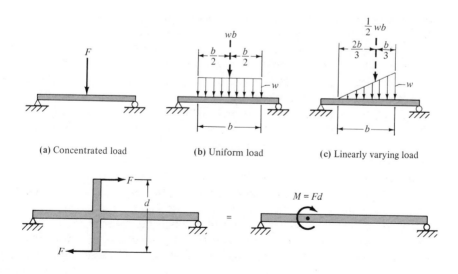

(a) Concentrated load   (b) Uniform load   (c) Linearly varying load

(d) Concentrated moment

FIGURE 6–5

may be replaced by its equivalent force. The equivalent force of a uniform load is equal to the load intensity w multiplied by the length of distribution b, and the line of action of the equivalent force passes through the midpoint of the length b, as shown in Fig. 6–5(b). The weight of the beam is an example of a uniformly distributed load.

Linearly Varying Loads

A *linearly varying load* is a distributed load with a uniform variation of intensity. Such a load condition occurs on a vertical or inclined wall due to liquid pressure. Figure 6–5(c) shows a linearly varying load with intensity varying uniformly from zero to a maximum value w.

A distributed force may be replaced by an equivalent concentrated force having a magnitude equal to the area of the load diagram and a line of action passing through the centroid of that area. For the linearly varying load in Figure 6–5(c), the equivalent concentrated force has a magnitude equal to the area of the load triangle, $\frac{1}{2}wb$, and a line of action passing through the centroid of the load triangle at a distance $b/3$ from the point with maximum load intensity w.

Concentrated Moments

A concentrated moment is a couple produced by two equal and opposite forces applied to the beam at a section. Figure 6–5(d) shows two equivalent representations of a concentrated moment.

6–4
TYPES OF BEAMS

Beams can be classified into the types shown in Fig. 6–6 according to the kind of support used. A beam supported at its ends with a pin and a roller, as shown in Fig. 6–6(a), is called a *simple beam*. A simply supported beam with over-hanging ends as shown in Fig. 6–6(b), is called an *overhanging beam*. A beam that is fixed at one end and free at the other, as shown in Fig. 6–6(c), is called a *cantilever beam*. Figure 6–6(d) shows a *propped beam* that is fixed at one end and simply supported at the other end. When both ends of a beam are fixed to supports as shown in Fig. 6–6(e), it is called a *fixed beam*. A *continuous beam* is a beam supported on a pin support and two or more roller supports, as shown in Fig. 6–6(f).

In the first three types of beams shown in Fig. 6–6(a), (b), and (c), there are three unknown reaction components; thus the reactions can be determined from the static equilibrium equations. Such beams are said to be *statically determinate*. When the number of unknown reaction components exceeds three, as in those beams shown in Fig. 6–6(d), (e), and (f), the three equilibrium equations are insufficient for determining the unknown reaction components. Such beams are said to be *statically indeterminate*. Calculation of reactions of statically determinate beams is reviewed in Section 6–5. The treatment of statically indeterminate beams is presented in Chapter 9.

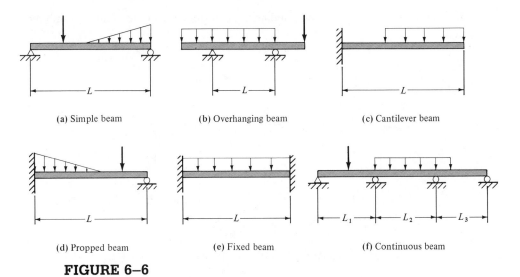

(a) Simple beam   (b) Overhanging beam   (c) Cantilever beam

(d) Propped beam   (e) Fixed beam   (f) Continuous beam

FIGURE 6–6

6–5
CALCULATION OF BEAM REACTIONS

Since the forces applied to a beam are in one plane, there are three equilibrium equations available for the determination of reactions. The equations are

$$\Sigma F_x = 0 \qquad \Sigma F_y = 0 \qquad \Sigma M_A = 0 \qquad\qquad (6\text{–}1)$$

where A is any point in the plane of loading.

Since subsequent computation of internal forces and stresses depends on beam reactions, it is important that the beam reactions be determined accurately.

——— **EXAMPLE 6–1** ———————————————————————

Determine the external reactions in the overhanging beam for the loading shown.

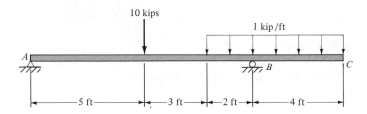

SOLUTION

The free-body diagram of the beam is shown in the following figure, with all the externally applied forces and the unknown reaction components drawn. Note that the uniform load is replaced by its resultant force in calculating reactions. After the free-body diagram is drawn, the static equilibrium equations are written and the unknown reaction components are determined from the equilibrium equations.

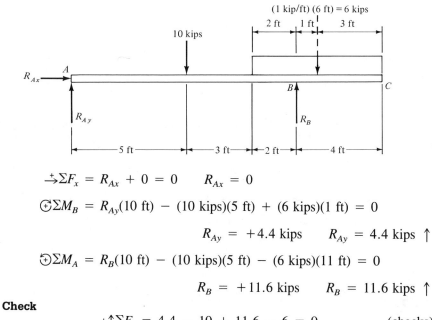

$$\overset{+}{\rightarrow}\Sigma F_x = R_{Ax} + 0 = 0 \qquad R_{Ax} = 0$$

$$\overset{+}{\circlearrowleft}\Sigma M_B = R_{Ay}(10 \text{ ft}) - (10 \text{ kips})(5 \text{ ft}) + (6 \text{ kips})(1 \text{ ft}) = 0$$

$$R_{Ay} = +4.4 \text{ kips} \qquad R_{Ay} = 4.4 \text{ kips} \uparrow$$

$$\overset{+}{\circlearrowright}\Sigma M_A = R_B(10 \text{ ft}) - (10 \text{ kips})(5 \text{ ft}) - (6 \text{ kips})(11 \text{ ft}) = 0$$

$$R_B = +11.6 \text{ kips} \qquad R_B = 11.6 \text{ kips} \uparrow$$

Check

$$+\uparrow\Sigma F_y = 4.4 - 10 + 11.6 - 6 = 0 \qquad \text{(checks)} \quad\blacksquare$$

─────── **EXAMPLE 6–2** ───────────────────────────────

Determine the external reactions of the cantilever beam for the loading shown.

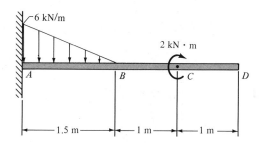

SOLUTION

The free-body diagram of the beam is first drawn and the unknown reactions are determined from the equilibrium equations.

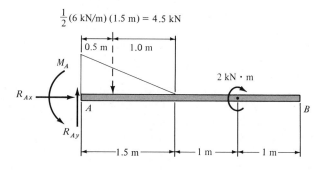

$$\xrightarrow{+}\Sigma F_x = R_{Ax} + 0 = 0 \qquad R_{Ax} = 0$$

$$^+\uparrow\Sigma F_y = R_{Ay} - 4.5 \text{ kN} = 0 \qquad R_{Ay} = +4.5 \text{ kN} \qquad R_{Ay} = 4.5 \text{ kN} \uparrow$$

$$\overset{+}{\circlearrowleft}\Sigma M_A = M_A - (4.5 \text{ kN})(0.5 \text{ m}) - 2 \text{ kN} \cdot \text{m} = 0$$

$$M_A = +4.25 \text{ kN} \cdot \text{m} \qquad M_A = 4.25 \text{ kN} \cdot \text{m}\circlearrowright$$

Note that the concentrated moment applied at C and reaction component M_A at the fixed support appear in the moment equations without multiplying by a moment arm, since these quantities are moments. ∎

PROBLEMS

In Problems 6–1 to 6–6, determine the external reactions on each beam due to the loading shown.

6–1

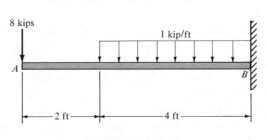

FIGURE P6–1

6–2

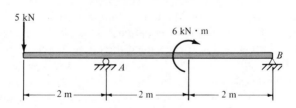

FIGURE P6–2

6–3

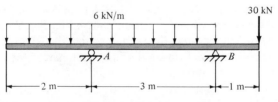

FIGURE P6–3

6–4

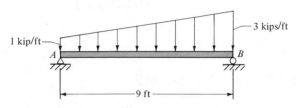

FIGURE P6–4

6–5

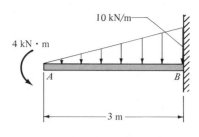

FIGURE P6–5

6–6

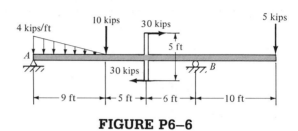

FIGURE P6–6

6–6
INTERNAL FORCES IN BEAMS

Stresses and deflections in beams are functions of internal forces. The internal forces in a beam section are those forces and moment required to resist the external forces and maintain equilibrium. Consider the beam of Fig. 6–7(a) that is subjected to the external forces and reactions shown. To find the internal forces at section 1–1, pass a plane through the section so that the beam is separated into two parts. Since the entire beam is in equilibrium, each part of the beam separated by section 1–1 is also in equilibrium.

Figure 6–7(b) shows the free-body diagram of the beam to the left of section 1–1. The internal forces of the section consist of the axial force P, the shear force V, and the bending moment M. The directions of the internal forces P, V, and M are arbitrarily assigned. Magnitudes and directions of P, V, and M can be determined from the equilibrium equations. A negative value for any internal force or moment indicates that the assumed direction should be reversed.

If the beam to the right of the section 1–1 is isolated as a free body and the equilibrium equations written [Fig. 6–7(c)], the internal forces and moment at the section are found to have the same magnitude but are opposite in direction to those determined in Fig. 6–7(b). Therefore, either side of the beam can be used to determine the internal forces. The side involving a fewer number of forces is usually chosen.

Since the internal forces in Fig. 6–7(b) and (c) for the same section have opposite directions, it is obvious that the algebraic sign conventions do not apply in this situation. Consequently, sign conventions for internal forces and moments in beam must be specified as presented in Section 6–7.

Although the internal forces include an axial force, the subsequent sections of this chapter will not include an analysis of the axial force. Members subjected to axial forces alone have been discussed in Chapter 2. The combined effect of axial force and bending moment is treated in Chapter 10.

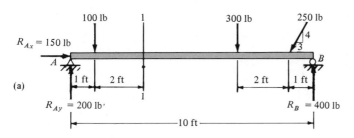

(a)

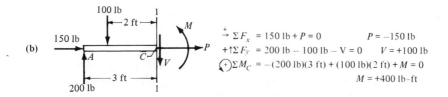

(b)

$$\xrightarrow{+} \Sigma F_x = 150 \text{ lb} + P = 0 \qquad P = -150 \text{ lb}$$

$$+\uparrow\Sigma F_y = 200 \text{ lb} - 100 \text{ lb} - V = 0 \qquad V = +100 \text{ lb}$$

$$(+)\Sigma M_C = -(200 \text{ lb})(3 \text{ ft}) + (100 \text{ lb})(2 \text{ ft}) + M = 0$$

$$M = +400 \text{ lb–ft}$$

(c)

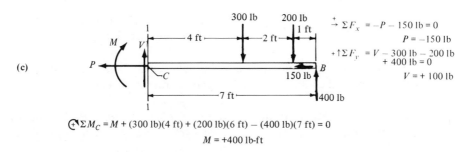

$$\xrightarrow{+} \Sigma F_x = -P - 150 \text{ lb} = 0$$

$$P = -150 \text{ lb}$$

$$+\uparrow\Sigma F_y = V - 300 \text{ lb} - 200 \text{ lb}$$
$$+ 400 \text{ lb} = 0$$

$$V = +100 \text{ lb}$$

$$(+)\Sigma M_C = M + (300 \text{ lb})(4 \text{ ft}) + (200 \text{ lb})(6 \text{ ft}) - (400 \text{ lb})(7 \text{ ft}) = 0$$

$$M = +400 \text{ lb-ft}$$

FIGURE 6–7

6–7
COMPUTATION OF SHEAR FORCE
AND BENDING MOMENT IN BEAMS

The shear force at a section tends to shear the section so that the left-hand side of the beam tends to move either

upward (⌷⎯⎯⌷) or downward (⌷⎯⎯⌷)

relative to the right-hand side. The bending moment at a section tends to bend the beam near the section into a curve either

concave upward (⌣) or concave downward (⌢).

The signs for shear force and bending moment will therefore be based on the effects produced by the shear force or by the bending moment.

A *positive shear force* at a section tends to make the left-hand side of the beam move upward relative to the right-hand side. Figure 6–8 shows the direction of positive internal shear force.

Thus, if the beam to the left of a section is analyzed, upward external forces cause positive internal resisting shear forces. On the other hand, if the beam to the right of the section is considered, downward external forces cause positive internal resisting shear force. A general rule can be stated:

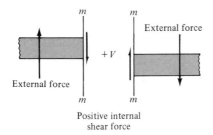

Positive internal
shear force

FIGURE 6-8

The internal shear force at any section of a beam is equal to the algebraic sum of the vertical components of external forces on the beam to the left of the section, treating upward forces as positive. That is,

$$\left[V \text{ (at any section)} = \left(\begin{array}{l} \text{sum of external forces to the} \\ \text{left of the section } (+\uparrow) \end{array} \right) \right] \qquad (6\text{-}2)$$

The internal shear force can also be obtained by considering the part of the beam to the right of the section. The general rule in this case is

The internal shear force at any section of a beam is equal to the algebraic sum of the vertical components of external forces on the beam to the right of the section, treating downward forces as positive. That is,

$$\left[V \text{ (at any section)} = \left(\begin{array}{l} \text{sum of external forces to the} \\ \text{right of the section } (+\downarrow) \end{array} \right) \right] \qquad (6\text{-}3)$$

A positive bending moment at a section tends to bend the beam concave upward near the section. Figure 6–9 shows the positive internal bending moment.

For the beam either to the left or to the right of a section, upward external forces always cause a positive internal resisting moment at a section. Thus a general rule can be stated as follows:

The internal bending moment at any section of a beam is equal to the algebraic sum of the moments of the external forces on the beam at either side of the section about the centroid of the cross section, treating moments produced by upward forces as positive. That is,

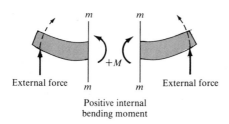

Positive internal
bending moment

FIGURE 6-9

$$\left[M \text{ (at any section)} = \left(\begin{array}{l} \text{sum of moments of external forces on} \\ \text{the beam at either side of the section} \\ \text{about the centroid of the section} \\ \text{(moment due to upward force as } +) \end{array} \right) \right] \quad (6\text{--}4)$$

Methods discussed above can be used to determine shear force and bending moment at any section along a beam, as illustrated in the following examples.

───── **EXAMPLE 6–3** ─────────────────────────────

Calculate the shear forces and bending moments at sections C and D of the beam shown.

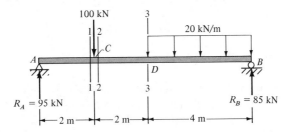

SOLUTION

If we consider the equilibrium of the entire beam, the reactions are found to be $R_A = 95$ kN and $R_B = 85$ kN.

[Method I]   Using the Method of Sections
and Equilibrium Equations

To determine the internal forces at section C, pass a section through C. Because the concentrated load is applied at C, a choice must be made as to whether the section is just to the left or just to the right of C. Let section 1–1 be just to the left and section 2–2 be just to the right of C. To find the internal forces at section 1–1 (denoted by V_{C-} and M_{C-}), consider the equilibrium of the beam to the left of section 1–1. The shear force and bending moment are both assumed to act in the positive direction according to the beam sign conventions. A minus sign in the result indicates simply that the shear or moment in the section is negative according to the beam sign conventions.

$$+\uparrow\Sigma F_y = +95 - V_{C-} = 0 \qquad V_{C-} = +95 \text{ kN}$$

$$\circlearrowleft\Sigma M_C = -95(2) + M_{C-} = 0 \qquad M_{C-} = +95(2) = +190 \text{ kN} \cdot \text{m}$$

To find the internal forces at section 2–2 (denoted by V_{C+} and M_{C+}), consider the equilibrium of the beam to the left of section 2–2. Note that now the free-body diagram would include the concentrated load acting at C.

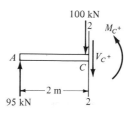

$$+\uparrow\Sigma F_y = +95 - 100 - V_{C+} = 0 \qquad V_C = +95 - 100 = -5 \text{ kN}$$

$$\circlearrowleft\Sigma M_C = -95(2) + 100(0) + M_{C+} = 0$$

$$M_{C+} = +95(2) = +190 \text{ kN} \cdot \text{m}$$

Although the internal bending moment is the same on the sections at either side of C, the internal shear force abruptly decreases from $+95$ kN on section 1–1 just to the left of C to -5 kN on section 2–2 just to the right of C. It is always true that at the section where a concentrated load is applied, the internal shear force abruptly changes. The amount of change is equal to the concentrated load at the section. An upward load (or reaction) causes an abrupt increase in shear force, and a downward load (or reaction) causes an abrupt decrease in shear force.

To determine the internal forces at D, pass section 3–3 through D. Although a uniform load is applied to the right of D, it does not make any difference whether section 3–3 is a little to the left or a little to the right of D because the amount of distributed load between the two sections is too small to cause any change in shear force. Both free-body diagrams of the beam to the left and to the right of section 3–3 are analyzed to demonstrate that identical results are obtained.

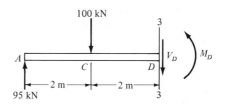

$$+\uparrow\Sigma F_y = +95 - 100 - V_D = 0$$

$$V_D = 95 - 100 = -5 \text{ kN}$$

$$\circlearrowleft\Sigma M_D = -95(4) + 100(2) + M_D = 0$$

$$M_D = 95(4) - 100(2) = +180 \text{ kN} \cdot \text{m}$$

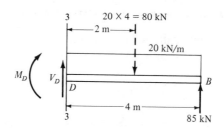

$$+\uparrow\Sigma F_y = V_D - 80 + 85 = 0$$

$$V_D = 80 - 85 = -5 \text{ kN}$$

$$\oplus\Sigma M_D = M_D + 80(2) - 85(4) = 0$$

$$M_D = -80(2) + 85(4) = +180 \text{ kN} \cdot \text{m}$$

[Method II]   Using the Rules in Eqs. (6–2) to (6–4)

Equation (6–2) is used to sum up vertical external forces to the left of the respective sections, treating upward forces as positive:

$$V_{C-} = V_{1-1} = +95 \text{ kN}$$

$$V_{C+} = V_{2-2} = +95 - 100 = -5 \text{ kN}$$

$$V_D = V_{3-3} = +95 - 100 = -5 \text{ kN}$$

We can also use Eq. (6–3) to sum up vertical external forces to the right of the respective section, treating downward forces as positive:

$$V_{C-} = V_{1-1} = -85 + 20(4) + 100 = +95 \text{ kN}$$

$$V_{C+} = V_{2-2} = -85 + 20(4) = -5 \text{ kN}$$

$$V_D = V_{3-3} = -85 + 20(4) = -5 \text{ kN}$$

Equation (6–4) is used to sum up moments of external forces to the left of respective section about the centroid of the section, treating moment produced by upward forces as positive:

$$M_C = +95(2) = +190 \text{ kN} \cdot \text{m}$$

$$M_D = +95(4) - 100(2) = +180 \text{ kN} \cdot \text{m}$$

We can also use Eq. (6–4) to sum up moments of external forces to the right of the respective section about the centroid of the section, treating moment produced by upward forces as positive:

$$M_C = +85(6) - (20 \times 4)(4) = +190 \text{ kN} \cdot \text{m}$$

$$M_D = +85(4) - (20 \times 4)(2) = +180 \text{ kN} \cdot \text{m}$$

—— **EXAMPLE 6–4** ——

Calculate the internal shear force and bending moment at sections A, B, C, D, E, and F of the simple beam shown.

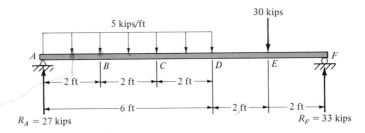

SOLUTION

If we consider the equilibrium of the entire beam, the reactions are found to be $R_A = 27$ kips and $R_F = 33$ kips.

Equation (6–2) is used to sum up vertical external forces to the left of the respective sections, treating upward forces as positive:

$$V_{A-} = 0$$

$$V_{A+} = +27 \text{ kips}$$

$$V_B = +27 - 5 \times 2 = +17 \text{ kips}$$

$$V_C = +27 - 5 \times 4 = +7 \text{ kips}$$

$$V_D = +27 - 5 \times 6 = -3 \text{ kips}$$

Equation (6–3) is used to sum up vertical external forces to the right of the respective sections, treating downward forces as positive:

$$V_{E-} = -33 + 30 = -3 \text{ kips}$$

$$V_{E+} = -33 \text{ kips}$$

$$V_{F-} = -33 \text{ kips}$$

$$V_{F+} = 0$$

Equation (6–4) is used to compute the internal bending moments at various sections, treating moment produced by upward external forces as positive:

$$M_A = 0$$

$$M_B = +27(2) - (5 \times 2)(1) = +44 \text{ kip-ft}$$

$$M_C = +27(4) - (5 \times 4)(2) = +68 \text{ kip-ft}$$

$$M_D = +27(6) - (5 \times 6)(3) = +72 \text{ kip-ft}$$

$$M_E = +33(2) = +66 \text{ kip-ft (from right)}$$

$$M_F = 0 \text{ (from right)}$$

PROBLEMS

In Problems 6–7 to 6–9, use the method of sections and the equilibrium equations to determine the shear forces and the bending moments at sections 1–1, 2–2, and 3–3.

6–7

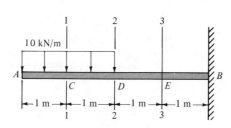

FIGURE P6–7

6–8

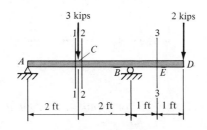

FIGURE P6–8

6–9

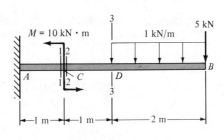

FIGURE P6–9

In Problems 6–10 to 6–12, use the rules in Eqs. (6–2) to (6–4) to determine the shear forces and bending moments at sections 1–1, 2–2, and 3–3.

6–10

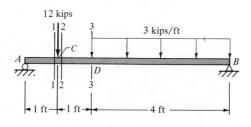

FIGURE P6–10

6–11

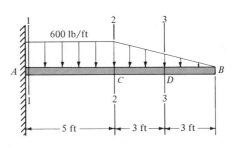

FIGURE P6–11

6–12

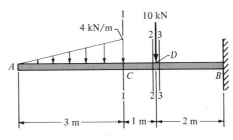

FIGURE P6–12

In Problems **6–13** *to* **6–15**, *determine the shear forces and bending moments at sections A, B, C, D, E, and F.*

6–13

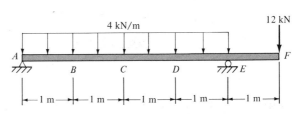

FIGURE P6–13

6–14

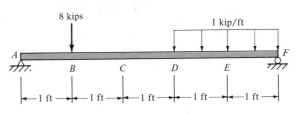

FIGURE P6–14

6–15

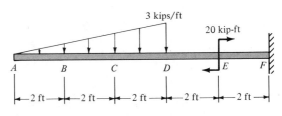

FIGURE P6–15

6–8
SHEAR FORCE AND BENDING MOMENT DIAGRAMS

Shear force and bending moment diagrams depict the variation of shear force and bending moment along a beam. To construct such diagrams, points with ordinates equal to the computed values of shear forces or bending moments of a beam are plotted from a baseline equal to the length of the beam. Beam sign conventions are used for plotting the shear forces and bending moment diagrams. A positive shear or moment is plotted above the baseline; a negative shear or moment is plotted below the baseline. When a series of points are plotted and interconnected, a shear force or bending moment diagram results. It is convenient to make the baselines of the diagrams directly below the beam, using the same horizontal scale.

Shear force and bending moment diagrams are important in beam design. With the aid of these diagrams, the magnitudes and locations of the maximum shear forces and bending moments become apparent.

―――――― **EXAMPLE 6–5** ――――――――――――――――――――――――――

Draw shear force and bending moment diagrams of the simple beam in Example 6–4.

SOLUTION
The values of shear forces and bending moments at sections A, B, C, D, E, and F computed in Example 6–4 are used as ordinates for the shear force and bending moment diagrams at the corresponding sections. When points are connected by lines, the following diagrams are obtained. Figure (a) is the loading diagram, Fig. (b) is the shear force diagram, and Fig. (c) is the bending moment

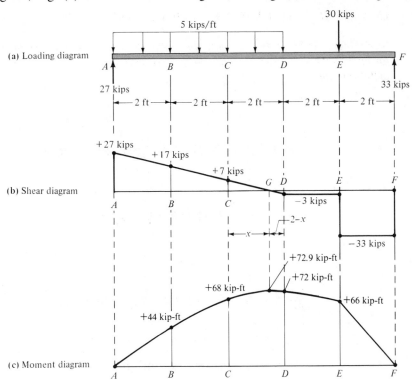

diagram. The three diagrams should be plotted in the same horizontal scale and the corresponding horizontal positions should lie on the same vertical line.

The point at which the shear force is zero is labeled as G and the distance CG is denoted by x. From the similar triangles in the shear diagram, we have

$$\frac{x}{7} = \frac{2 - x}{3}$$

from which

$$x = 1.4 \text{ ft}$$

As will be shown in Section 6–10, the moment is either a maximum or a minimum at the section where the shear stress is zero. Computed from the left, the maximum moment at G is

$$M_G = M_{max} = +27 \times 5.4 - (5 \times 5.4)\left(\frac{5.4}{2}\right) = +72.9 \text{ kip-ft}$$

■

PROBLEMS

*In Problems **6–16** to **6–18**, draw the shear force and bending moment diagrams for each beam.*

6–16   The simple beam in Problem 6–13 (on p. 156)

6–17   The overhanging beam in Problem 6–14 (on p. 156)

6–18   The cantilever beam in Problem 6–15 (on p. 156)

*In Problems **6–19** to **6–21**, draw the shear force and bending moment diagrams for each beam. Locate the section with zero shear force and determine the moment at the section.*

6–19

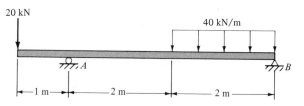

FIGURE P6–19

6–20

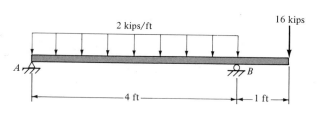

FIGURE P6–20

6–21

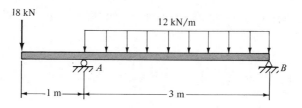

FIGURE P6–21

6–9
SHEAR AND MOMENT EQUATIONS

Shear forces and bending moments at sections along a beam can be expressed as functions of a horizontal distance, x. These functions can be plotted into shear and moment diagrams, as demonstrated in the following examples.

EXAMPLE 6–6

Rework Example 6–5 by first finding equations expressing shear and moment as functions of x, and then plotting the equations into shear and moment diagrams.

SOLUTION

The loading diagram of Example 6–5 is shown in Fig. (a). The x-axis is along the beam with the origin at A. Sections 1–1, 2–2, and 3–3 represents general sections for segments AD, DE, and EF, respectively. Each section is located by distance x from the left end A to the section.

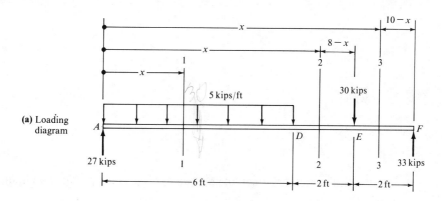

(a) Loading diagram

Equations for shear forces at the three sections can be obtained by using Eq. (6–2) or (6–3). Thus

$$V_{1-1} = V_{AD} = 27 - 5x$$

$$V_{2-2} = V_{DE} = -33 + 30 = -3 \quad \text{(from right)}$$

$$V_{3-3} = V_{EF} = -33 \quad \text{(from right)}$$

Equations for bending moment at the three sections can be obtained by using Eq. (6–3). Thus

$$M_{1-1} = M_{AD} = 27x - \underbrace{(5x)}_{\text{(resultant)}}\underbrace{\left(\frac{x}{2}\right)}_{\text{(arm)}} = 27x - 2.5x^2$$

$$M_{2-2} = M_{DE} = 33(10 - x) - 30(8 - x) = 90 - 3x \quad \text{(from right)}$$

$$M_{3-3} = M_{EF} = 33(10 - x) \quad \text{(from right)}$$

These equations can be plotted into shear force and bending moment diagrams, as shown in Figs. (b) and (c). In Fig. (b), x_G is determined by equating V_{AD} to zero, we have

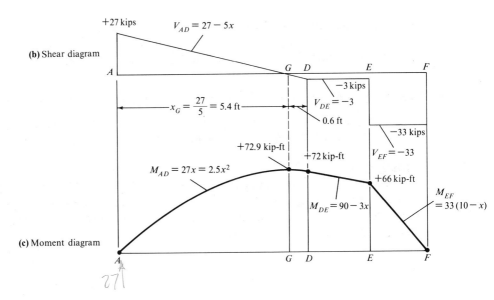

(b) Shear diagram

(c) Moment diagram

$$V_G = 27 - 5x_G = 0$$

from which

$$x_G = \frac{27}{5} = 5.4 \text{ ft}$$

In Fig. (c), the moment diagram between AD is a parabola. The maximum moment occurs at

$$\frac{dM}{dx} = 27 - 2.5(2x) = 27 - 5x = 0$$

which gives $x = 27/5 = 5.4$ ft, the same value as x_G. Thus

$$M_{\max} = M_G = 27(5.4) - 2.5(5.4)^2 = +72.9 \text{ kip-ft}$$

It is generally true that the maximum or minimum moment occurs at the section where the shear force is zero or where the shear force changes sign (see Section 6–10 for detailed relationships between the shear and moment diagrams).

───── **EXAMPLE 6–7** ─────────────────────────────────────

Find the shear force and bending moment at section 1–1 as a function of x and plot the shear and moment diagrams of the cantilever beam subjected to the triangular loading shown in Fig. (a).

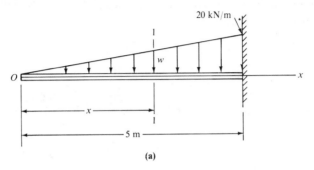

20 kN/m

O

w

x

5 m

(a)

SOLUTION

Since the distributed load varies linearly, we have

$$\frac{w}{x} = \frac{20}{5}$$

or

$$w = 4x$$

The linearly varying load to the left of section 1–1 is replaced by its resultant, as shown in Fig. (b).

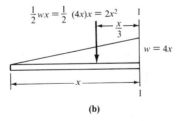

$\frac{1}{2}wx = \frac{1}{2}(4x)x = 2x^2$

$\frac{x}{3}$

$w = 4x$

x

(b)

Using Eq. (6–2), the internal shear force at section 1–1 is

$$V = -2x^2$$

Using Eq. (6–4), the internal bending moment at section 1–1 is

$$M = -(2x^2)\left(\frac{x}{3}\right) = -\frac{2}{3}x^3$$

From these equations the values of shear force and bending moment at every 1-m section are tabulated in the following table.

x (m)	0	1	2	3	4	5
$V = -2x^2$ (kN)	0	-2	-8	-18	-32	-50
$M = -\dfrac{2}{3}x^3$ (kN · m)	0	-0.67	-5.33	-18.0	-42.7	-83.3

These values are used to draw the shear force and bending moment diagrams shown in the following figures.

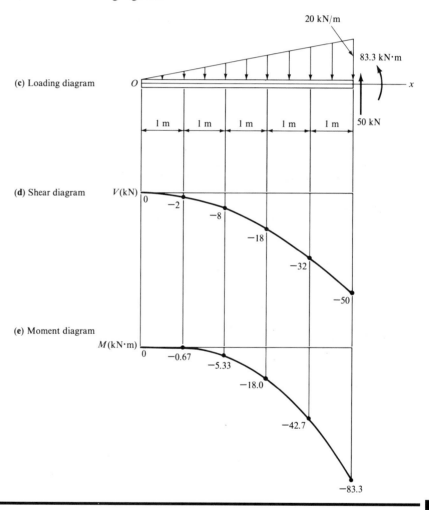

(c) Loading diagram

(d) Shear diagram

(e) Moment diagram

PROBLEMS

In Problems **6–22** *to* **6–24,** *find equations that express the shear force and bending moment at section* 1–1 *as functions of x and plot the shear force and bending moment diagrams.*

6–22

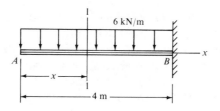

FIGURE P6–22

6–23

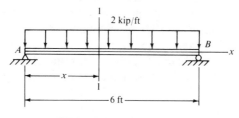

FIGURE P6–23

6–24

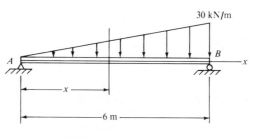

FIGURE P6–24

In Problems **6–25** *to* **6–27,** *plot the shear force and bending moment diagrams by using shear and moment equations.*

6–25   Use the beam in Problem 6–19 (on p. 158).

6–26   Use the beam in Problem 6–20 (on p. 158).

6–27   Use the beam in Problem 6–21 (on p. 159).

6–10
SHEAR AND MOMENT DIAGRAMS
BY THE SUMMATION METHOD

Several methods can be used to draw the shear and moment diagrams. By the method discussed in previous sections, we plot points from section to section along the beam, using the shear or moment calculated at the sections as ordinates. The points are then joined by straight or curved lines to produce shear and moment diagrams. Although this process is simple, it is not very efficient.

In this section relationships that exist between load, shear, and moment diagrams will be established. An alternative procedure that uses these relationships, referred to as the *summation method,* can be used conveniently and efficiently to construct shear and moment diagrams.

Consider an element of differential length dx between two sections 1–1 and 2–2 of a beam that is subjected to an arbitrary loading, as shown in Fig. 6–10(a). The element is isolated from the beam, and the free-body diagram of the element is shown in Fig. 6–10(b). The shear force and bending moment on section 1–1 are denoted by V and M. Both V and M are shown in a positive sense in the free-body diagram according to the beam sign conventions. Let dV and dM represent the change in shear force and bending moment, respectively, between sections 1–1 and 2–2. Then the shear force and bending moment on section 2–2 are equal to $V + dV$ and $M + dM$, respectively, both of which are shown in a positive sense in the free-body diagram. Since the differential length dx is an infinitesimal quantity, the distributed load acting on the element can be regarded as having uniform intensity w. Upward loads are considered positive.

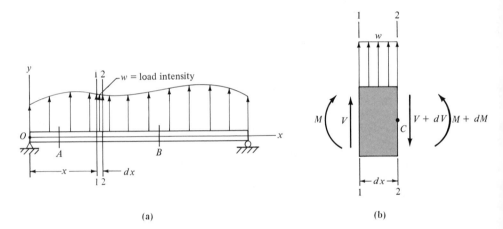

(a)   (b)

FIGURE 6–10

The equilibrium equation $\Sigma F_y = 0$ for the element gives

$$+\uparrow \Sigma F_y = V + w\,dx - (V + dV) = 0$$

from which

$$dV = w\,dx \qquad (6\text{–}5)$$

or

$$\frac{dV}{dx} = w \qquad (6\text{–}6)$$

Equation (6–6) means that *the slope of the shear diagram (the rate of change of the shear force per unit length of beam) at a given section is equal to the load intensity at that section.*

Integrating Eq. (6–5) from section A to section B [Fig. 6–10(a)] gives

$$\int_A^B dV = \int_{x_A}^{x_B} w \, dx$$

which gives

$$V_B - V_A = \begin{pmatrix} \text{total external load} \\ \text{between } A \text{ and } B \end{pmatrix}$$

or

$$\left[V_B = V_A + \begin{pmatrix} \text{total external load} \\ \text{between } A \text{ and } B \end{pmatrix} (+\uparrow) \right] \qquad (6\text{–}7)$$

Equation (6–7) means that *the shear force at a section is equal to the shear force at a section to the left plus the total external load between the two sections. Upward loads are considered positive.*

The equilibrium equation $\Sigma M_C = 0$ for the element gives

$$\circlearrowleft\Sigma M_C = -M - V \, dx - (w \, dx)\left(\frac{dx}{2}\right) + (M + dM) = 0$$

from which

$$dM = V \, dx + \tfrac{1}{2}w(dx)^2$$

or

$$\frac{dM}{dx} = V + \tfrac{1}{2}w \, dx$$

Since dx is an infinitesimal quantity, the term $\tfrac{1}{2}w \, dx$ approaches zero. We have

$$\frac{dM}{dx} = V \qquad (6\text{–}8)$$

Equation (6–8) means that *the slope of the moment diagram (the rate of change of moment per unit length of beam) at any section is equal to the value of the shear force at that section.*

The maximum or minimum moment occurs where the moment diagram either has a zero slope or the slope changes sign. Therefore, from Eq. (6–8), we see that *the maximum or minimum moment occurs at the section where the shear force is either zero* (see Example 6–8) *or the sign of shear force changes* (see Example 6–9).

Writing Eq. (6–8) in differential form and integrating from section A to section B, we have

$$dM = V \, dx$$

$$\int_A^B dM = \int_{x_A}^{x_B} V \, dx$$

Since $V\,dx$ is the differential area of a vertical strip in the shear diagram, the integral $\int_{x_A}^{x_B} V\,dx$ gives the total area of the shear diagram between sections A and B. Thus the equation above can be written as

$$M_B - M_A = \begin{pmatrix} \text{total area under the shear} \\ \text{diagram between } A \text{ and } B \end{pmatrix}$$

or

$$\left[M_B = M_A + \begin{pmatrix} \text{total area under the shear} \\ \text{diagram between } A \text{ and } B \end{pmatrix} \begin{pmatrix} \text{area above the} \\ \text{base line as } + \end{pmatrix} \right] \quad (6\text{–}9)$$

Equation (6–9) means that *the moment at a section is equal to the moment at a section to the left plus the area under the shear diagram between the two sections. The area above the base line is considered positive.*

EXAMPLE 6-8 ──

Rework Example 6–5 by using the summation method. The loading diagram of the beam is shown in Fig. (a).

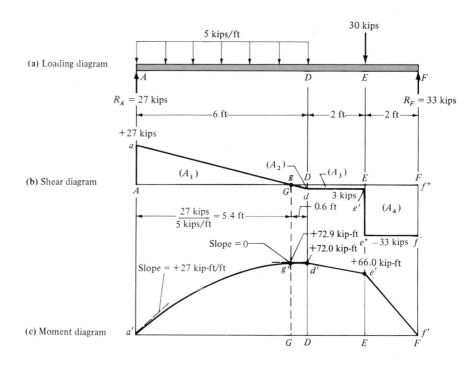

(a) Loading diagram

(b) Shear diagram

(c) Moment diagram

SOLUTION

The shear diagram is shown in Fig. (b). To draw the shear diagram, start at point a with a shear force equal to the upward reaction $R_A = 27$ kips. From a to d the shear diagram is a straight line because the uniform load has a constant intensity, and therefore the shear diagram has a constant slope. The total downward load between A and D is $(-5 \text{ kips/ft})(6 \text{ ft}) = -30$ kips, which causes the shear force to decrease from $+27$ kips at a to $+27 - 30 = -3$ kips at d. From d to e^- (just to the left of E) there is no load to cause any change of shear force; thus line de^- is horizontal. The concentrated load of 30 kips acting downward causes the shear force to drop abruptly from -3 kips at e^- to $-3 - 30 = -33$ kips at e^+ (just to the right of E). From e^+ to f^- there is no load to cause a change in shear force; hence the line e^+f^- is horizontal. The upward reaction $R_F = 33$ kips causes the shear force to increase abruptly from -33 kips at f^- to zero at f^+. This provides a useful check of the solution.

The distance between AG can be determined by using Eq. (6–7); thus

$$V_G = V_A + [\text{external load between } A \text{ and } G]$$

or

$$0 = +27 \text{ kips} + (-5 \text{ kips/ft}) \quad AG$$

from which

$$AG = \frac{27 \text{ kips}}{5 \text{ kips/ft}} = 5.4 \text{ ft}$$

Then

$$GD = 6 \text{ ft} - 5.4 \text{ ft} = 0.6 \text{ ft}$$

The areas of the shear diagram labeled as A_1, A_2, A_3, and A_4 are useful in plotting the moment diagram; hence the areas are calculated as follows:

$$A_1 = \tfrac{1}{2}(+27 \text{ kips})(5.4 \text{ ft}) = +72.9 \text{ kip-ft}$$

$$A_2 = \tfrac{1}{2}(-3 \text{ kips})(0.6 \text{ ft}) = -0.9 \text{ kip-ft}$$

$$A_3 = (-3 \text{ kips})(2 \text{ ft}) = -6.0 \text{ kip-ft}$$

$$A_4 = (-33 \text{ kips})(2 \text{ ft}) = -66.0 \text{ kip-ft}$$

To draw the moment diagram as shown in Fig. (c), start at a' with zero moment at the simple support. From Eq. (6–9), the moment at g' is $M_G = M_A + A_1 = 0 + 72.9 = 72.9$ kip-ft. The moment at d' is $M_D = M_G + A_2 = +72.9 - 0.9 = +72.0$ kip-ft. The slope at a' is equal to $V_A = +27$ kip-ft/ft. The slope at g' is zero because $V_G = 0$. The slope at d' is equal to $V_D = -3$ kip-ft/ft. The curve $a'g'd'$ is parabolic with its vertex at g'. The moment diagram has a maximum value at g'.

Since the shear force has a constant value of -3 kips between DE, the moment diagram from d' to e' decreases at a uniform rate of -3 kip-ft/ft. The moment at e' is $M_E = M_D + A_3 = +72 - 6 = +66$ kip-ft. From e' to f' the moment again decreases uniformly, but the rate is equal to the constant shear force of -33 kip-ft/ft in this region. The moment at f' is $M_F = M_E + A_4 = +66 - 66 = 0$, which provides a useful check, since the moment at the simple support or free end of a beam is equal to zero unless a concentrated moment is applied at the end.

The computations for shear force and bending moment discussed above can be organized as follows:

For Shear Force	For Bending Moment
$0 = V_{A-}$	$0 = M_A$
$+)\qquad R_A = +27$	$+)\quad A_1 = +72.9$
$+27 = V_{A+}$	$+72.9 = M_G = M_{max}$
$+)\quad -5 \times 6 = -30$	$+)\quad A_2 = -0.9$
$-3 = V_D = V_{E-}$	$+72.0 = M_D$
$+)\qquad\qquad -30$	$+)\quad A_3 = -6.0$
$-33 = V_{E+} = V_{F-}$	$+66.0 = M_E$
$+)\qquad R_F = +33$	$+)\quad A_4 = -66.0$
$0 = V_{F+}$ (checks)	$0 = M_D$ (checks)

■

EXAMPLE 6–9

Draw the shear force and bending moment diagrams for the beam subjected to the loading as shown in Fig. (a).

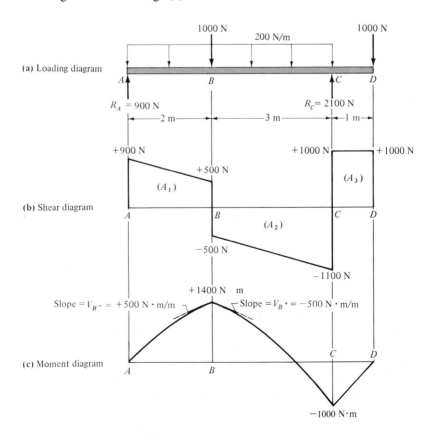

(a) Loading diagram

(b) Shear diagram

(c) Moment diagram

SOLUTION

The shear diagram shown in Fig. (b) starts with the value of the upward reaction R_A, or $+900$ N. Subtract the total downward load between A and B^- of $(200 \text{ N/m})(2 \text{ m})$ or 400 N from $+900$ N to get $V_{B^-} = +500$ N. The shear diagram between A and B^- is a straight line because the load is uniform. Subtract the magnitude of the downward concentrated load of 1000 N from $+500$ N to get $V_{B^+} = -500$ N. Between B^+ and C^- the shear force decreases at a uniform rate. Subtract (200 N/m) (3 m) or 600 N from -500 N to get $V_{C^-} = -1100$ N. Add the upward reaction $R_C = 2100$ N to -1100 N to get $V_{C^+} = +1000$ N. There is no change of shear force between C^+ and D^-; thus $V_{D^-} = V_{C^+} = +1000$ N, and the line is horizontal. Subtract the downward load of 1000 N from $+1000$ N to get $V_{D^+} = 0$, which provides a useful check.

The areas of the shear diagrams are calculated as follows:

$$A_1 = \tfrac{1}{2}(900 \text{ N} + 500 \text{ N})(2 \text{ m}) = +1400 \text{ N} \cdot \text{m}$$

$$A_2 = \tfrac{1}{2}(-500 \text{ N} - 1100 \text{ N})(3 \text{ m}) = -2400 \text{ N} \cdot \text{m}$$

$$A_3 = (+1000 \text{ N})(1 \text{ m}) = +1000 \text{ N} \cdot \text{m}$$

The moment diagram shown in Fig. (c) starts with zero at the simple support A. Add $A_1 = +1400$ N $\cdot$ m to zero to get $M_B = +1400$ N $\cdot$ m. The curve between A and B is parabolic because its slope (which is equal to the shear force at a section) changes at a uniform rate. Add $A_2 = -2400$ N $\cdot$ m to M_B to get $M_C = -1000$ N $\cdot$ m. The curve between B and C, is also parabolic because its slope also changes at a uniform rate. But the two parabolas meeting at B have different slopes at B. The slope at B^- is equal to $V_{B^-} = +500$ N $\cdot$ m/m, while the slope at B^+ is equal to $V_{B^+} = -500$ N $\cdot$ m/m, as shown in the moment diagram in Fig. (c).

Add $A_3 = +1000$ N $\cdot$ m to M_C to get $M_D = 0$, which is a useful check of the solution. The moment diagram between C and D is a straight line because of the constant shear force in the region.

The computations for shear forces and bending moments discussed above can be organized as follows:

For Shear Force		For Bending Moment	
	$0 = V_{A^-}$		$0 = M_A$
$+)$ $\quad R_A = \quad +900$		$+)$ $\quad A_1 = +1400$	
	$+900 = V_{A^+}$		$+1400 = M_B$
$+) -200 \times 2 = \quad -400$		$+)$ $\quad A_2 = -2400$	
	$+500 = V_{B^-}$		$-1000 = M_C$
$+)$ $\quad -1000$		$+)$ $\quad A_3 = +1000$	
	$-500 = V_{B^+}$		$0 = M_D$ (checks)
$+) -200 \times 3 = \quad -600$			
	$-1100 = V_{C^-}$		
$+)$ $\quad R_C = +2100$			
	$+1000 = V_{C^+} = V_{D^-}$		
$+)$ $\quad -1000$			
	$0 = V_{D^+}$ (checks)		

━━━━ **EXAMPLE 6–10** ━━

Draw the shear force and bending moment diagram of the beam subjected to the loading shown in Fig. (a). The reactions are indicated in the figure.

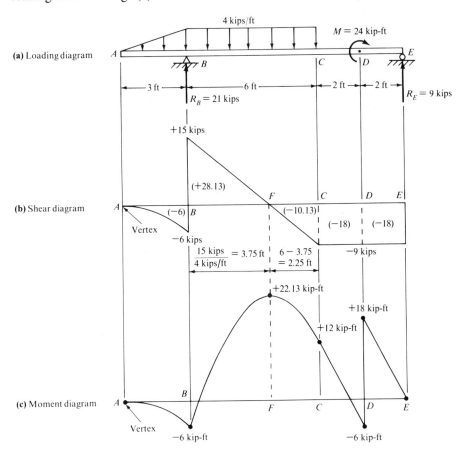

(a) Loading diagram

(b) Shear diagram

(c) Moment diagram

SOLUTION

The shear diagram is plotted as shown in Fig. (b). The diagram starts from zero at A. The total linearly varying load between A and B is $\frac{1}{2}(-4 \text{ kips/ft})(3 \text{ ft}) = -6$ kips, which causes the shear force to decrease to -6 kips at the section just to the left of B. The curve between A and B is parabolic with vertex at A, because the slope at A, which is equal to the load intensity at A, is zero. The slope is decreasing at a uniform rate due to the linearly varying downward load. Adding $R_B = 21$ kips to -6 kips, we get $V_{B+} = +15$ kips. Subtracting the uniform load between B and C of $(4 \text{ kips/ft})(6 \text{ ft}) = 24$ kips from $+15$ kips, we get $V_C = -9$ kips. The shear diagram between B and C is a straight line because the load in this region is uniform. From C to E^- there is no vertical force; therefore, the shear diagram is a horizontal line in this region. At section D the couple does not have any force component along the vertical direction (or along any direction, for that matter). Therefore, the couple does not cause any change in the shear diagram at the section. The upward reaction $R_E = 9$ kips causes the shear diagram to return to zero at E^+, which is a useful check.

Before the moment diagram is plotted, the areas of the shear diagram must be calculated. The resulting value of each area is indicated inside the area in parentheses in Fig. (b).

The moment diagram is plotted as shown in Fig. (c). The diagram starts from zero at the free end A. Adding the area (-6) of the shear diagram between A and B, we get $M_B = -6$ kip-ft. The curve between A and B is a cubic parabola with its vertex at A. Adding the area $(+28.13)$ of the shear diagram between B and F we get $M_F = 22.13$ kip-ft, which is the maximum positive moment. Adding the area (-10.13) of the shear diagram between F and C, we get $M_C = +12$ kip-ft. The moment diagram between B and C is a parabola with its vertex at F, where the shear force is zero. Adding the area (-18) of the shear diagram between C and D, we get $M_{D-} = -6$ kip-ft. The moment diagram between C and D is a straight line because the shear force in this region is a constant.

The clockwise concentrated moment of 24 kip-ft applied at E causes an abrupt increase in the value of the moment at the section. Thus the moment just to the right of E is

$$M_{D+} = M_{D-} + M = -6 + 24 = +18 \text{ kip-ft}$$

At the section where the concentrated moment is applied, a clockwise concentrated moment always causes an abrupt increase in the value of the moment, while a counterclockwise concentrated moment always causes an abrupt decrease in the value of the moment.

Adding the area (-18) of the shear diagram between D and E, we get $M_E = 0$, which is a useful check.

The computations for shear force and bending moment discussed above can be organized as follows:

For Shear Force			For Bending Moment		
		$0 = V_A$			$0 = M_A$
$+)$	$-\frac{1}{2}(4)(3) =$	-6	$+) [A]_A^B =$		-6
		$-6 = V_{B-}$			$-6 = M_B$
$+)$	$R_B = $	$+21$	$+) [A]_B^F =$		$+28.13$
		$+15 = V_{B+}$			$+22.13 = M_F$
$+)$	$-(4)(6) =$	-24	$+) [A]_F^C =$		-10.13
		$-9 = V_C = V_E$			$+12 = M_C$
$+)$	$R_E = $	$+9$	$+) [A]_C^D =$		-18
		$0 = V_{E+}$ (checks)			$-6 = M_{D-}$
			$+)$	$M =$	$+24$
					$+18 = M_{D+}$
			$+) [A]_D^E =$		-18
					$0 = M_E$ (checks)

PROBLEMS

In Problems **6–28** *to* **6–47,** *draw the shear force and bending moment diagrams for each beam by the summation method.*

6–28

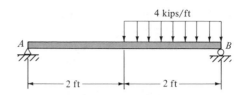

FIGURE P6–28

6–29

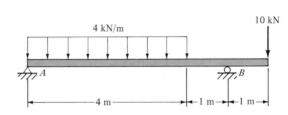

FIGURE P6–29

6–30

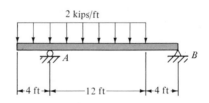

FIGURE P6–30

6–31

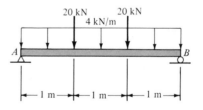

FIGURE P6–31

6–32

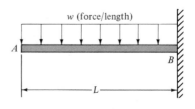

FIGURE P6–32

6–33

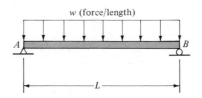

FIGURE P6–33

6–34

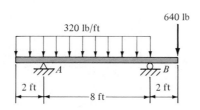

FIGURE P6–34

6–35

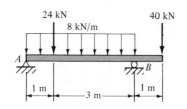

FIGURE P6–35

6–36

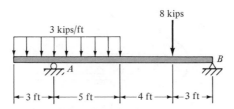

FIGURE P6–36

6–37

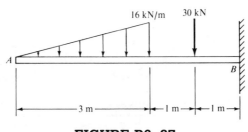

FIGURE P6–37

6–38

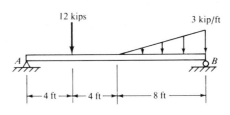

FIGURE P6–38

6–39

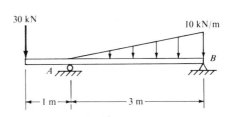

FIGURE P6–39

6–40

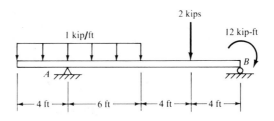

FIGURE P6–40

6–41

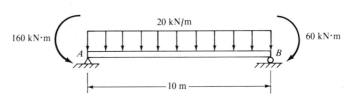

FIGURE P6–41

6–42

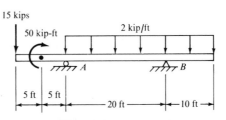

FIGURE P6–42

6–43

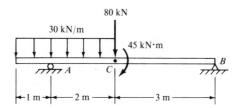

FIGURE P6–43

6–44

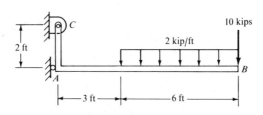

FIGURE P6–44

6–45

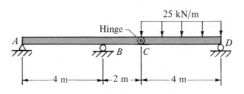

FIGURE P6–45

6–46

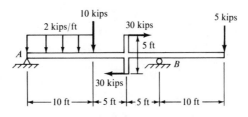

FIGURE P6–46

6–47

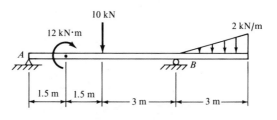

FIGURE P6–47

6–48   A uniform beam is to be lifted by a crane, with cables attached at A and B, as shown in Fig. P6–48. Determine the distance b if the absolute value of the maximum moment is to be as small as possible. (**HINT:** Draw the bending moment diagram in terms of w, b, and L, and then equate the maximum positive moment and the absolute value of the maximum negative moment.)

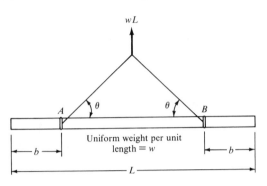

FIGURE P6–48

CHAPTER 7

Stresses in Beams

7–1
INTRODUCTION

In this chapter the stresses caused by bending moments and shear forces are considered. Normal stresses along the longitudinal direction are caused by bending moments, and shear stresses are caused by shear forces. It is the purpose of this chapter to study the distribution of normal and shear stresses in a beam and relate these stresses to the bending moment and shear force in the beam.

Normal stresses in beams due to bending (also called *flexural stresses*) are discussed first, followed by a study of shear stresses in beams. The design of beams for strength is discussed next. Finally, shear connectors in beams and beams of composite materials are studied.

7–2
DISTRIBUTION OF NORMAL STRESSES IN BEAMS

For a straight beam having constant cross-sectional area with a vertical axis of symmetry, as shown in Fig. 7–1, a line through the centroid of all cross sections is referred to as the *axis of the beam*. Consider two cross sections *ab* and *cd* in the beam. Before application of loads, the cross sections are in the vertical direction. Assume that the beam segment between the two sections is subjected to a positive bending moment $+M$. The beam bends and the cross sections *ab* and *cd* tilt slightly, but the sections remain planar and perpendicular to the axis of the beam, as shown in Fig. 7–2.

Imagine that the beam is composed of an infinite number of fibers along the longitudinal direction. The fibers along *bd* become longer and those along *ac* become shorter. Hence the fibers along *bd* are subjected to tension and those along *ac* are subjected to compression. The fibers along the beam axis *mn* do not undergo any change of length due to bending; hence these fibers are not subjected to any normal stresses. The fibers along *mn* form a surface called the *neutral surface*. The intersection of the neutral surface with a cross section is called a *neutral axis*. *For a beam subjected to pure bending (no axial force), the neutral axis passes through the centroid of the cross-sectional area.*

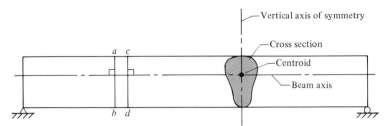

FIGURE 7–1

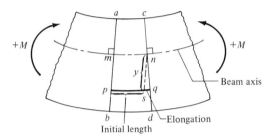

FIGURE 7–2

In Fig. 7–2, consider a typical fiber pq parallel to the neutral surface and located at a distance y from it. From point n, draw a line ns parallel to mp; then

$$ps = mn = \text{initial undeformed length of the fiber}$$

and the fiber pq elongates by an amount sq. From triangle nsq, we see that the elongation (or compression) of a fiber varies linearly as the distance y to the neutral surface. Since the initial length of all fibers between the sections is the same, we conclude that:

Linear strain of a longitudinal beam fiber due to bending varies linearly as the distance of the fiber from the neutral surface.

For elastic bending of the beam, Hooke's law applies; that is, stress is proportional to strain. For most materials, the moduli of elasticity in tension and in compression are equal. Under these conditions, we conclude further that:

Flexural stress of a point on beam section varies linearly as the distance of the point from the neutral axis.

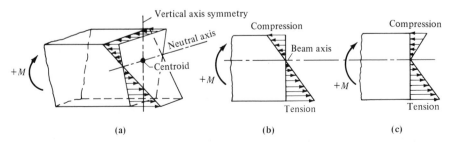

FIGURE 7–3

Figure 7–3(a) shows the flexural stress distribution in a beam section. Points located at equal distance from the neutral axis have the same flexural stress. The maximum flexural stress occurs at the points most remote from the neutral axis. Two alternative ways of representing the stress distribution are shown in Fig. 7–3(b) and (c).

7–3
FLEXURE FORMULA

Consider a beam segment subjected to a positive bending moment $+M$ as shown in Fig. 7–4(a). At section m–m the applied moment is resisted by flexural stresses that vary linearly from the neutral axis passing through the centroid of the section. The maximum flexural stresses occur at points on the bottom of the section, as these points are located at the greatest distance from the neutral axis. Denote the maximum normal stress by σ_{max} and the distance from the neutral axis to the bottom of the section by c. Then the normal stress σ at the narrow strip of area dA located at distance y from the neutral axis is, by proportion,

$$\sigma = \frac{y}{c} \sigma_{max} \qquad (7\text{–}1)$$

The force on the differential area dA is $\sigma \, dA$. The differential moment of this force about the neutral axis is

$$dM = (\sigma \, dA) \, y$$

The total resisting moment developed by the flexural stresses can be obtained by integrating this expression over the entire section. This resisting moment must be equal to the external moment M to satisfy the equilibrium condition. We have

$$M = \int_A dM = \int_A (\sigma \, dA)y = \int_A \left(\frac{y}{c} \sigma_{max} \, dA \right) y = \frac{\sigma_{max}}{c} \int_A y^2 \, dA \qquad (a)$$

where the integral $\int_A y^2 \, dA$ is, by the definition given in Chapter 4, the moment of inertia, I, of the cross-sectional area with respect to the neutral axis. The value

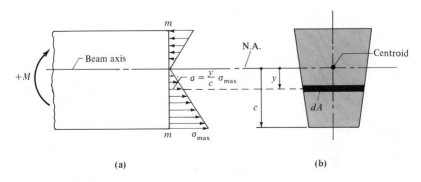

(a)   (b)

FIGURE 7–4

of I depends on the size and shape of the cross-sectional area; it is a constant for a given section. Equation (a) can be written as

$$M = \frac{\sigma_{max}}{c} I$$

from which

$$\sigma_{max} = \frac{Mc}{I} \qquad (7\text{--}2)$$

Substituting in Eq. (7–1), we have

$$\sigma = \frac{My}{I} \qquad (7\text{--}3)$$

Equations (7–2) and (7–3) are two forms of the famous and widely used *flexure formula*.

From Table 4–2, the moment of inertia of a rectangular section of width d and height h with respect to the neutral axis is

$$I = \frac{bh^3}{12} \qquad (7\text{--}4)$$

and the moment of inertia of circular section of radius R or diameter d with respect to the neutral axis is

$$I = \frac{\pi R^4}{4} = \frac{\pi d^4}{64} \qquad (7\text{--}5)$$

The moments of inertia of sections with composite areas can be determined by the method discussed in Section 4–8.

The sketches in Fig. 7–5(a) and (b) are helpful for determining whether a fiber is in tension or in compression due to a given bending moment.

Since I and c are both constants for a given section, the quotient I/c is also a constant. We denote I/c by S; that is,

$$S = \frac{I}{c} \qquad (7\text{--}6)$$

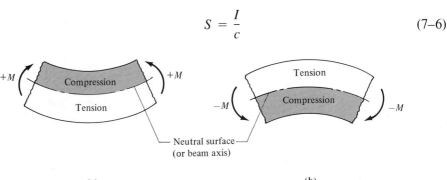

(a)   (b)

FIGURE 7–5

The constant S is called the *section modulus*. Expressed in terms of the section modulus, Eq. (7–2) becomes

$$\sigma_{max} = \frac{Mc}{I} = \frac{M}{I/c} = \frac{M}{S} \tag{7–7}$$

which means that the maximum flexural stress at a section can be determined simply by dividing the bending moment at the section by the section modulus.

Equation (7–7) is widely used in engineering practice because of its simplicity. The section moduli for rectangular and circular sections are shown in Fig. 7–6.

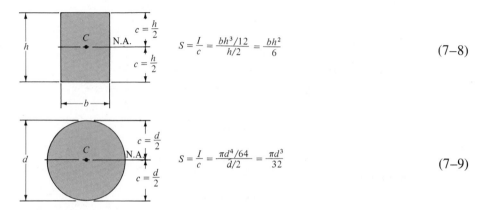

$$S = \frac{I}{c} = \frac{bh^3/12}{h/2} = \frac{bh^2}{6} \tag{7–8}$$

$$S = \frac{I}{c} = \frac{\pi d^4/64}{d/2} = \frac{\pi d^3}{32} \tag{7–9}$$

FIGURE 7–6

To facilitate computations, section moduli for manufactured sections are tabulated in handbooks. Values of section moduli for selected structural steel shapes are given in the Appendix Tables.

For sections that are not symmetrical with respect to their neutral axis, such as the inverted T-section in Fig. 7–7, the maximum tensile and compressive stresses are not equal. Expressions for the maximum flexure stresses are shown in the figure for a positive bending moment.

The moment of inertia has the units in.^{4} or m^4. The section modulus has the units in.^{3} or m^3. It should be emphasized that when numerical values are substituted in Eq. (7–2), (7–3), or (7–7), consistent units must be used for each quantity. To avoid using incorrect units, units must be written in each quantity in the flexure formula.

Since the bending moments usually vary along a beam, the maximum flexural stress along a beam occurs at the extreme fibers of the section where the absolute

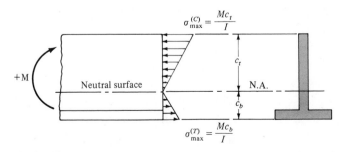

FIGURE 7–7

value of the bending moment is a maximum. For sections that are not symmetrical with respect to the neutral axis, the maximum flexural stresses must be calculated at both the section with the maximum positive moment and the section with the maximum negative moment, as illustrated in Example 7–3.

───── **EXAMPLE 7–1** ─────

A 4-m cantilever beam with a circular cross section of 100 mm diameter is subjected to a concentrated load $P = 2$ kN, as shown. Determine the maximum flexural stress in the beam due to the load.

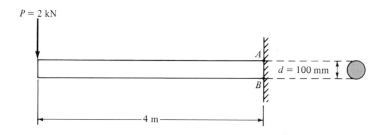

SOLUTION

The maximum moment at the fixed end is

$$M_{max} = -(2 \text{ kN})(4 \text{ m}) = -8 \text{ kN} \cdot \text{m}$$

The moment of inertia of the circular section about the neutral axis is

$$I = \frac{\pi d^4}{64} = \frac{\pi (0.1 \text{ m})^4}{64} = 4.91 \times 10^{-6} \text{ m}^4$$

From the flexure formula, the maximum value of flexural stress in the beam is

$$\sigma_{max} = \pm \frac{M_{max}\, c}{I} = \pm \frac{(8 \text{ kN} \cdot \text{m})(0.05 \text{ m})}{4.91 \times 10^{-6} \text{ m}^4} = \pm 81.5 \times 10^3 \text{ kN/m}^2$$

$$= \pm 81.5 \text{ MPa}$$

The maximum tension occurs at A and the maximum compression occurs at B. These stresses have the same magnitude because the section is symmetrical with respect to the neutral axis.

If we use Eqs. (7–9) and (7–7), we get the same result, as shown in the following.

$$\frac{r}{c} = \frac{\pi d}{}$$

$$S = \frac{\pi d^3}{32} = \frac{\pi (0.1 \text{ m})^3}{32} = 9.82 \times 10^{-5} \text{ m}^3$$

$$\sigma_{max} = \pm \frac{M_{max}}{S} = \pm \frac{8 \text{ kN} \cdot \text{m}}{9.82 \times 10^{-5} \text{ m}^3} = \pm 81.5 \times 10^3 \text{ kN/m}^2$$

$$= \pm 81.5 \text{ MPa}$$

———— EXAMPLE 7–2 ————

Determine the maximum intensity of a uniform load that a structural steel W14 × 38 beam can carry over a simple span of 12 ft without exceeding an allowable flexural stress of 24 000 psi.

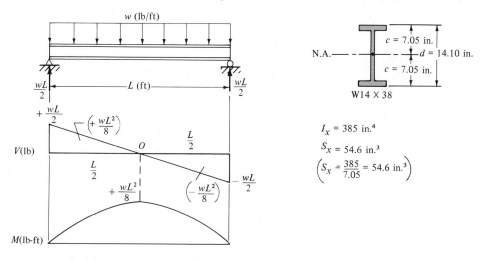

$I_x = 385$ in.^{4}

$S_x = 54.6$ in.^{3}

$\left(S_x = \dfrac{385}{7.05} = 54.6 \text{ in.}^3 \right)$

SOLUTION

The shear force and bending moment diagrams of a simple beam of span length L (ft) subjected to a uniform load w (lb/ft) are shown in the figure above. The properties of W14 × 38 section are obtained from Table A–1 in the Appendix Tables.

The maximum flexure stress occurs at the midspan of the beam, where the bending moment is a maximum. We have

$$M_{\text{max}} = \frac{wL^2}{8} = \frac{(w,\ \text{lb/ft})(12\ \text{ft})^2}{8} = 18w\ (\text{lb-ft})$$

$$\sigma_{\text{max}} = \frac{M_{\text{max}}}{S} = \frac{18w \times 12\ (\text{lb-in.})}{54.6\ \text{in.}^3} = 3.96w\ (\text{psi}) \leq 24\ 000\ \text{psi}$$

from which we get

$$w \leq \frac{24\ 000}{3.96} = 6070\ \text{lb/ft}$$

The dead weight of the beam (38 lb/ft) must be subtracted from the load. Thus the maximum uniform load that the beam can carry is

$$6070 - 38 = 6032\ \text{lb/ft}$$

———— EXAMPLE 7–3 ————

The overhanging beam is built up with two full-sized timber planks 2 in. by 6 in. glued together to form a T section as shown in Fig. (a). The beam is subjected to a uniform load of 400 lb/ft, which includes the weight of the beam. Determine the maximum tensile and compressive flexural stresses in the beam.

SOLUTION

The reactions are first determined from the equilibrium conditions of the beam. The shear force and bending moment diagrams are then drawn, as shown in Figs. (b) and (c).

The centroid of the section is determined to be 5 in. above the bottom of the section. The neutral axis passes through the centroid. The moment of inertia of the section about the neutral axis is determined as shown.

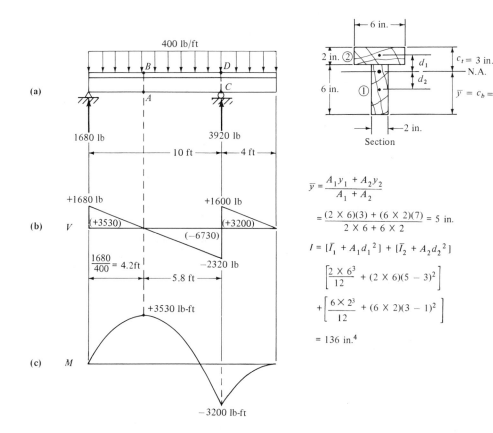

$$\bar{y} = \frac{A_1 y_1 + A_2 y_2}{A_1 + A_2}$$

$$= \frac{(2 \times 6)(3) + (6 \times 2)(7)}{2 \times 6 + 6 \times 2} = 5 \text{ in.}$$

$$I = [\bar{I}_1 + A_1 d_1{}^2] + [\bar{I}_2 + A_2 d_2{}^2]$$

$$\left[\frac{2 \times 6^3}{12} + (2 \times 6)(5 - 3)^2 \right]$$

$$+ \left[\frac{6 \times 2^3}{12} + (6 \times 2)(3 - 1)^2 \right]$$

$$= 136 \text{ in.}^4$$

At the section where the maximum positive moment occurs, the maximum tensile stress at point A and the maximum compressive stress at point B are, respectively,

$$\sigma_A = \frac{M_{\max}^{(+)} c_b}{I} = \frac{(3530 \times 12 \text{ lb-in.})(5 \text{ in.})}{136 \text{ in.}^4} = 1560 \text{ psi (T)}$$

$$\sigma_B = \frac{M_{\max}^{(+)} c_t}{I} = \frac{(3530 \times 12 \text{ lb-in.})(3 \text{ in.})}{136 \text{ in.}^4} = 934 \text{ psi (C)}$$

At the section over the roller support where the maximum negative moment occurs, the maximum tensile stress at D and the maximum compressive stress at C are, respectively,

$$\sigma_D' = \frac{M_{max}^{(-)} c_t}{I} = \frac{(3200 \times 12 \text{ lb-in.})(3 \text{ in.})}{136 \text{ in.}^4} = 953 \text{ psi (T)}$$

$$\sigma_C = \frac{M_{max}^{(-)} c_b}{I} = \frac{(3200 \times 12 \text{ lb-in.})(5 \text{ in.})}{136 \text{ in.}^4} = 1410 \text{ psi (C)}$$

Thus the maximum tensile stress in the beam is

$$\sigma_{max}^{(T)} = \sigma_A = 1560 \text{ psi}$$

and the maximum compressive stress in the beam is

$$\sigma_{max}^{(C)} = \sigma_C = 1410 \text{ psi}$$

PROBLEMS

In Problems **7–1** *to* **7–6**, *determine the flexural stresses of the longitudinal fibers at the points A, B, and C in each section shown subjected to the bending moment indicated.*

7–1

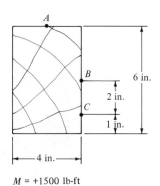

6 in.

2 in.

1 in.

4 in.

$M = +1500$ lb-ft

FIGURE P7–1

7–2

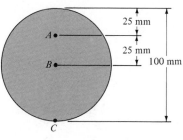

25 mm

25 mm

100 mm

$M = -10$ kN $\cdot$ m

FIGURE P7–2

7–3

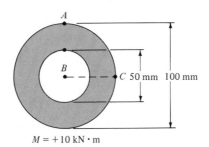

$M = +10 \text{ kN} \cdot \text{m}$

FIGURE P7–3

7–4

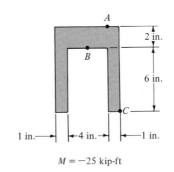

$M = -25 \text{ kip-ft}$

FIGURE P7–4

7–5

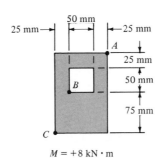

$M = +8 \text{ kN} \cdot \text{m}$

FIGURE P7–5

7–6

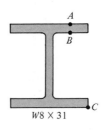

$W8 \times 31$

$M = +50 \text{ kip-ft}$

FIGURE P7–6

7–7 Verify the section moduli tabulated in the Appendix Tables for the following sections.

(a) S_x for W18 × 35
(b) S_y for W10 × 49
(c) S_x and S_y for S12 × 31.8
(d) S_x for C10 × 30

In Problems **7–8** *to* **7–12**, *determine the moment capacity (the maximum moment that a beam can resist) for each cross section shown about the horizontal neutral axis based on the allowable flexural stress indicated.*

7–8

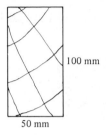

100 mm

50 mm

$\sigma_{\text{allow}} = 10$ MPa

FIGURE P7–8

7–9

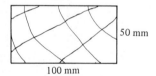

50 mm

100 mm

$\sigma_{\text{allow}} = 10$ MPa

FIGURE P7–9

7–10

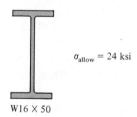

$\sigma_{\text{allow}} = 24$ ksi

W16 × 50

FIGURE P7–10

7–11

$\sigma_{\text{allow}} = 24$ ksi

S12 × 50

FIGURE P7–11

7–12

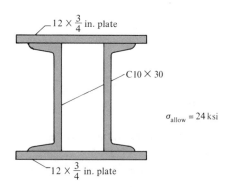

$12 \times \frac{3}{4}$ in. plate

C 10 × 30

$\sigma_{\text{allow}} = 24 \, \text{ksi}$

$12 \times \frac{3}{4}$ in. plate

FIGURE P7–12

7–13 A cast-iron machine part has a channel section, as shown in Fig. P7–13. Determine the maximum positive moment about the horizontal neutral axis that the section can resist without exceeding the allowable stress of 21 MPa in tension and 84 MPa in compression.

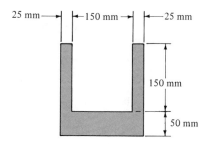

25 mm — 150 mm — 25 mm

150 mm

50 mm

FIGURE P7–13

7–14 In Problem 7–13, determine the maximum negative moment that the section can resist without exceeding the given allowable stresses in tension and in compression.

7–15 In Fig. P7–15, determine the maximum load P that can be applied to the midspan of the simply supported structural steel W14 × 82 beam shown without exceeding an allowable flexural stress of 33 ksi. Neglect the weight of the beam.

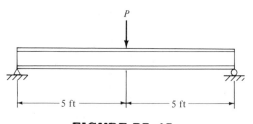

P

5 ft — 5 ft

FIGURE P7–15

7–16   In Fig. P7–16, determine the maximum uniform load w (lb/ft) that the structural steel S15 × 50 cantilever beam shown can carry without exceeding an allowable flexural stress of 24 ksi.

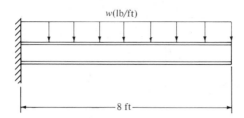

FIGURE P7–16

7–17   Determine the maximum tensile and compressive stresses in the inverted T-beam subjected to the two concentrated loads shown in Fig. P7–17. Neglect the weight of the beam.

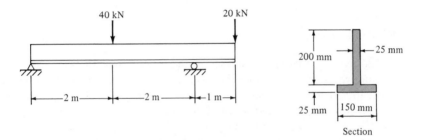

FIGURE P7–17

7–18   Determine the maximum tensile and compressive stresses in the beam shown in Fig. P7–18.

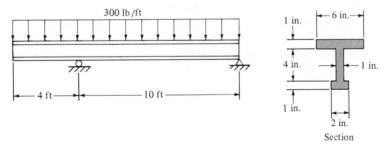

FIGURE P7–18

7–4
SHEAR STRESS FORMULA FOR BEAMS

Internal shear forces exist in cross sections of a beam. The shear forces cause shear stresses in the cross sections. Since equal shear stresses exist on mutually perpendicular planes at a point (see Section 5–5), shear stresses will also exist in the longitudinal sections of a beam.

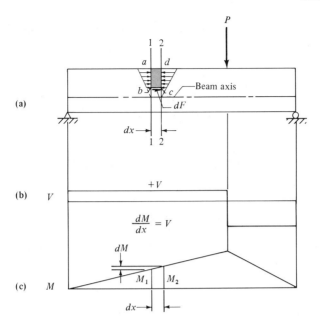

FIGURE 7–8

The existence of shear stresses in the longitudinal sections can be seen from the following consideration. Figure 7–8(a) shows a simple beam subjected to a concentrated load. The shear force and bending moment diagrams of the beam are shown in parts (b) and (c) of the figure. Consider the forces acting on element *abcd*, which is bounded between two adjoining cross sections 1–1 and 2–2 at a differential distance *dx* apart, and the longitudinal section *bc*. Since M_2 (the moment at section 2–2) is greater than M_1 (the moment at section 1–1), the resultant of the normal stresses on *cd* is greater than the resultant of the normal stresses on *ab*. The difference of the resultant forces on the two sides is resisted by the shear force acting on the longitudinal section *bc*.

The shear stress formula for beam may be obtained by considering the equilibrium of element *abcd* in Fig. 7–8(a). The free-body diagram of the element is shown in Fig. 7–9(a). The cross section of the beam is shown in Fig. 7–9(c).

The resultant F_1 of the flexural stresses that act on the shaded area A' of section 1–1 due to the moment M_1 is

$$F_1 = \int_{A'} \sigma \, dA = \int_{A'} \frac{M_1 y}{I} \, dA = \frac{M_1}{I} \int_{A'} y \, dA$$

Similarly, the resultant F_2 of the flexural stresses on the shaded area A' of section 2–2 due to the moment M_2 is

$$F_2 = \int_{A'} \sigma \, dA = \int_{A'} \frac{M_2 y}{I} \, dA = \frac{M_2}{I} \int_{A'} y \, dA$$

Equilibrium of element *abcd* along the horizontal direction requires that the shear force *dF* acting on side *bc* be equal to

$$dF = F_2 - F_1 = \frac{M_2 - M_1}{I} \int_{A'} y \, dA = \frac{dM}{I} \int_{A'} y \, dA$$

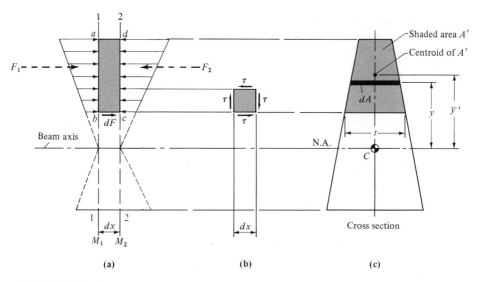

FIGURE 7–9

Assume that the shear stress τ is uniformly distributed across side bc of width t. Then the shear stress in the area may be obtained by dividing dF by the area $t\,dx$. This gives the horizontal shear stress τ. It has been shown in Section 5–5 that for a small element, numerically equal shear stresses act on the mutually perpendicular planes, as indicated in Fig. 7–9(b). Hence the shear stress τ in both the longitudinal plane and the vertical section is

$$\tau = \frac{dF}{t\,dx} = \frac{dM}{dx}\frac{\int_{A'} y\,dA}{It} \tag{a}$$

This equation can be simplified. From Section 6–8, $dM/dx = V$, and the integral can be written as

$$\int_{A'} y\,dA = A'\bar{y}' = Q$$

where $Q = A'\bar{y}'$ represents the first moment of the shaded area A' [the cross-sectional area above (or below) the level at which shear stress is to be determined] about the neutral axis, and $\bar{y}'$ is the distance from the neutral axis to the centroid of the shaded area A'. Thus Eq. (a) becomes

$$\tau = \frac{VQ}{It} \tag{7–10}$$

This is the shear formula for beams. This formula can be used to calculate the shear stresses either on the vertical section or on the longitudinal planes.

In Eq. (7–10), the shear force V and the moment of inertia I are constant for a given section; it follows that the shear stresses in a section vary in accordance with the variation of Q/t. In a rectangular section (Fig. 7–10), since $t = b$ is a constant and the maximum value of Q occurs at the neutral axis, the maximum shear stress occurs at the neutral axis and is equal to

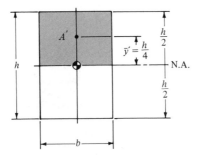

FIGURE 7–10

$$\tau_{max} = \frac{VQ}{It} = \frac{VA'\bar{y}'}{It} = \frac{V(bh/2)(h/4)}{(bh^3/12)(b)} = \frac{3V}{2bh} = 1.5\frac{V}{A} \qquad (7\text{–}11)$$

where $A = bh$ is the area of the entire section. Thus the maximum shear stress in a rectangular section is 1.5 times the average shear stress in the section.

Since beams of rectangular cross section are frequently used in engineering practice, especially timber beams, Eq. (7–11) is very useful. Timber beams have a tendency to split along the neutral surface where the shear stress is a maximum because the shear strength of wood in longitudinal planes parallel to the grain is weaker than the shear strength perpendicular to the grain.

──── **EXAMPLE 7–4** ────────────────────────────────────

A simple beam shown is subjected to a concentrated load at midspan. The beam has a rectangular section with dimensions indicated. Determine the shear stresses at points along lines 1, 2, 3, and 4. Sketch the shear stresses distribution in the section.

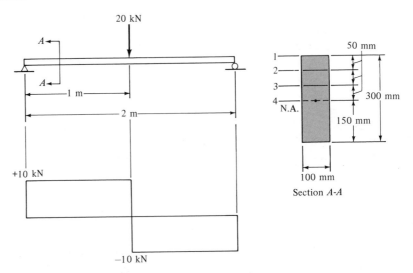

SOLUTION

From the shear diagram shown, the shear force at section A–A is

$$V = +10 \text{ kN}$$

The moment of inertia of the rectangular section with respect to the neutral axis is

$$I = \frac{bh^3}{12} = \frac{(0.1 \text{ m})(0.3 \text{ m})^3}{12} = 2.25 \times 10^{-4} \text{ m}^4$$

The value of $V/(It)$ in the shear formula is a constant and is equal to

$$\frac{V}{It} = \frac{10 \text{ kN}}{(2.25 \times 10^{-4} \text{ m}^4)(0.1 \text{ m})} = 4.44 \times 10^5 \text{ kN/m}^5$$

At points along line 1:

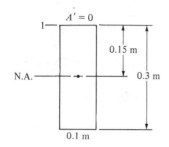

$$Q = A'\bar{y}' = (0)(0.15) = 0$$

$$\tau_1 = \frac{VQ}{It} = (4.44 \times 10^5)(0) = 0$$

At points along line 2:

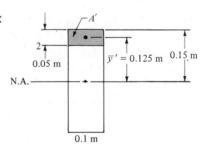

$$Q = A'\bar{y}' = (0.1 \text{ m} \times 0.05 \text{ m})(0.125 \text{ m}) = 6.25 \times 10^{-4} \text{ m}^3$$

$$\tau_2 = \frac{VQ}{It} = (4.44 \times 10^5 \text{ kN/m}^5)(6.25 \times 10^{-4} \text{ m}^3)$$

$$= 278 \text{ kN/m}^2 = 278 \text{ kPa}$$

At points along line 3:

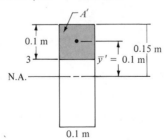

$$Q = A'\bar{y}' = (0.1 \text{ m} \times 0.1 \text{ m})(0.1 \text{ m}) = 1.0 \times 10^{-3} \text{ m}^3$$

$$\tau_3 = \frac{VQ}{It} = (4.44 \times 10^5 \text{ kN/m}^5)(1.0 \times 10^{-3} \text{ m}^3)$$

$$= 444 \text{ kN/m}^2 = 444 \text{ kPa}$$

At points along line 4 (the neutral axis):

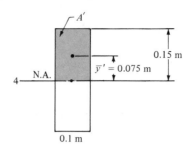

$$Q = A'\bar{y}' = (0.1 \text{ m} \times 0.15 \text{ m})(0.075 \text{ m}) = 1.125 \times 10^{-3} \text{ m}^3$$

$$\tau_4 = \frac{VQ}{It} = (4.44 \times 10^5 \text{ kN/m}^5)(1.125 \times 10^{-3} \text{ m}^3)$$

$$= 500 \text{ kN/m}^2 = 500 \text{ kPa}$$

The shear stress at the neutral axis (level 4) can also be computed from Eq. (7–11), which gives

$$\tau_{max} = \tau_4 = 1.5 \frac{V}{A} = 1.5 \frac{10 \text{ kN}}{0.1 \times 0.3 \text{ m}^2} = 500 \text{ kPa}$$

The shear stresses at the levels below the neutral axis can be calculated in the same way, except that for convenience the area A' is taken below the level where the shear stress is to be computed. The magnitudes of the shear stresses are symmetrical with respect to the neutral axis. The distribution of the shear stresses in the section is shown in the following figure. Note that the sense of the shear stresses coincides with the sense of the shear force on the section.

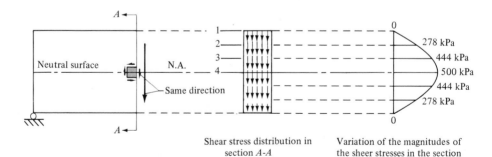

Shear stress distribution in
section *A-A*

Variation of the magnitudes of
the sheer stresses in the section

EXAMPLE 7–5

Determine the maximum shear stress in the beam of Example 7–3 (on p. 183).

SOLUTION
The shear diagram plotted in Example 7–3 is shown as follows:

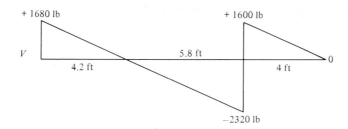

From the shear diagram, the maximum absolute value of the shear force is

$$|V_{max}| = 2320 \text{ lb}$$

The section properties of the T-section calculated in Example 7–3 are shown in the following figure. The moment of inertia of the section about the neutral axis is $I = 136$ in.[4]

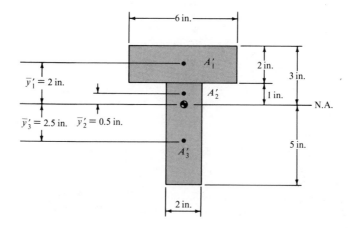

The maximum shear stress occurs at the neutral axis of the section, where Q is a maximum and t is a minimum.

If the area above the neutral axis is taken as A', the first moment of the area about the neutral axis is

$$Q = A_1'\bar{y}_1' + A_2'\bar{y}_2' = (6 \times 2)(2) + (2 \times 1)(0.5) = 25 \text{ in.}^3$$

Or if the area below the neutral axis is taken as A', the first moment of the area about the neutral axis is

$$Q = A_3'\bar{y}_3' = (2 \times 5)(2.5) = 25 \text{ in.}^3$$

Thus we see that either way we get the same value of Q.

The maximum shear stress can now be determined by the shear stress formula:

$$\tau_{max} = \frac{V_{max}Q}{It} = \frac{(2320 \text{ lb})(25 \text{ in.}^3)}{(136 \text{ in.}^4)(2 \text{ in.})} = 213 \text{ psi}$$

━━━ EXAMPLE 7–6 ━━━

Determine the maximum shear stress and the maximum tensile stress in the wide-flange beam subjected to the uniform load shown and indicate the maximum stresses on rectangular elements at the points where they occur.

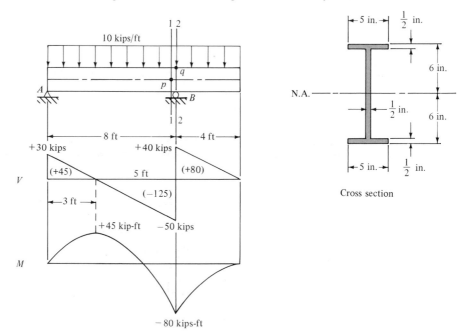

SOLUTION

Treating the cross-sectional area to be a rectangle 5 in. by 12 in. minus a rectangle 4.5 in. by 11 in., the moment of inertia of the section about the neutral axis is

$$I = \frac{(5 \text{ in.})(12 \text{ in.})^3}{12} - \frac{(4.5 \text{ in.})(11 \text{ in.})^3}{12} = 221 \text{ in.}^4$$

The shear force and bending moment diagrams are first plotted as shown. From the shear diagram, the maximum absolute value of the shear force is 50 kips at section 1–1 just to the left of B. The maximum shear stress occurs at the neutral axis of section 1–1. Thus

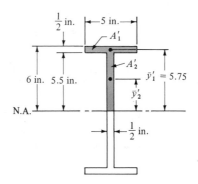

$$Q = A_1'\bar{y}_1' + A_2'\bar{y}_2'$$

$$= \left(5 \times \frac{1}{2} \text{ in.}^2\right)(5.75 \text{ in.}) + \left(\frac{1}{2} \times 5.5 \text{ in.}^2\right)\left(\frac{5.5 \text{ in.}}{2}\right)$$

$$= 21.94 \text{ in.}^3$$

$$\tau_{max} = \frac{VQ}{It} = \frac{(50 \text{ kips})(21.91 \text{ in.}^3)}{(221 \text{ in.}^4)(\frac{1}{2} \text{ in.})} = 9.91 \text{ ksi}$$

The maximum shear stress is indicated on the element at p in the following figure.

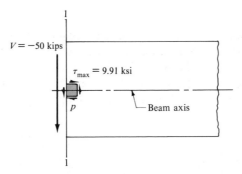

Note that the element is at the neutral surface, where the flexural stress is zero; the element is thus under pure shear.

From the bending moment diagram, the magnitude of the maximum moment is 80 kip-ft at section 2–2 directly above B. Since this is a negative moment, the maximum tensile stress occurs at the top of the section. Thus

$$\sigma_{max}^{(T)} = \frac{Mc}{I} = \frac{(80 \times 12 \text{ kip-in.})(6 \text{ in.})}{221 \text{ in.}^4} = 26.1 \text{ ksi}$$

The maximum tensile stress is indicated on the element at q in the following figure.

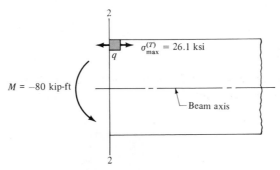

Note that the element is at the top of the section, where the shear stress is zero; the element is thus subjected to tensile stress only.

The minimum shear stress in the web occurs at the junction of the flange and the web and is equal to

$$(\tau_{web})_{min} = \frac{VQ}{It} = \frac{(50 \text{ lb})(5 \times \frac{1}{2} \text{ in.}^2)(5.75 \text{ in.})}{(221 \text{ in.}^4)(\frac{1}{2} \text{ in.})} = 6.51 \text{ ksi}$$

Thus the vertical shear stresses throughout the web of a wide-flange section are distributed as shown in the following figure.

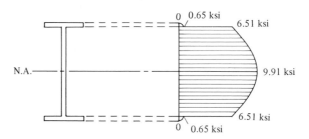

Note that the shear stresses in the flange are very small because at the flanges $t = 5$ in., which is 10 times the thickness of the web. Therefore, at the junction of the flange and the web, the shear stress at the flange is one-tenth the shear stress in the web. Hence the shear force at the section is carried mainly by the web. The maximum shear stress can be approximated by

$$(\tau_{max})_{approx} = \frac{V_{max}}{A_{web}} \tag{7-12}$$

where A_{web} stands for the area of the web. In this example

$$(\tau_{max})_{approx} = \frac{50 \text{ kips}}{(\frac{1}{2} \text{ in.})(11 \text{ in.})} = 9.09 \text{ ksi}$$

which is about 8 percent off from $\tau_{max} = 9.91$ ksi calculated by the shear formula. This approximation is widely used in engineering practice because, in most beams, the maximum shear stress is well within the allowable shear strength of the material. ∎

PROBLEMS

*In Problems **7–19** to **7–21**, determine the shear stresses at points A, B, and C in each section subjected to the shear force indicated.*

7–19

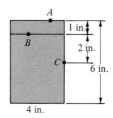

$V = 1900$ lb

FIGURE P7–19

7–20

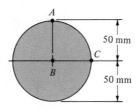

$V = 15$ kN

FIGURE P7–20

7–21

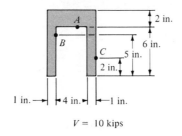

$V = 10$ kips

FIGURE P7–21

7–22 A rectangular timber beam is supported as shown in Fig. P7–22. Determine the maximum value of P if the shear stress in the bar may not exceed 800 kPa.

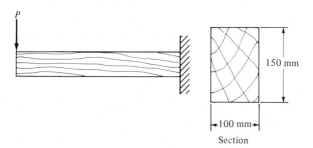

Section

FIGURE P7–22

7–23 A simple beam of rectangular section carries a uniform load, as shown in Fig. P7–23. The beam is made of oak with allowable flexural stress of 1900 psi and allowable longitudinal shear stress (parallel to the grain) of 145 psi. Determine the maximum intensity of the uniform load w in lb/ft that can be applied to the beam.

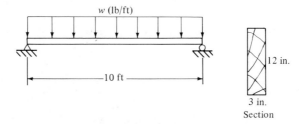

Section

FIGURE P7–23

7–24 Show that the maximum shear stress at the neutral axis in a beam having a solid circular cross-sectional area A is

$$\tau_{max} = \frac{4}{3}\frac{V}{A}$$

7–25 Determine the maximum load P in kN that can be applied to the beam of circular section shown in Fig. P7–25. The beam has an allowable flexural stress of 9 MPa and an allowable shear stress parallel to the grain of 850 kPa. (**HINT:** Use the formula derived in Problem 7–24.)

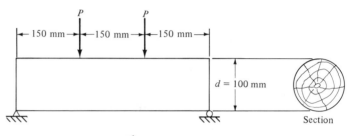

FIGURE P7–25

7–26 Determine the maximum value of shear stress in the beam shown in Fig. P7–17 (on p. 189). (**HINT:** The maximum shear stress occurs at the neutral axis of the section where the absolute value of the shear force is a maximum.)

7–27 Determine the maximum value of shear stress in the beam of Fig. P7–18 (on p. 189). (See the hint in Problem 7–26.)

7–28 An overhanging beam having a T section ($I = 136$ in.^{4}) is subjected to a concentrated load as shown in Fig. P7–28. Determine the shear stresses in section A–A at the levels indicated. Show the distribution of shear stresses in the section with figures similar to those in Example 7–4.

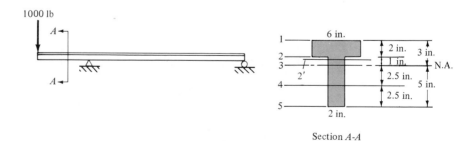

Section A-A

FIGURE P7–28

7–29   Determine the maximum value of the shear stress and the maximum value of the flexural stress in the beam with box section shown in Fig. P7–29.

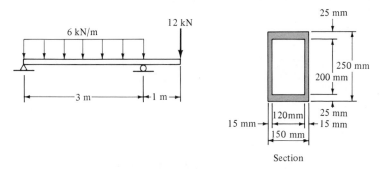

FIGURE P7–29

7–30   A beam on simple supports has a box section with the dimensions indicated in Fig. P7–30. If the moment diagram of the beam is shown, determine the maximum flexural stress and the maximum shear stress in the beam. (**HINT:** Find shear forces by $V = dM/dx$.)

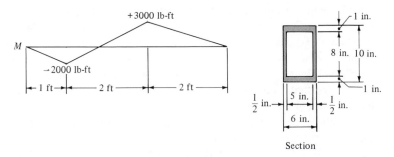

FIGURE P7–30

7–31   The cast-iron beam shown in Fig. P7–31 has the inverted T-section shown. Determine the values of the maximum shear stress, the maximum tensile stress, and the maximum compressive stress in the beam.

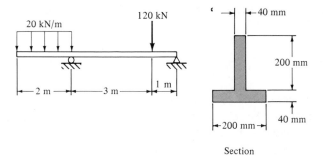

FIGURE P7–31

7–32   The W14 × 38 beam shown in Fig. P7–32 supports a uniform load of 3 kips/ft, including the weight of the beam. Determine the normal and shear stresses acting on the elements at A and B. Show the senses of these stresses on the elements.

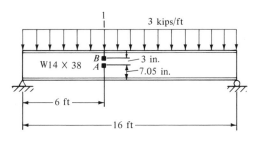

FIGURE P7–32

7–5
DESIGN OF BEAMS FOR STRENGTH

In beam design problems, generally the span length, the supporting conditions, and the loading are given; the proper size of the beam section is required. The size of the section should be such that the maximum flexural stress and the maximum shear stress at the critical sections are within the allowable limits. One critical section occurs where the absolute value of bending moment is a maximum; the other critical section occurs where the absolute value of shear force is a maximum. To determine the locations of these critical sections along a beam, the shear force and bending moment diagrams of the given beam are very useful. For simple loadings, however, construction of complete shear and moment diagrams may be omitted. Handbooks provide formulas for maximum shear force and maximum moment for many different loading conditions.

The allowable stresses used in the design are often specified by the design code. The code frequently used for structural steel design is the American Institute of Steel Construction (AISC) code. Under the most favorable conditions, the AISC code prescribes that the allowable bending stress be $0.66\sigma_y$, and the allowable shear stress be $0.4\sigma_y$. For the most commonly used structural steel, the yield strength σ_y is 36 ksi. Thus the allowable bending stress is $\sigma_{allow} = 0.66 \times 36 = 24$ ksi, and the allowable shear stress is $\tau_{allow} = 0.4 \times 36 = 14.5$ ksi.

In ordinary practices, beams are designed for bending and then checked for shear. In some cases, the deflection of a beam may govern the size of the beam. This topic is discussed in Chapter 8.

For the steel beam design, the minimum section modulus required is calculated by

$$S_{req} = \frac{M_{max}}{\sigma_{allow}} \tag{7–13}$$

Then a suitable wide-flange steel beam (W shape) or a standard steel I-beam (S shape) may be selected from Tables A–1 or A–2 in the Appendix Tables. The beam selected is checked for shear stress by using the approximate formula, Eq. (7–12); that is,

$$(\tau_{max})_{approx} = \frac{V_{max}}{A_{web}} < \tau_{allow} \tag{7–14}$$

where A_{web} is the area of the web and is approximately equal to the depth of the section d times the thickness of the web t_w.

In the timber beam design, because of the small allowable shear stress parallel to the grain, the shear stress frequently controls the dimensions of the cross section. Timber beams are usually available in rectangular sections for which the maximum shear stress is 1.5 times the average shear stress [Eq. (7–11)]. Therefore, in addition to the minimum required section modulus calculated from Eq. (7–13), the minimum rectangular cross-sectional area required must be calculated from

$$A_{req} = \frac{1.5V_{max}}{\tau_{allow}} \tag{7-15}$$

The proper size of timber section is selected from Table A–6 in the Appendix Tables. Rough-sawed timber is full-sized. Dressed or surfaced timber is $\frac{3}{8}$ in. or $\frac{1}{2}$ in. smaller in dimension than rough-sawed timber. For example, a rough-sawed plank 4 in. by 8 in. is $3\frac{5}{8}$ in. by $7\frac{1}{2}$ in. dressed. In Table A–6, the cross-sectional areas and the sectional moduli are computed using the dressed sizes.

Narrow, deep timber beams are more effective than wide, shallow beams in resisting bending moments. In engineering practice, the depth of a timber beam is usually $1\frac{1}{2}$ to 3 times of its width.

EXAMPLE 7–7

Select a Douglas fir beam of rectangular cross section to carry two concentrated loads shown. The allowable stresses are 1300 psi in bending and 85 psi in shear parallel to the grain.

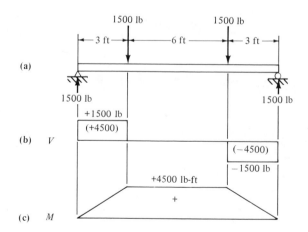

SOLUTION

Shear force and bending moment diagrams are plotted as shown in Figs. (b) and (c). From the moment diagram the maximum moment is

$$M_{max} = (4500 \text{ lb-ft})\left(\frac{12 \text{ in.}}{1 \text{ ft}}\right) = 54\,000 \text{ lb-in.}$$

From Eq. (7–13), the minimum section modulus required is

$$S_{req} = \frac{M_{max}}{\sigma_{allow}} = \frac{54\ 000\ \text{lb-in.}}{1300\ \text{lb/in.}^2} = 41.5\ \text{in.}^3$$

From the shear diagram, the maximum shear force is

$$V_{max} = 1500\ \text{lb}$$

From Eq. (7–15), the minimum cross-sectional area required is

$$A_{req} = \frac{1.5V_{max}}{\tau_{allow}} = \frac{1.5(1500\ \text{lb})}{85\ \text{lb/in.}^2} = 26.5\ \text{in.}^2$$

From Table A–6, the following nominal size rectangular sections fulfill the above requirements:

$$3 \times 12: \quad A = 30.2\ \text{in.}^2, \quad S = 57.9\ \text{in.}^3$$
$$4 \times 10: \quad A = 34.4\ \text{in.}^2, \quad S = 54.5\ \text{in.}^3$$
$$6 \times 8: \quad A = 41.3\ \text{in.}^2, \quad S = 51.6\ \text{in.}^3$$

The 3×12 section has the smallest cross-sectional area and is hence the lightest section, but its depth may be too large for some applications. The 4×10 section has a depth-to-width ratio of $2\frac{1}{2}$, which is a reasonable ratio; therefore, this section is selected. From Table A–6, the weight of the section is 9.57 lb/ft. This uniform load causes a maximum shear force at the support equal to

$$\frac{wL}{2} = \frac{(9.57\ \text{lb/ft})(12\ \text{ft})}{2} = 57\ \text{lb}$$

and a maximum moment at the midspan equal to

$$\frac{wL^2}{8} = \frac{(9.57\ \text{lb/ft})(12\ \text{ft})^2}{8} = 172\ \text{lb-ft}$$

Thus the maximum stresses are

$$\tau_{max} = 1.5\frac{V_{max}}{A} = 1.5 \times \frac{(1500 + 57)\ \text{lb}}{34.4\ \text{in.}^2} = 67.9\ \text{psi} < \tau_{allow} = 85\ \text{psi}$$

$$M_{max} = (4500\ \text{lb-ft} + 172\ \text{lb-ft})(12\ \text{in.}/1\ \text{ft}) = 56\ 060\ \text{lb-in.}$$

$$\sigma_{max} = \frac{M_{max}}{S} = \frac{56\ 060\ \text{lb-in.}}{54.5\ \text{in.}^3} = 1029\ \text{psi} < \sigma_{allow} = 1300\ \text{psi}$$

Hence the selected 4×10 section (dressed size $3\frac{5}{8}$ in. by $9\frac{1}{2}$ in.) is satisfactory. ■

──────── **EXAMPLE 7–8** ───

Select a wide-flange steel beam or a standard steel I-beam for a girder subjected to the loads shown. The uniform load does not include the weight of the beam. Allowable stresses are 24 ksi in bending and 14.5 ksi in shear.

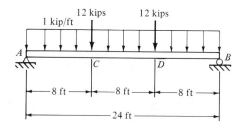

SOLUTION

Assume the weight of the beam to be 50 lb/ft (0.05 kip/ft). Then the total uniform load is 1.05 kips/ft. Due to the concentrated loads, the maximum moment between CD is

$$(12 \text{ kips})(8 \text{ ft}) = 96 \text{ kip-ft}$$

Due to the uniform load, the maximum moment at the midspan is

$$\frac{wL^2}{8} = \frac{(1.05 \text{ kips/ft})(24 \text{ ft})^2}{8} = 75.6 \text{ kip-ft}$$

Therefore, the maximum bending moment at the midspan of the beam is equal to

$$M_{\text{max}} = 96 + 75.6 = (171.6 \text{ kip-ft})\left(\frac{12 \text{ in.}}{1 \text{ ft}}\right) = 2060 \text{ kip-in.}$$

From Eq. (7–13), the minimum section modulus required is

$$S_{\text{req}} = \frac{M_{\text{max}}}{\sigma_{\text{allow}}} = \frac{2060 \text{ kip-in.}}{24 \text{ kips/in.}^2} = 85.8 \text{ in.}^3$$

From Tables A–1 and A–2, the following steel sections were found to fulfill the requirement on the minimum section modulus:

$$\text{W18} \times \text{50:} \quad S = 88.9 \text{ in.}^3$$
$$\text{W16} \times \text{57:} \quad S = 92.2 \text{ in.}^3$$
$$\text{W14} \times \text{61:} \quad S = 92.2 \text{ in.}^3$$
$$\text{S18} \times \text{54.7:} \quad S = 89.4 \text{ in.}^3$$

If an 18-in. depth is permissible, W18 × 50 is the lightest section and hence it is the most economical one.

To check the shear stress, the maximum shear force in the beam is calculated as follows:

$$V_{max} = 12 \text{ kips} + \tfrac{1}{2}(1.05 \text{ kips/ft})(24 \text{ ft}) = 24.6 \text{ kips}$$

From Eq. (7–14), the approximate value of the maximum shear stress in the W18 × 50 beam is

$$(\tau_{max})_{approx} = \frac{V_{max}}{dt_w} = \frac{24.6 \text{ kips}}{17.99 \times 0.355 \text{ in.}^2} = 3.85 \text{ ksi} < \tau_{allow} = 14.5 \text{ ksi}$$

The weight of the beam happens to be the same as assumed; no revision needs to be made. Thus the steel beam W18 × 50 is satisfactory and is therefore selected.

If the 18-in. depth is not allowed due to the limitation on the vertical space, a shallower W section should be used. On the other hand, if the width of the beam is limited, then S18 × 54.7 may be an alternative choice because it is narrower. ■

PROBLEMS

7–33 Select the lightest oak beam of rectangular section for a simple beam of 12-ft span subjected to a concentrated load of 10 kips at the midspan. Given: $\sigma_{allow} = 1900$ psi and $\tau_{allow} = 145$ psi.

7–34 Select the most economical hemlock beam to be used as a simple beam of 16-ft span. The beam carries a uniform load of 800 lb/ft, which includes the weight of the beam. Keep the ratio of depth to width no more than 3, and use allowable stresses of 1100 psi in bending and 90 psi in shear parallel to the grain.

In Problems 7–35 to 7–38, select the most economical rectangular timber section for the beam and loading shown. The allowable stresses are 1200 psi in bending and 100 psi in shear parallel to the grain. Keep the ratio of depth to width no more than 3. The weight of the beam is already included in the uniform load.

7–35

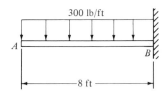

FIGURE P7–35

7–36

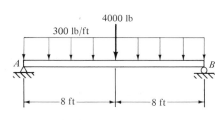

FIGURE P7–36

7–37

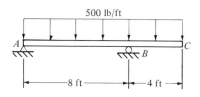

500 lb/ft

A   B   C

—8 ft— —4 ft—

FIGURE P7–37

7–38

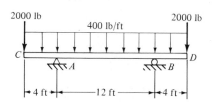

2000 lb   2000 lb

400 lb/ft

C   A   B   D

—4 ft— —12 ft— —4 ft—

FIGURE P7–38

7–39 Select the most economical wide-flange steel beam to be used as a simple beam of 20-ft span. The beam carries a uniform load of 4000 lb/ft, which does not include the weight of the beam. Given: σ_{allow} = 24 000 psi and τ_{allow} = 14 500 psi.

7–40 Select the most economical standard steel I-beam for Problem 7–39.

*In Problems **7–41** to **7–44**, select the most economical wide-flange steel beam or the standard steel I-beam for the beam and loading shown. The allowable stresses are 24 ksi in bending and 14.5 ksi in shear. The weight of the beam is already included in the uniform load.*

7–41

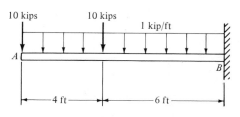

10 kips   10 kips

1 kip/ft

A   B

—4 ft— —6 ft—

FIGURE P7–41

7–42

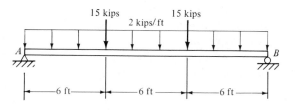

15 kips   15 kips

2 kips/ft

A   B

—6 ft— —6 ft— —6 ft—

FIGURE P7–42

7-43

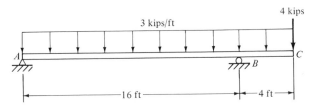

FIGURE P7-43

7-44

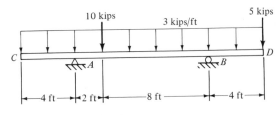

FIGURE P7-44

*7-6
SHEAR FLOW

Beams are sometimes fabricated by joining several component parts to form a single section. Figure 7–11 gives three typical examples of built-up beams. Figure 7–11(a) shows a T-beam fabricated by nailing two timber planks. Figure 7–11(b) shows plywood and boards nailed together to form a box section. Figure 7–11(c) shows a fabricated steel beam consisting of a steel channel bolted to the top flange of a wide-flange steel beam.

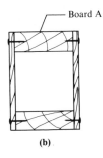

FIGURE 7-11

The nails or bolts that connect the components of a built-up beam must be spaced at a proper pitch (the spacing of nails or bolts along the length of the beam) so that the connectors would carry adequately the longitudinal shear force on the contact surface. Consider the T-beam in Fig. 7–12 subjected to a concentrated load at the midspan. Since the shear force is constant throughout the beam, the nails used to connect the two planks are spaced at a constant pitch p. Each nail is required to carry the shear force on the contact surface for each pitch length. The longitudinal shear stress at the contact surface (level 1–1) is

$$\tau_{1-1} = \frac{VQ}{It}$$

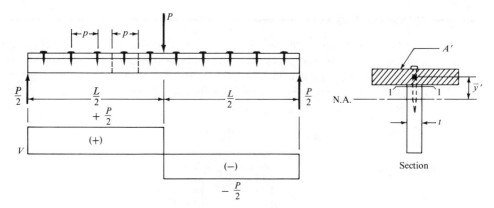

FIGURE 7-12

The quantity $Q = A'\bar{y}'$ represents the first moment of the area A' (area of the plank on the top) about the neutral axis. The shear stress is uniformly distributed in the contact surface, which has an area equal to the product of p and t. Thus the total shear force in the area is

$$F_s = (p \cdot t)\tau_{1-1} = p\frac{VQ}{I} \tag{7-16}$$

which is the shear force required to be carried by each nail. The quantity VQ/I in Eq. (7-16) represents the shear force in a longitudinal section per unit of beam length; it is referred to as the *shear flow* and is denoted here by q; that is,

$$q = \frac{VQ}{I} \tag{7-17}$$

If the maximum allowable shear force of each nail is $(F_s)_{\text{allow}}$, then from Eq. (7-16) the maximum pitch is

$$p_{\text{max}} = \frac{(F_s)_{\text{allow}}}{q} \tag{7-18}$$

Equation (7-18) can also be used to determine the pitch of the nails in Fig. 7-11(b) and the pitch of the bolts in Fig. 7-11(c). In the case of Fig. 7-11(b), the quantities Q and $(F_s)_{\text{allow}}$ are

Q = first moment of the area of board A about the neutral axis
$(F_s)_{\text{allow}}$ = allowable shear force of two nails

and in the case of Fig. 7-11(c), the quantities Q and $(F_s)_{\text{allow}}$ are

Q = first moment of the area of steel channel about the neutral axis
$(F_s)_{\text{allow}}$ = allowable shear force of two bolts

━━━━━ **EXAMPLE 7–9** ━━━━━

A beam is made up of four full-sized, 50-mm-thick boards nailed together to form a box section as shown. Each nail is capable of resisting 400 N of shear force. If this beam transmits a constant vertical shear force of 6000 N, determine the pitch (a) of the nails connecting board ① to boards ② and ③, (b) of the nails connecting board ④ to boards ② and ③.

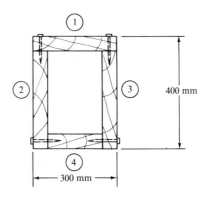

SOLUTION

The moment of inertia of the section about the neutral axis is first calculated as follows:

$$I = \frac{0.3 \times 0.4^3}{12} - \frac{0.2 \times 0.3^3}{12} = 0.001\ 15 \text{ m}^4$$

(a) The maximum pitch of the nails connecting board ① to boards ② and ③ is

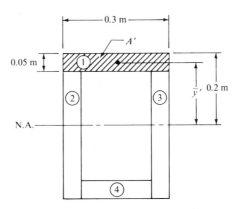

$$Q = A'\bar{y}' = (0.3 \times 0.05)(0.2 - 0.025)$$

$$= 0.002\ 625 \text{ m}^3$$

$$q = \frac{VQ}{I} = \frac{(6000 \text{ N})(0.002\ 625 \text{ m}^3)}{0.001\ 15 \text{ m}^4}$$

$$= 13\ 700 \text{ N/m}$$

$$p_{max} = \frac{(F_s)_{allow}}{q} = \frac{2 \times 400 \text{ N}}{13\ 700 \text{ N/m}}$$

$$= 0.0584 \text{ m} = 58.4 \text{ mm}$$

Thus a pitch of 50 mm may be used for the nails on the top.

(b) The maximum pitch of the nails connecting board ④ to boards ② and ③ is

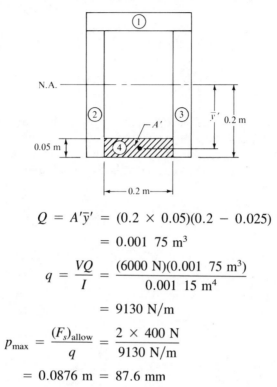

$$Q = A'\bar{y}' = (0.2 \times 0.05)(0.2 - 0.025)$$

$$= 0.001\ 75\ \text{m}^3$$

$$q = \frac{VQ}{I} = \frac{(6000\ \text{N})(0.001\ 75\ \text{m}^3)}{0.001\ 15\ \text{m}^4}$$

$$= 9130\ \text{N/m}$$

$$p_{\text{max}} = \frac{(F_s)_{\text{allow}}}{q} = \frac{2 \times 400\ \text{N}}{9130\ \text{N/m}}$$

$$= 0.0876\ \text{m} = 87.6\ \text{mm}$$

Thus a pitch of 80 mm may be used for the nails on the bottom. ∎

EXAMPLE 7–10

A girder is subjected to the loads shown. It is fabricated with four steel angles riveted to a web plate, as shown in the section. Determine the pitch of the rivets if they have a diameter of $\frac{3}{4}$ in. and an allowable shear stress of 15 ksi.

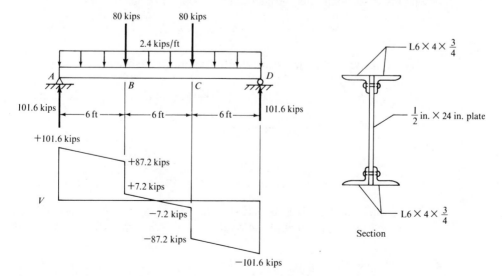

SOLUTION

The neutral axis is located at the middepth of the section due to symmetry. The moment of inertia of the entire section about the neutral axis is

$$I = I_{web} + 4I_{angle} = \frac{bh^3}{12} + 4[\bar{I} + Ad^2]_{angle}$$

$$= \frac{(\frac{1}{2})(24)^3}{12} + 4[8.68 + 6.94(12 - 1.08)^2] = 3920 \text{ in.}^4$$

By inspecting the shear diagram plotted, the maximum pitch for regions AB and CD is calculated using the maximum shear force $V_{max} = 101.6$ kips. The value of Q is determined by calculating the first moment of the areas of both angles at the top (the shaded area shown in the following figure) about the neutral axis. Thus

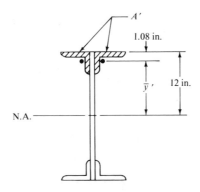

$$A' = 2A_{angle} = 2(6.94) = 13.88 \text{ in.}^2$$

$$Q = A'\bar{y}' = (13.88)(12 - 1.08) = 151.6 \text{ in.}^3$$

$$q = \frac{VQ}{I} = \frac{(101.6 \text{ kips})(151.6 \text{ in.}^3)}{3920 \text{ in.}^4} = 3.93 \text{ kips/in.}$$

The cross-sectional area of each rivet is

$$A = \frac{1}{4}\pi(0.75 \text{ in.})^2 = 0.442 \text{ in.}^2$$

The rivets are in double shear. Hence the allowable shear force in each rivet is

$$(F_s)_{allow} = \tau_{allow}[2A] = (15 \text{ ksi})[2(0.442 \text{ in.}^2)] = 13.3 \text{ kips}$$

From Eq. (7–18), the maximum pitch is

$$p_{max} = \frac{(F_s)_{allow}}{q} = \frac{13.3 \text{ kips}}{3.93 \text{ kips/in.}} = 3.37 \text{ in.}$$

Thus a pitch of 3 in. may be used for the rivets in regions AB and CD.

To determine the pitch of the rivets in region BC, we use $V = 7.2$ kips. Thus

$$q = \frac{VQ}{I} = \frac{(7.2 \text{ kips})(151.6 \text{ in.}^3)}{3920 \text{ in.}^4} = 0.278 \text{ kip/in.}$$

$$p_{\max} = \frac{(F_s)_{\text{allow}}}{q} = \frac{13.3 \text{ kips}}{0.278 \text{ kips/in.}} = 47.8 \text{ in.}$$

In this case the maximum pitch permissible by the code will govern for region *BC*. A pitch of 12 in. may be used conveniently unless the code requires closer spacing. ∎

PROBLEMS

7–45   Four 25 mm by 150 mm full-sized boards are nailed together to form a box section, as shown in Fig. P7–45. If the beam is subjected to a constant vertical shear force of 4.5 kN and the allowable shear force per nail is 800 N, determine the maximum pitch of the nails.

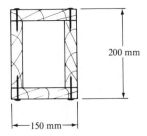

200 mm

150 mm

FIGURE P7–45

7–46   Two "two by four" pine studs with a dressed size of $1\frac{3}{4}$ in. $\times$ $3\frac{3}{4}$ in. are nailed together to make a beam of approximately square cross section, as shown in Fig. P7–46. Two 0.192-in. diameter nails are used in each spacing at a pitch of $2\frac{1}{2}$ in. If the allowable flexural stress for pine is 1100 psi and the allowable shear stress for the nails is 12 000 psi, determine the maximum uniform load that the beam can carry over a 4-ft simple span.

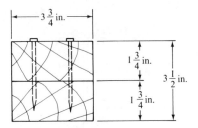

$3\frac{3}{4}$ in.

$1\frac{3}{4}$ in.

$1\frac{3}{4}$ in.

$3\frac{1}{2}$ in.

FIGURE P7–46

7–47   Three 50 mm by 100 mm full-sized timbers are fastened by 5-mm-diameter bolts spaced at a pitch of 40 mm, as shown in Fig. P7–47. If the allowable flexural stress in timber is 8 MPa and the allowable shear stress in the bolts is 100 MPa, determine the maximum concentrated load that can be applied at the midpoint of a 3-m simple span. Neglect the weight of the beam.

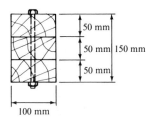

FIGURE P7–47

7–48   A simple beam of 12 ft span is subjected to a uniform load of 1200 lb/ft. The cross section of the beam is fabricated by two pieces of full-sized $\frac{1}{2}$ in. by 24 in. plywood lagged to two full-sized 2 in. by 8 in. timber planks, as shown in Fig. P7–48. If the screws are $\frac{1}{8}$ in. in diameter with an allowable shear stress of 8000 psi, determine the required pitch of the screws near the supports.

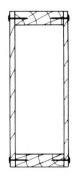

FIGURE P7–48

7–49   A built-up beam consists of cover plates fastened to the steel wide-flange beam by $\frac{3}{4}$-in.-diameter rivets, as shown in Fig. P7–49. If the beam is subjected to a constant vertical shear force of 120 kips and the allowable shear stress in the rivets is 15 ksi, determine the required pitch of the rivets.

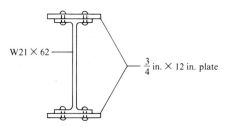

FIGURE P7–49

7–50   An overhanging beam subjected to a concentrated load is fabricated from two steel channels and two steel cover plates, as shown in Fig. P7–50. If the bolts are $\frac{1}{2}$ in. in diameter with an allowable shear stress of 15 ksi, determine the required pitches of the rivets in segments AB and BC.

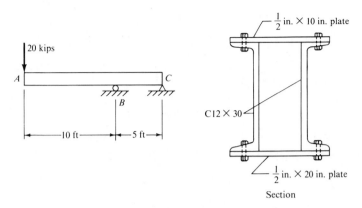

FIGURE P7–50

7–51   A simple beam of 16 ft span has a cross section consisting of a steel channel and a wide-flange steel section fastened together by 1-in. bolts, as shown in Fig. P7–51. The beam is subjected to a vertical downward load of 50 kips at the midspan. Consider the weight of the beam as a uniform load. The moment of inertia of the entire section about the neutral axis is 1250 in.[4] (see Example 4–11 on p. 109). If the allowable shear stress in the rivets is 15 ksi, determine the required pitch.

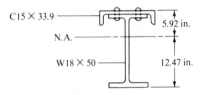

FIGURE P7–51

*7–7
BEAMS OF COMPOSITE MATERIALS

Beams previously discussed were of homogeneous material. In engineering practice, however, beams made of several materials, especially those made of two materials, are in common use. For example, timber beams are often strengthened by metal plates, and concrete beams are almost always reinforced with steel bars.

Consider the rectangular wooden beam reinforced with a steel plate, as shown in Fig. 7–13(a). Assume that the steel plate is properly fastened to the wooden section so that there is no sliding between steel and wood during bending. The basic deformation assumption used in deriving the flexure formula remains valid. That is, plane sections at right angle to the axis remain plane. Therefore, the strains of the longitudinal fibers vary linearly from the neutral axis, as shown in Fig. 7–13(b). For elastic deformations, stress is proportional to strain. Thus when the section is subjected to a positive moment, the flexural stress is distributed as shown in Fig. 7–13(c). Note that since $E_{st} > E_{wd}$, the flexural stress in the steel plate is much greater than the flexural stress in the wooden section. At the interface (level 3), the stress in steel is

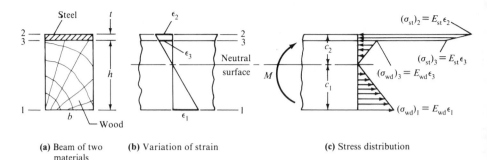

(a) Beam of two materials　　**(b)** Variation of strain　　　　　　　　**(c)** Stress distribution

(d) Tranformed section in steel

(e) Stress distribution in the transformed section

FIGURE 7–13

$$(\sigma_{st})_3 = E_{st}\epsilon_3 = \frac{E_{st}}{E_{wd}}E_{wd}\epsilon_3 = \frac{E_{st}}{E_{wd}}(\sigma_{wd})_3$$

Let the ratio of the moduli of elasticity of steel and wood, E_{st}/E_{wd}, be denoted by n. Then

$$(\sigma_{st})_3 = n(\sigma_{wd})_3$$

The flexural stresses in the section can be determined by using the *transformed section* technique, which consists of constructing a section of one material on which the resisting forces are the same as on the composite section. A choice must be made as to the material in the section to be transformed. It can be either one of the component materials. To obtain a transformed section in steel, as shown in Fig. 7-13(d), the dimensions of steel section remain unchanged. The width b of the wooden section is reduced to b/n, while its depth remains unchanged. The flexural stresses vary linearly from the neutral axis on the transformed section. Therefore, the flexure formula applies. The stress distribution in the transformed section is shown in Fig. 7–13(e), where the stress at level 1 is

$$(\sigma_{st})_1 = E_{st}\epsilon_1 = nE_{wd}\epsilon_1 = n(\sigma_{wd})_1$$

Thus it is seen that while the width of the wooden section is reduced to b/n in the transformed section, the stress is increased by n times. Therefore, the resisting force of the composite section is equivalent to that of the transformed section. Since there is no change in the vertical dimensions, the resisting moment of the original section is equivalent to that of the transformed section.

The maximum flexural stresses in steel and in wood are, respectively,

$$(\sigma_{st})_{max} = (\sigma_{st})_2 = \frac{Mc_2}{I_{st}} \tag{7–19}$$

$$(\sigma_{wd})_{max} = (\sigma_{wd})_1 = \frac{(\sigma_{st})_1}{n} = \frac{1}{n}\left[\frac{Mc_1}{I_{st}}\right] \tag{7–20}$$

where I_{st} is the moment of inertia of the entire transformed section in steel [Fig. 7–13(d)] with respect to the neutral axis.

An alternative solution is to transform the section into wood, as shown in Fig. 7–14(a). The stress distribution in this transformed section is shown in Fig. 7–14(b).

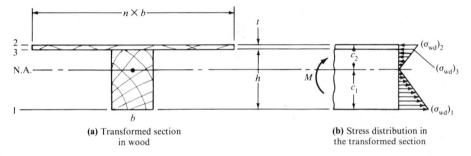

(a) Transformed section in wood   (b) Stress distribution in the transformed section

FIGURE 7–14

In this case the maximum flexural stresses in wood and in steel are, respectively,

$$(\sigma_{wd})_{max} = (\sigma_{wd})_1 = \frac{Mc_1}{I_{wd}} \tag{7–21}$$

$$(\sigma_{st})_{max} = (\sigma_{st})_2 = n(\sigma_{wd})_2 = n\left[\frac{Mc_2}{I_{wd}}\right] \tag{7–22}$$

where I_{wd} is the moment of inertia of the entire transformed section in wood [Fig. 7–14(a)] with respect to the neutral axis.

━━━━ **EXAMPLE 7–11** ━━━━

A composite beam consists of a wooden section reinforced with steel plate at the bottom as shown. If the beam is subjected to a maximum positive moment of 50 kN · m, determine the maximum absolute value of the flexural stresses in steel and in wood. Given: $E_{st} = 210$ GPa and $E_{wd} = 10.5$ GPa.

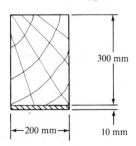

300 mm

←200 mm→

10 mm

SOLUTION

The ratio of the moduli of elasticity is

$$n = \frac{E_{st}}{E_{wd}} = \frac{210 \text{ GPa}}{10.5 \text{ GPa}} = 20$$

The transformed section in steel and its moment of inertia about the neutral axis are

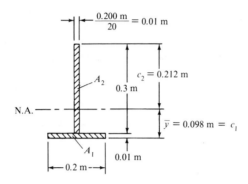

$$\bar{y} = \frac{A_1 y_1 + A_2 y_2}{A_1 + A_2}$$

$$= \frac{(0.2 \times 0.01)(0.005) + (0.01 \times 0.3)(0.01 + 0.15)}{0.2 \times 0.01 + 0.01 \times 0.3} = 0.098 \text{ m}$$

$$I_{st} = [\bar{I}_1 + A_1 d_1^2] + [\bar{I}_2 + A_2 d_2^2]$$

$$= \left[\frac{0.2 \times 0.01^3}{12} + (0.2 \times 0.01)(0.098 - 0.005)^2 \right]$$

$$+ \left[\frac{0.01 \times 0.3^3}{12} + (0.01 \times 0.3)(0.212 - 0.15)^2 \right] = 5.14 \times 10^{-5} \text{ m}^4$$

From Eqs. (7–19) and (7–20), the maximum flexural stresses in steel and in wood are, respectively,

$$(\sigma_{st})_{max} = \frac{Mc_1}{I_{st}} = \frac{(50 \text{ kN} \cdot \text{m})(0.098 \text{ m})}{5.14 \times 10^{-5} \text{ m}^4} = 95.3 \times 10^3 \text{ kN/m}^2$$

$$= 95.3 \text{ MPa (T)}$$

$$(\sigma_{wd})_{max} = \frac{1}{n} \left[\frac{Mc_2}{I_{st}} \right] = \frac{1}{20} \left[\frac{(50)(0.212)}{5.14 \times 10^{-5}} \right] = 10.3 \times 10^3 \text{ kN/m}^2$$

$$= 10.3 \text{ MPa (C)}$$

■

PROBLEMS

7–52 Rework Example 7–11 by using a transformed section in wood.

7–53   A timber beam reinforced with steel plates has a cross-sectional dimensions as shown in Fig. P7–53. Determine the maximum bending stress in each material due to a bending moment of 100 kN · m. E_{st} = 210 MPa and E_{wd} = 10.5 MPa.

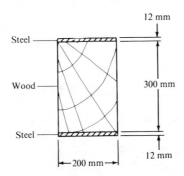

FIGURE P7–53

7–54   A composite beam of steel and aluminum having the cross-sectional dimensions as shown in Fig. P7–54 is subjected to a positive bending moment of 300 kip-in. Determine the maximum bending stresses in steel and aluminum. E_{st} = 30 000 ksi and E_{al} = 10 000 ksi.

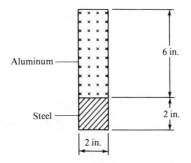

FIGURE P7–54

*In Problems **7–55** and **7–56**, determine the allowable bending moment about horizontal neutral axis for the composite beam of wood and steel having cross-sectional dimensions as shown. The moduli of elasticity are E_{st} = 210 GPa and E_{wd} = 10.5 GPa. The allowable flexural stresses are $(\sigma_{st})_{allow}$ = 160 MPa and $(\sigma_{wd})_{allow}$ = 8.5 MPa.*

7–55

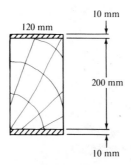

FIGURE P7–55

7–56

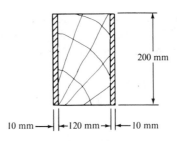

FIGURE P7–56

7–57 A beam is built up with standard steel I-beam S 12 × 31.8 and timbers, as shown in Fig. P7–57. Materials are properly fastened so that they act as one unit. Determine the allowable uniform load in lb/ft that the beam can carry over a simple span of 20 ft. Given: $E_{st} = 30 \times 10^6$ psi, $E_{wd} = 1.5 \times 10^6$ psi, $(\sigma_{st})_{allow} = 24\,000$ psi, and $(\sigma_{wd})_{allow} = 1200$ psi.

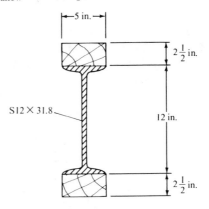

FIGURE P7–57

*7–8
REINFORCED CONCRETE BEAMS

The transformed section method used in Section 7–7 can be applied to the reinforced concrete beams shown in Fig. 7–15(a). The reinforced concrete section is usually transformed into equivalent concrete area.

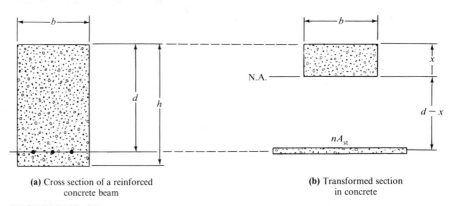

(a) Cross section of a reinforced concrete beam

(b) Transformed section in concrete

FIGURE 7–15

Concrete is very weak in resisting tension and it may crack in the tension zone. Therefore, concrete is not reliable in resisting tension. The transformed section takes the shape shown in Fig. 7–15(b), where the area of concrete in the compression zone remains unchanged, but it does not include any concrete area in the tension zone. The tensile force is assumed to be carried by steel reinforcing bars alone. The transformed area of steel bars into concrete is nA_{st}, where $n = E_{st}/E_{cn}$ and A_{st} is the total cross-sectional area of the bars in the section. For computation purpose, the area nA_{st} is assumed to be concentrated at the centerline of the steel bars.

Since the neutral axis must pass through the centroid of the transformed section, the first moment of the concrete area in the compression zone about the neutral axis must be equal to the first moment of the transformed steel area about the same axis. Thus

$$(bx)\left(\frac{x}{2}\right) = nA_s(d - x) \tag{7–23}$$

from which x can be solved.

The moment of inertia of the transformed section about the neutral axis is then

$$I = \frac{bx^3}{3} + nA_{st}(d - x)^2 \tag{7–24}$$

The maximum compressive stress in concrete and the tensile stress in the steel bars are, respectively,

$$(\sigma_{cn})_{max} = \frac{M_{max}x}{I} \tag{7–25}$$

$$(\sigma_{st})_{max} = n\left[\frac{M_{max}(d - x)}{I}\right] \tag{7–26}$$

EXAMPLE 7–12

Determine the maximum compressive stress in concrete and the maximum tensile stress in the steel bars of a simply supported reinforced concrete beam shown in Figs. (a) and (b). Assume that $n = E_{st}/E_{cn} = 10$ and the concrete weighs 150 lb/ft³.

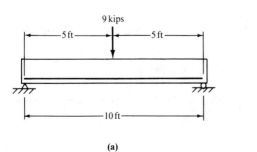

(a)

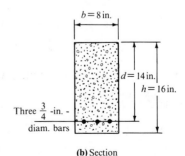

(b) Section

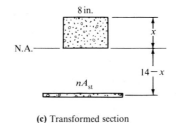

(c) Transformed section

SOLUTION

The transformed section of the given beam is shown in Fig. (c). The total cross-sectional area of three $\frac{3}{4}$-in.-diameter bars is

$$A_{st} = 3[\tfrac{1}{4}\pi d^2] = 3[\tfrac{1}{4}\pi(0.75 \text{ in.})^2] = 1.32 \text{ in.}^2$$

Thus the transformed area of steel (into concrete) is

$$nA_{st} = 10 \times 1.32 = 13.2 \text{ in.}^2$$

The neutral axis of the transformed section can be located by solving Eq. (7–23). We have

$$8x\left(\frac{x}{2}\right) = 13.2(14 - x)$$

or

$$4x^2 + 13.2x - 185 = 0$$

From the quadratic formula, the solution is

$$x = \frac{-13.2 \pm \sqrt{(13.2)^2 - 4(4)(-185)}}{2(4)} = \frac{-13.2 \pm 56.0}{8}$$

$$= 5.35 \text{ in.} \qquad \text{(the negative solution is dropped)}$$

From Eq. (7–24), the moment of inertia of the transformed section about the neutral axis is

$$I = \frac{8(5.35)^3}{3} + 13.2(14 - 5.35)^2 = 1400 \text{ in.}^4$$

The weight of the beam per foot of length is

$$w = \left[\frac{8 \times 16}{144} \text{ ft}^2\right](150 \text{ lb/ft}^3) = 133 \text{ lb/ft} = 0.133 \text{ kip/ft}$$

The maximum moment at the midspan of the beam is

$$M_{max} = +\frac{PL}{4} + \frac{wL^2}{8} = +\frac{(9 \text{ kips})(10 \text{ ft})}{4} + \frac{(0.133 \text{ kip/ft})(10 \text{ ft})^2}{8}$$

$$= +24.2 \text{ kip-ft}$$

From Eq. (7–25), the maximum compressive stress in concrete is

$$(\sigma_{cn})_{max} = \frac{M_{max}x}{I} = \frac{(24.2 \times 12 \text{ kip-in.})(5.36 \text{ in.})}{1400 \text{ in.}^4} = 1.11 \text{ ksi}$$

From Eq. (7–26), the maximum tensile stress in the steel bars is

$$(\sigma_{st})_{max} = n\left[\frac{M_{max}(d - x)}{I}\right] = 10\left[\frac{(24.2 \times 12 \text{ kip-in.})(14 \text{ in.} - 5.35 \text{ in.})}{1400 \text{ in.}^4}\right]$$

$$= 18.0 \text{ ksi}$$

■

PROBLEMS

7–58 A rectangular concrete beam is reinforced with three $\frac{3}{4}$-in.-diameter steel bars ($A_{st} = 1.32 \text{ in.}^2$), as shown in Fig. P7–58. Locate the neutral axis of the section. Assume that $n = E_{st}/E_{cn} = 10$.

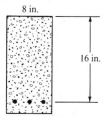

8 in.

16 in.

FIGURE P7–58

7–59 Determine the maximum stresses in concrete and steel for a reinforced concrete beam with the section shown in Fig. P7–59. The beam is subjected to a maximum positive bending moment of 70 kip-ft. Assume that $n = E_{st}/E_{cn} = 9$.

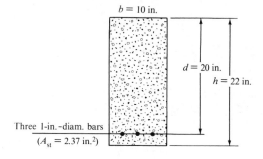

$b = 10$ in.

$d = 20$ in.

$h = 22$ in.

Three 1-in.-diam. bars
($A_{st} = 2.37 \text{ in.}^2$)

FIGURE P7–59

7–60 A reinforced concrete beam has the cross section shown in Fig. P7–60. Determine the allowable positive moment that the beam can resist. Given $n = E_{st}/E_{cn} = 8$, $(\sigma_{st})_{allow} = 20\ 000$ psi, and $(\sigma_{cn})_{allow} = 1800$ psi.

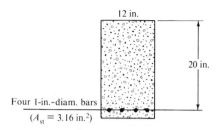

FIGURE P7–60

7–61   A 16-ft simply supported reinforced concrete beam has the cross section shown
in Fig. 7–61. Knowing that $n = E_{st}/E_{cn} = 10$, determine **(a)** the location of the
neutral axis if both the concrete and the steel are stressed to their maximum

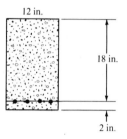

FIGURE P7–61

allowable values of $(\sigma_{cn})_{allow} = 1000$ psi and $(\sigma_{st})_{allow} = 20\ 000$ psi, **(b)** the
required area of steel A_{st}, **(c)** the allowable uniform load w lb/ft (including its
own weight) over the entire span.
[**HINTS:** (a) The transformed section and the stress distribution in the transformed
section are shown in the following:

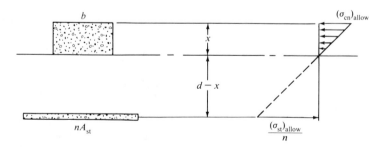

The value of x can be determined by proportion. (b) Solve A_{st} from Eq. (7–23).
(c) Solve M_{max} from Eq. (7–25) or (7–26) and then equate it to $wL^2/8$.]

CHAPTER 8

Deflections of Beams

8–1
INTRODUCTION

Under the action of transverse loads, a beam deflects from its unloaded position. The movement of a point normal to the axis of the beam is called the *deflection* of the beam at the point. If the maximum flexural stress in the beam is within the elastic limit of the beam material, the beam undergoes elastic deflection. In this chapter we discuss the computation of elastic beam deflections.

In structural or machine design, deformation analysis is as important as strength analysis. Accurate values of beam deflections are needed in beam design. The deflection of beams to which the plastered ceiling is attached must be limited so that the beam will not crack the plaster. Power-transmission shafts carrying gears must be rigid enough to ensure proper meshing of the gear teeth. Consideration of deflections of a beam is also needed in solving statically indeterminate beam problems.

There are many methods that can be used to solve beam deflection problems. The double integration method, the superposition method, and the moment-area method are studied in this chapter.

The relationship between the moment and the radius of curvature are derived first. Then the general rules for sketching the beam deflection curve are discussed. The double integration method is discussed next. Beam deflection formulas for some simple cases can be derived by the double integration method, which is followed by the method of superposition using the beam deflection formulas. Finally, the theorem of moment area method is developed and the computation of beam deflection by the moment-area method is discussed.

8–2
RELATION BETWEEN BENDING MOMENT
AND RADIUS OF CURVATURE

Consider a beam segment bent into a concave upward curvature due to a positive bending moment as shown in Fig. 8–1. Two adjacent sections *ab* and *cd* remain

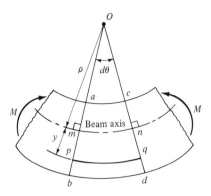

FIGURE 8–1

planar and normal to the axis of the beam. The point of intersection O of lines ab and cd is called the *center of curvature,* and the distance ρ (the Greek lowercase letter rho) from O to the beam axis is called the *radius of curvature.* The length of fiber mn along the axis of the beam remains unchanged, while that of fiber pq is elongated. Let the angle $d\theta$ at O be measured in radians. Then the length of mn is $\rho\, d\theta$ and the elongated length of pq is $(\rho + y)\, d\theta$. By definition, the strain ε of the fiber pq is

$$\varepsilon = \frac{pq - mn}{mn} = \frac{(\rho + y)\, d\theta - \rho\, d\theta}{\rho\, d\theta} = \frac{y}{\rho} \tag{a}$$

For elastic bending, the stress is proportional to strain. Thus we have

$$\sigma = E\varepsilon = \frac{Ey}{\rho} \tag{b}$$

When the stress $\sigma = My/I$ from the flexure formula is substituted in Eq. (b), we have

$$\frac{My}{I} = \frac{Ey}{\rho}$$

or

$$\frac{1}{\rho} = \frac{M}{EI} \tag{8–1}$$

This equation relates the bending moment M at any section of a beam to the radius of curvature of the elastic deflection curve of the beam. It states that for a beam with constant EI, the radius of curvature of the deflection curve is inversely proportional to the moment. This relationship is fundamental to the methods for determining beam deflections.

8–3
SKETCH OF BEAM DEFLECTION CURVE

Before the magnitudes of beam deflections are calculated, it is important to know the general shape of the deflection curve. To sketch the beam deflection curve, the following rules concerning beam deflections must be observed.

1. There is no vertical deflection at a simple unyielding support (roller or pinned support). The deflection curve must pass through such a support.
2. There can be no vertical deflection rotation of the tangent at a fixed support. The deflection curve must pass through the point of support and be tangent to the undeformed axis of the beam.
3. The moment diagram of the beam gives an indication of the shape of the deflection curve. For the part of beam where the moment is positive, the deflection curve is concave upward (⌣), and for the part of beam where the moment is negative, the deflection curve is concave downward (⌢). At the point where the bending moment changes sign, the concavity of the beam changes. This point is called the *point of inflection*.
4. As established in Section 8–2, the radius of curvature of the deflection curve is inversely proportional to the magnitude of the moment. The curve is "sharper" (i.e., has a shorter radius of curvature) where the magnitude of the moment is larger, and the curve is "flatter" (i.e., has a longer radius of curvature) where the magnitude of the moment is smaller. The deflection curve is a straight line (the radius of curvature is infinity) for the part of beam where the bending moment is zero.

──────── **EXAMPLE 8–1** ────────────────────────────────────

Sketch the deflection curve of the beam subjected to the loading shown in Fig. (a).

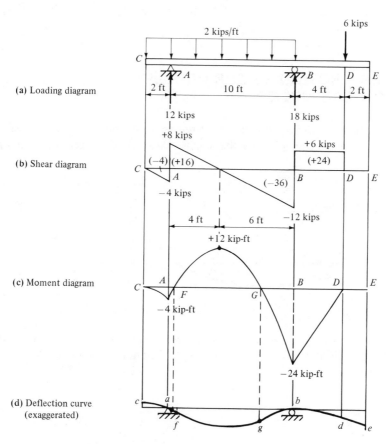

(a) Loading diagram

(b) Shear diagram

(c) Moment diagram

(d) Deflection curve
    (exaggerated)

SOLUTION

The reactions at the supports are determined by considering the free-body diagram of the entire beam. The results are indicated in Fig. (a). The shear and moment diagrams are plotted by using the summation method, as shown in Figs. (b) and (c).

The exaggerated deflection curve is sketched in Fig. (d). The curve passes through the simple supports at points a and b. Points f and g are points of inflection, since at the corresponding points F and G the moment changes sign. Parts cf and gd of the deflection curve are concave downward, since the moment at these parts are negative. Part fg of the deflection curve is concave upward, since the moment at each section in this region is positive. Part de is a straight line, since the moment is zero in this part of the beam.

The magnitude of deflection at each point along the beam can be determined by the methods to be discussed later in this chapter. ∎

PROBLEMS

8–1   Sketch the deflection curve of the beam in Example 6–9 (on p. 168).

8–2   Sketch the deflection curve of the beam in Problem 6–36 (on p. 173).

8–3   Sketch the deflection curve of the beam in Problem 6–39 (on p. 174).

8–4   Sketch the deflection curve of the beam in Problem 6–42 (on p. 174).

8–4
BEAM DEFLECTION BY THE DOUBLE INTEGRATION METHOD

Consider the deflected axis AB of the beam shown in Fig. 8–2. The undeformed axis of the beam is along the x-axis. Point O is the center of curvature of the

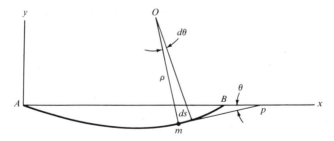

FIGURE 8–2

deflection curve at m. The radius of curvature Om is denoted by ρ. The angle $d\theta$ is the central angle at O subtended by a differential length ds along the beam. We have

$$d\theta = \frac{ds}{\rho} \tag{a}$$

From Eq. (8–1),

$$\frac{1}{\rho} = \frac{M}{EI}$$

Substituting in Eq. (a) gives

$$d\theta = \frac{M \, ds}{EI}$$

or

$$EI \, d\theta = M \, ds \qquad \text{(b)}$$

Since deflection is always small for engineering structures, we have, approximately,

$$ds \approx dx$$

Thus Eq. (b) becomes

$$EI \, d\theta = M \, dx \qquad \text{(8–2)}$$

In Fig. 8–2 the line mp is tangent to the curve at the point m. The angle between the tangent line and the x-axis is denoted by θ. The slope of the deflection curve at the point m is

$$\tan \theta = \frac{dy}{dx} \qquad \text{(c)}$$

For small deflection, the angle θ is also very small; thus

$$\tan \theta \approx \theta \text{ (in radians)}$$

and Eq. (c) becomes

$$\frac{dy}{dx} = \theta \qquad \text{(d)}$$

The general expression for the slope θ can be obtained by integrating Eq. (8–2); we have

$$EI\theta = \int M \, dx \qquad \text{(8–3)}$$

where EI is a constant. To carry out the integration, the bending moment M must be expressed as a function of x.

Expressing Eq. (d) in differential form and multiplying both sides by EI we write

$$EI \, dy = EI\theta \, dx \qquad \text{(8–4)}$$

Integrating this expression gives the general expression for the deflection as a function of x. That is,

$$EIy = \int EI\theta \, dx \qquad (8\text{–}5)$$

The resulting function of x for $EI\theta$ from Eq. (8–3) must be substituted in Eq. (8–5) before performing the integration.

From Eqs. (8–3) and (8–5), we see that by integrating the moment function $M(x)$ twice, a general expression of beam deflection is obtained. This method is therefore called the *double integration method*.

Recall from the discussion in Section 1–8 that each time an indefinite integral is integrated, a constant of integration is introduced. A beam deflection problem usually involves two or more constants of integration. These constants are evaluated by boundary conditions of the beam, as illustrated in the following examples.

EXAMPLE 8–2

Determine the slope θ_B and the deflection y_B at the free end B of the cantilever timber beam subjected to a uniform load shown. For timber, $E = 10$ GPa.

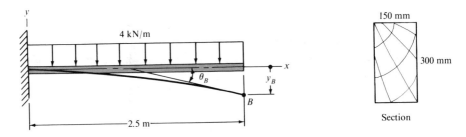

Section

SOLUTION

The reactions at the fixed support are first determined by considering the equilibrium of the entire beam. The results are indicated in the following diagram of the beam.

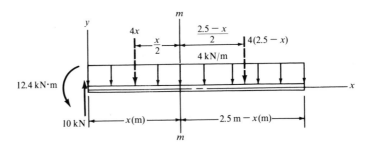

Recall from Section 6–7 that the moment at a section is the summation of the external moments about the section due to all the forces either to the left or to the right of the section. Computed from the left, the moment at section m–m is

$$M = -12.5 + 10x - (4x)\left(\frac{x}{2}\right) = -12.5 + 10x - 2x^2$$

Computed from the right, the moment at section m–m is

$$M = -[4(2.5 - x)]\frac{2.5 - x}{2} = -2(2.5 - x)^2 = -12.5 + 10x - 2x^2$$

From Eq. (8–3), integrating the moment function, we get

$$EI\theta = \int M\, dx = \int (-12.5 + 10x - 2x^2)\, dx \tag{a}$$
$$= -12.5x + 5x^2 - \tfrac{2}{3}x^3 + C_1$$

From Eq. (8–5), integrating the $EI\theta$ function, we get

$$EIy = \int EI\theta\, dx = \int (-12.5x + 5x^2 - \tfrac{2}{3}x^3 + C_1)\, dx \tag{b}$$
$$= -6.25x^2 + \tfrac{5}{3}x^3 - \tfrac{1}{6}x^4 + C_1x + C_2$$

The two constants of integration can be evaluated by the following boundary conditions:

1. The slope is zero at the fixed support, that is, $(\theta)_{x=0} = 0$. From Eq. (a),
$$0 = 0 + C_1 \qquad C_1 = 0$$

2. The deflection is zero at the fixed support, that is, $(y)_{x=0} = 0$. From Eq. (b),
$$0 = 0 + C_2 \qquad C_2 = 0$$

Therefore, the general expressions for the slope and deflection of the beam are

$$\theta = \frac{1}{EI}\left(-12.5x + 5x^2 - \frac{2}{3}x^3\right) \tag{c}$$

$$y = \frac{1}{EI}\left(-6.25x + \frac{5}{3}x^3 - \frac{1}{6}x^4\right) \tag{d}$$

These expressions can be used to compute the slope and deflection of any point along the beam by plugging in the proper values of EI and x.
The value of EI for the beam is

$$EI = (10 \times 10^6\ \text{kN/m}^2)\left(\frac{0.15 \times 0.3^3}{12}\ \text{m}^4\right) = 3375\ \text{kN} \cdot \text{m}^2$$

Note that if consistent units are used, the angle θ will be in radians and the deflection y will be in the length unit. In this problem, kN and m units are used for all quantities.

For the free end B, $x = 2.5$ m, and from Eqs. (c) and (d) we get

$$\theta_B = \frac{1}{3375}\left[-12.5(2.5) + 5(2.5)^2 - \frac{2}{3}(2.5)^3\right]$$

$$= -0.003\ 086\ \text{rad} = (-0.003\ 086\ \text{rad})\left(\frac{180°}{\pi\ \text{rad}}\right) = -0.177°$$

$$y_B = \frac{1}{3375}\left[-6.25(2.5)^2 + -\frac{5}{3}(2.5)^3 - \frac{1}{6}(2.5)^4\right]$$

$$= -0.005\ 79\ \text{m} = -5.79\ \text{mm}$$

The minus sign for θ indicates that the angle is measured clockwise from the x-axis to the tangent at B. The minus sign for y indicates that the deflection at B is measured downward from the x-axis. ■

EXAMPLE 8-3

Determine the deflection at points C and D of the beam of a wide-flange steel section W10 × 22 due to the loads shown.

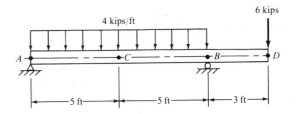

SOLUTION

The reactions of the beam are determined first; the results are shown in the following figure.

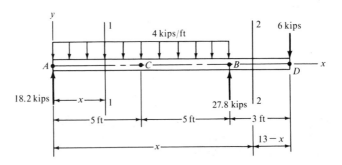

Computed from the left end, the bending moment at section 1–1 is

$$M_{1-1} = 18.2x - (4x)\left(\frac{x}{2}\right) = (18.2x - 2x^2)\ \text{kip-ft}$$

which is valid for any section between A and B. Computed from the right end, the bending moment at section 2–2 is

$$M_{2-2} = -6(13 - x) = (6x - 78) \text{ kip-ft}$$

which is valid for any section between B and D.

For segment AB:

$$EI\theta_{AB} = \int M_{1-1} \, dx = 9.1x^2 - \frac{2}{3}x^3 + C_1 \tag{a}$$

$$EIy_{AB} = \int EI\theta_{AB} \, dx = \frac{9.1}{3}x^3 - \frac{1}{6}x^4 + C_1x + C_2 \tag{b}$$

For segment BD:

$$EI\theta_{BD} = \int M_{2-2} \, dx = 3x^2 - 78x + C_3 \tag{c}$$

$$EIy_{BD} = \int EI\theta_{BD} \, dx = x^3 - 39x^2 + C_3x + C_4 \tag{d}$$

The four constants of integration can be determined by the following boundary conditions:

1. The deflection is zero at the left support A, that is, $(y_{AB})_{x=0} = 0$. From Eq. (b),

$$0 = 0 + C_2 \qquad C_2 = 0$$

2. The deflection is zero at the right support B (applied to y_{AB}), that is, $(y_{AB})_{x=10} = 0$. From Eq. (b),

$$0 = \frac{9.1}{3}(10)^3 - \frac{1}{6}(10)^4 + C_1(10) \qquad C_1 = -136.7$$

3. The slope computed from Eq. (a) and Eq. (b) must be the same, that is, $(\theta_{AB})_{x=10} = (\theta_{BD})_{x=10}$. From Eqs. (a) and (c),

$$9.1(10)^2 - \tfrac{2}{3}(10)^3 - 136.7 = 3(10)^2 - 78(10) + C_3 \qquad C_3 = 586.6$$

4. The deflection is zero at the left support B (applied to y_{BD}), that is, $(y_{BD})_{x=10} = 0$. From Eq. (d),

$$0 = (10)^3 - 39(10)^2 + 586.6(10) + C_4 \qquad C_4 = -2966$$

Therefore, the general expressions for the beam deflection are

$$y_{AB} = \frac{1}{EI}\left(\frac{9.1}{3}x^3 - \frac{1}{6}x^4 - 136.7x \right) \tag{e}$$

$$y_{BD} = \frac{1}{EI}(x^3 - 39x^2 + 586.6x - 2966) \tag{f}$$

From Table A–1, for W10 × 22, $I = 118$ in.^{4}. The constant EI value is

$$EI = (30 \times 10^3 \text{ kips/in.}^2)(118 \text{ in.}^4) = 3.54 \times 10^6 \text{ kip-in.}^2$$
$$= 2.46 \times 10^4 \text{ kip-ft}^2$$

Note that consistent units of kips and feet must be used for all quantities.

The deflections at C and D are

$$y_C = (y_{AB})_{x=5} = \frac{1}{2.46 \times 10^4} \left[\frac{9.1}{3}(5)^3 - \frac{1}{6}(5)^4 - 136.7(5) \right]$$

$$= -0.0166 \text{ ft} = -0.199 \text{ in. } \downarrow$$

$$y_D = (y_{BD})_{x=13} = \frac{1}{2.46 \times 10^4} [(13)^3 - 39(13)^2 + 586.6(13) - 2966]$$

$$= +0.0108 \text{ ft} = +0.130 \text{ in. } \uparrow$$

■

EXAMPLE 8–4

Determine the deflection at quarter points D, B, and E of the simply supported beam of a wide-flange steel section W10 × 30 due to the load shown.

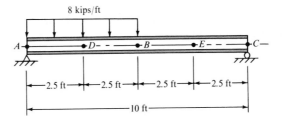

SOLUTION

The reactions of the beam are determined first; the results are shown in the following figure.

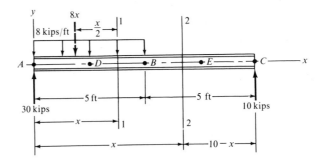

Computed from the left end, the bending moment at section 1–1 is

$$M_{1-1} = 30x - (8x)\left(\frac{x}{2}\right) = (30x - 4x^2) \text{ kip-ft}$$

which is valid for any section between A and B. Computed from the right end, the bending moment at section 2–2 is

$$M_{2-2} = 10(10 - x) = (100 - 10x) \text{ kip-ft}$$

which is valid for any section between B and C.

For segment AB:

$$EI\theta_{AB} = \int M_{1-1}\, dx = 15x^2 - \tfrac{4}{3}x^3 + C_1 \tag{a}$$

$$EIy_{AB} = \int EI\theta_{AB}\, dx = 5x^3 - \tfrac{1}{3}x^4 + C_1 x + C_2 \tag{b}$$

For segment BC:

$$EI\theta_{BC} = \int M_{2-2}\, dx = 100x - 5x^2 + C_3 \tag{c}$$

$$EIy_{BC} = \int EI\theta_{BC}\, dx = 50x^2 - \tfrac{5}{3}x^3 + C_3 x + C_4 \tag{d}$$

The four constants of integration can be determined by the following boundary conditions:

1. The deflection is zero at the left support, that is, $(y_{AB})_{x=0} = 0$. From Eq. (b),

$$0 = 0 + C_2 \qquad C_2 = 0$$

2. The slope at B computed from Eqs. (a) and (c) must be equal, that is, $(\theta_{AB})_{x=5} = (\theta_{BC})_{x=5}$. From Eqs. (a) and (c),

$$15(5)^2 - \tfrac{4}{3}(5)^3 + C_1 = 100(5) - 5(5)^2 + C_3 \qquad C_1 - C_3 = 166.7 \tag{e}$$

3. The deflection at B computed from Eqs. (b) and (d) must be equal, that is, $(y_{AB})_{x=5} = (y_{BC})_{x=5}$. From Eqs. (b) and (d),

$$5(5)^3 - \tfrac{1}{3}(5)^4 + C_1(5) + 0 = 50(5)^2 - \tfrac{5}{3}(5)^3 + C_3(5) + C_4$$

$$5C_1 - 5C_3 - C_4 = 625 \tag{f}$$

4. The deflection is zero at the right support, that is, $(y_{BC})_{x=10} = 0$. From Eq. (d),

$$0 = 50(10)^2 - \tfrac{5}{3}(10)^3 + C_3(10) + C_4 \qquad 10C_3 + C_4 = -3333 \tag{g}$$

Solving Eqs. (e), (f), and (g) simultaneously, we get

$$C_1 = -187.5$$

$$C_3 = -354.2$$

$$C_4 = +209$$

Therefore, the general expressions for the beam deflections are

$$y_{AB} = \frac{1}{EI}\left(5x^3 - \frac{1}{3}x^4 - 187.5x\right) \tag{i}$$

$$y_{BC} = \frac{1}{EI}\left(50x^2 - \frac{5}{3}x^3 - 354.2x + 209\right) \tag{j}$$

From Table A–1, for a W10 × 30, $I = 170$ in.^{4}, and for steel, $E = 30 \times 10^3$ ksi, the constant value of EI is

$$EI = (30 \times 10^3 \text{ ksi/in.}^2)(170 \text{ in.}^4) = 5.1 \times 10^6 \text{ kip-in.}^2$$
$$= 3.54 \times 10^4 \text{ kip-ft}^2$$

When consistent units in kips and feet are used, we get, from Eqs. (i) and (j), the following deflections at the quarter points:

$$y_D = (y_{AB})_{x=2.5} = \frac{1}{3.54 \times 10^4} [5(2.5)^3 - \tfrac{1}{3}(2.5)^4 - 187.5(2.5)]$$
$$= -0.0114 \text{ ft} = -0.1368 \text{ in. } \downarrow$$

$$y_B = (y_{AB})_{x=5} = \frac{1}{3.54 \times 10^4} [5(5)^3 - \tfrac{1}{3}(5)^4 - 187.5(5)]$$
$$= -0.0147 \text{ ft} = -0.176 \text{ in. } \downarrow$$

$$y_E = (y_{BC})_{x=7.5} = \frac{1}{3.54 \times 10^4} [50(7.5)^2 - \tfrac{5}{3}(7.5)^3 - 354.2(7.5) + 209]$$
$$= -0.009\ 55 \text{ ft} = -0.1146 \text{ in. } \downarrow$$

EXAMPLE 8–5

Derive the general expressions for the slope and deflection of a simple beam subjected to a uniform load over the entire span in terms of the parameters shown in the figure.

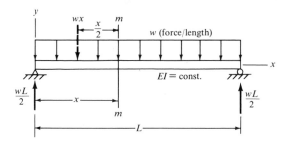

SOLUTION
The moment at an arbitrary section m–m is

$$M = \frac{wL}{2}x - (wx)\left(\frac{x}{2}\right) = \frac{1}{2}wLx - \frac{1}{2}wx^2$$

Integrating the moment function, we get

$$EI\theta = \int M\ dx = \tfrac{1}{4}wLx^2 - \tfrac{1}{6}wx^3 + C_1$$

Due to symmetry, the slope at the midspan is zero. Setting $(\theta)_{x=L/2} = 0$, we have

$$0 = \frac{1}{4}wL\left(\frac{L}{2}\right)^2 - \frac{1}{6}w\left(\frac{L}{2}\right)^3 + C_1 \qquad C_1 = -\frac{1}{24}wL^3$$

A second integration gives

$$EIy = \int EI\theta \, d\theta = \tfrac{1}{12}wLx^3 - \tfrac{1}{24}wx^4 - \tfrac{1}{24}wL^3x + C_2$$

The deflection at the left support must be zero, that is, $(y)_{x=0} = 0$, then C_2 must be zero. Therefore, the general expressions for the slope and deflection are

$$\theta = \frac{1}{EI}\left(\frac{1}{4}wLx^2 - \frac{1}{6}wx^3 - \frac{1}{24}wL^3\right)$$

or

$$\theta = -\frac{w}{24EI}(L^3 - 6Lx^2 + 4x^3)$$

and

$$y = \frac{1}{EI}\left(\frac{1}{12}wLx^3 - \frac{1}{24}wx^4 - \frac{1}{24}wL^3x\right)$$

or

$$y = -\frac{wx}{24EI}(L^3 - 2Lx^2 + x^3)$$

The deflection at the right support must be zero. This condition provides a useful check. Thus

$$(y)_{x=L} = -\frac{wL}{24EI}(L^3 - 2L^3 + L^3) = 0 \qquad \text{(checks)}$$

The maximum deflection occurs at the midspan with $x = L/2$; thus

$$y_{max} = y_{L/2} = -\frac{w(\tfrac{1}{2}L)}{24EI}\left[L^3 - 2L\left(\frac{L}{2}\right)^2 + \left(\frac{L}{2}\right)^3\right] = -\frac{5wL^4}{384EI}$$

where the minus sign indicates that the deflection is downward.   ■

PROBLEMS

In Problems **8–5** *to* **8–8**, *use the double integration method to determine the maximum deflection at the free end of the cantilever beam due to the loading shown. Use the EI value indicated in the figure.*

8–5

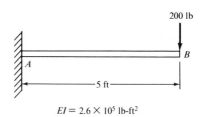

$EI = 2.6 \times 10^5$ lb-ft^2

FIGURE P8–5

8–6

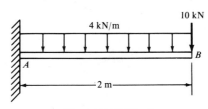

$EI = 4.5 \times 10^3$ kN·m^2

FIGURE P8–6

8–7

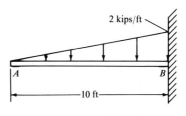

$EI = 1.73 \times 10^4$ kip-ft^2

FIGURE P8–7

8–8

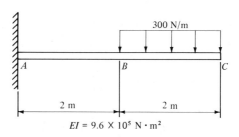

$EI = 9.6 \times 10^5$ N·m^2

FIGURE P8–8

In Problems 8–9 to 8–12, use the double integration method to determine the deflection at the quarter points of the simple beam due to the loading shown.

8–9

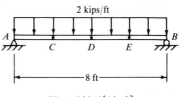

$EI = 1.2 \times 10^4$ kip-ft^2

FIGURE P8–9

8–10

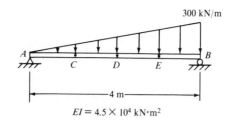

$EI = 4.5 \times 10^4$ kN·m²

FIGURE P8–10

8–11

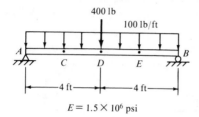

$E = 1.5 \times 10^6$ psi

FIGURE P8–11

(**HINT:** Due to symmetry, the deflection is symmetrical with respect to the midspan and the slope at the midspan is zero.)

8–12

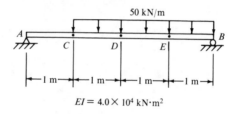

$EI = 4.0 \times 10^4$ kN·m²

FIGURE P8–12

In Problems **8–13** *to* **8–15**, *use the double integration method to determine the deflections at the midpoint C and the end D of the overhanging beam due to the loading shown.*

8–13

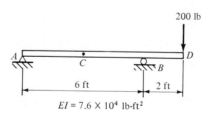

$EI = 7.6 \times 10^4$ lb-ft²

FIGURE P8–13

8–14

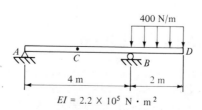

$EI = 2.2 \times 10^5$ N · m²

FIGURE P8–14

8–15

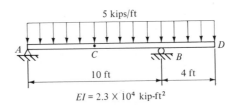

$EI = 2.3 \times 10^4$ kip-ft^2

FIGURE P8–15

In Problems **8–16** to **8–18**, *derive a general expression for the deflection of the beam subjected to the given loading in terms of the parameters shown in the figure.*

8–16

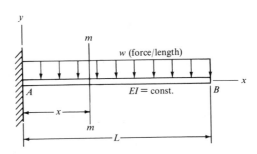

FIGURE P8–16

8–17

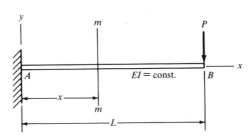

FIGURE P8–17

8–18

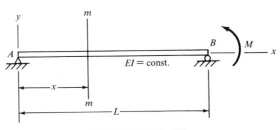

FIGURE P8–18

8–5
BEAM DEFLECTION FORMULAS

Formulas derived by using the double integration method, such as those in Example 8–5 and Problems 8–16, 8–17, and 8–18, are listed in Table 8–1. The table includes formulas for cantilever beams (cases 1 to 5) and simple beams (cases 6 to 10) for several different loading conditions. Each case in the table is given a case number in the first column for easy reference. The second column shows a sketch of the beam indicating the support conditions, the dimensions, the loading, and the meaning of notations used in the formulas. In each case, the location of the section where the maximum deflection occurs is indicated in the sketch. In the third column are the general formulas of beam deflection in terms of x, the distance from the left support to the section. In cases 2 and 7, two different formulas applicable to two different segments of the beam are listed. In the fourth column, formulas for maximum slopes, denoted by θ_{max} (measured in radians), are listed for cantilever beams at the free ends. For simple beams, formulas are listed for the slopes at the left and right ends of the beam, denoted by θ_l and θ_r. In the last column are the formulas for computing the maximum deflection along the beam. In simple beams with unsymmetrical loadings (cases 7, 9, and 10), the formulas for computing the midspan deflection are also included. The table is reproduced at the inside of the back cover for easy reference.

8–6
BEAM DEFLECTION BY THE METHOD OF SUPERPOSITION

The method of superposition can be used to determine small elastic beam deflections. In this method, deflection caused by each load is calculated separately; then the algebraic sum of the deflections gives the resultant deflection due to all the loads acting simultaneously.

The method of superposition is widely used in solving problems involving beam deflection. The method consists of utilizing the deflection formulas tabulated in Table 8–1. A variety of beam deflection problems can be solved this way. Ingenuity often plays an important role in using the formulas, as illustrated in the following examples.

──────── **EXAMPLE 8–6** ────────────────────────────────────

Use the method of superposition to find the maximum deflection at the free end C of the cantilever timber beam of rectangular section shown. For timber, $E = 10$ GPa.

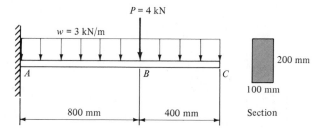

SOLUTION
The constant EI value of the beam is

$$EI = (10 \times 10^6 \text{ kN/m}^2)\left(\frac{0.1 \times 0.2^3}{12} \text{ m}^4\right) = 667 \text{ kN} \cdot \text{m}^2$$

The cantilever is subjected to a concentrated load and a uniform load; this loading can be considered to be the superposition of cases 2 and 3 of Table 8–1.

TABLE 8–1 Beam Deflection Formulas

Case	Beam	General Deflection Formula (y is downward as positive)	Slope at Ends	Maximum Deflection
1		$$y = \frac{Px^2}{6EI}(3L - x)$$	$$\theta_{max} = \frac{PL^2}{2EI}$$	$$y_{max} = \frac{PL^3}{3EI}$$
2		$$y_{AB} = \frac{Px^2}{6EI}(3a - x)$$ $$y_{BC} = \frac{Pa^2}{6EI}(3x - a)$$	$$\theta_{max} = \frac{Pa^2}{2EI}$$	$$y_{max} = \frac{Pa^2}{6EI}(3L - a)$$
3	w(force/length)	$$y = \frac{wx^2}{24EI}(x^2 + 6L^2 - 4Lx)$$	$$\theta_{max} = \frac{wL^3}{6EI}$$	$$y_{max} = \frac{wL^4}{8EI}$$
4	w (force/length)	$$y = \frac{wx^2}{120EIL}(10L^3 - 10L^2x + 5Lx^2 - x^3)$$	$$\theta_{max} = \frac{wL^3}{24EI}$$	$$y_{max} = \frac{wL^4}{30EI}$$
5		$$y = \frac{Mx^2}{2EI}$$	$$\theta_{max} = \frac{ML}{EI}$$	$$y_{max} = \frac{ML^2}{2EI}$$

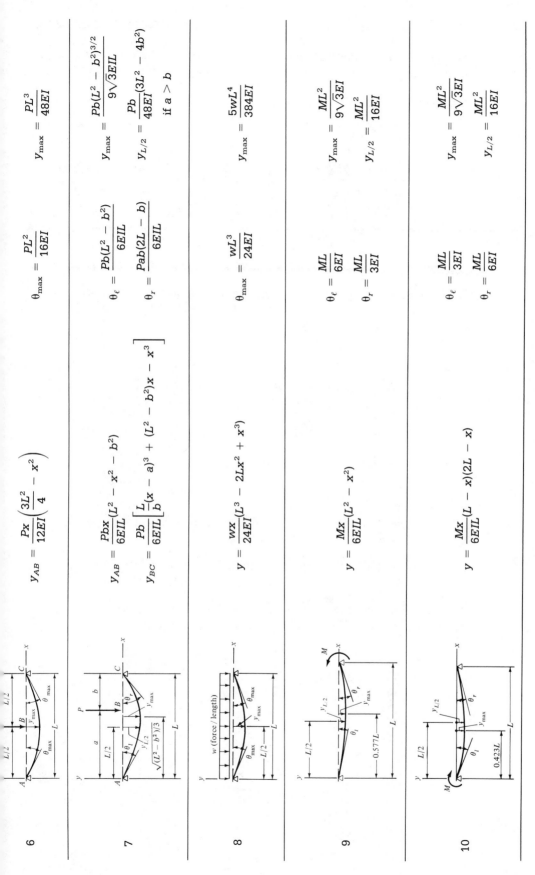

6

$$y_{AB} = \frac{Px}{12EI}\left(\frac{3L^2}{4} - x^2\right)$$

$$\theta_{max} = \frac{PL^2}{16EI}$$

$$y_{max} = \frac{PL^3}{48EI}$$

7

$$y_{AB} = \frac{Pbx}{6EIL}(L^2 - x^2 - b^2)$$

$$y_{BC} = \frac{Pb}{6EIL}\left[\frac{L}{b}(x - a)^3 + (L^2 - b^2)x - x^3\right]$$

$$\theta_\ell = \frac{Pb(L^2 - b^2)}{6EIL}$$

$$\theta_r = \frac{Pab(2L - b)}{6EIL}$$

$$y_{max} = \frac{Pb(L^2 - b^2)^{3/2}}{9\sqrt{3}EIL}$$

$$y_{L/2} = \frac{Pb}{48EI}(3L^2 - 4b^2)$$

if $a > b$

8

$$y = \frac{wx}{24EI}(L^3 - 2Lx^2 + x^3)$$

$$\theta_{max} = \frac{wL^3}{24EI}$$

$$y_{max} = \frac{5wL^4}{384EI}$$

9

$$y = \frac{Mx}{6EIL}(L^2 - x^2)$$

$$\theta_\ell = \frac{ML}{6EI}$$

$$\theta_r = \frac{ML}{3EI}$$

$$y_{max} = \frac{ML^2}{9\sqrt{3}EI}$$

$$y_{L/2} = \frac{ML^2}{16EI}$$

10

$$y = \frac{Mx}{6EIL}(L - x)(2L - x)$$

$$\theta_\ell = \frac{ML}{3EI}$$

$$\theta_r = \frac{ML}{6EI}$$

$$y_{max} = \frac{ML^2}{9\sqrt{3}EI}$$

$$y_{L/2} = \frac{ML^2}{16EI}$$

Therefore, the maximum deflection at the free end C of the beam is the sum of the maximum deflections of cases 2 and 3. Thus

$$y_{max}(+\downarrow) = [y_{max}]_{case\ 2}^{due\ to\ P} + [y_{max}]_{case\ 3}^{due\ to\ w}$$

$$= \frac{Pa^2}{6EI}(3L - a) + \frac{wL^4}{8EI}$$

$$= \frac{(4\ kN)(0.8\ m)^2}{6(667\ kN \cdot m^2)}(3 \times 1.2\ m - 0.8\ m) + \frac{(3\ kN/m)(1.2\ m)^4}{8(667\ kN \cdot m^2)}$$

$$= 0.00179\ m + 0.00117\ m = +0.00296\ m = +2.96\ mm \downarrow \quad \blacksquare$$

EXAMPLE 8–7

Use the method of superposition to find the maximum deflection at the free end C of the cantilever beam of the W12 $\times$ 40 steel section due to the uniform load over part of the beam, as shown.

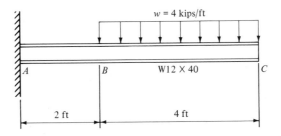

SOLUTION

From Table A–1, for a W12 $\times$ 40, $I = 310$ in.^{4}, and for steel, $E = 30 \times 10^3$ ksi. Then

$$EI = (30 \times 10^3\ kips/in.^2)(310\ in.^4) = 9.30 \times 10^6\ kip\text{-}in.^2$$
$$= 6.46 \times 10^4\ kip\text{-}ft^2$$

The uniform load over part of the beam may be considered to be the superposition of the two uniform loads shown in Figs. (a) and (b).

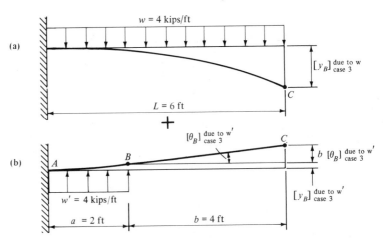

In Fig. (a), due to the downward uniform load w over the entire length, the downward deflection at the free end of the beam from case 3 of Table 8–1 is

$$[y_C]_{\text{case 3}}^{\text{due to } w} = \frac{wL^4}{8EI}$$

In Fig. (b), due to the upward uniform load w over the length AB, the upward deflection and the slope at B from case 3 of Table 8–1 are

$$[y_B]_{\text{case 3}}^{\text{due to } w'} = \frac{w'a^4}{8EI}$$

$$[\theta_B]_{\text{case 3}}^{\text{due to } w'} = \frac{w'a^3}{6EI}$$

Since θ_B in radians is very small and sine BC part of the beam is a straight line, the additional upward deflection at the free end C in Fig. (b) due to the slope θ_B is

$$b[\theta_B]_{\text{case 3}}^{\text{due to } w'} = b\frac{w'a^3}{6EI}$$

Therefore, the superposed deflection at C due to the given load is

$$y_{\max}(+\downarrow) = y_C = \frac{wL^4}{8EI} - \frac{w'a^4}{8EI} - b\frac{w'a^3}{6EI}$$

$$= \frac{4 \text{ kips/ft}}{6.47 \times 10^4 \text{ kip-ft}^2}\left[\frac{(6 \text{ ft})^4}{8} - \frac{(2 \text{ ft})^4}{8} - (4 \text{ ft})\frac{(2 \text{ ft})^3}{6}\right]$$

$$= +0.009\ 58 \text{ ft} = +0.115 \text{ in.} \downarrow$$

EXAMPLE 8–8

Use the method of superposition to find the deflection at the center C and point D of the simply supported timber beam of rectangular section shown. For timber, $E = 10$ GPa.

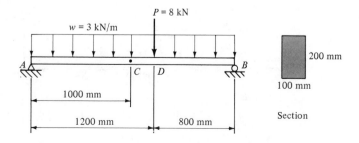

SOLUTION

The constant EI value of the beam is

$$EI = (10 \times 10^6 \text{ kN/m}^2) \frac{(0.1 \text{ m})(0.2 \text{ m})^3}{12} = 667 \text{ kN·m}^2$$

The simple beam is subjected to a concentrated load and a uniform load; this loading can be considered to be the superposition of cases 7 and 8 of Table 8–1. By the method of superposition, the deflection at the center C is

$$y_C(+\downarrow) = [y_{L/2}]^{\text{due to } P}_{\text{case } 7} + [y_{L/2}]^{\text{due to } w}_{\text{case } 8}$$

$$= \frac{Pb}{48EI}(3L^2 - 4b^2) + \frac{5wL^4}{384EI}$$

$$= \frac{(8 \text{ kN})(0.8 \text{ m})}{48(667 \text{ kN} \cdot \text{m}^2)}[3(2.0 \text{ m})^2 - 4(0.8 \text{ m})^2] + \frac{5(3 \text{ kN/m})(2.0 \text{ m})^4}{384(667 \text{ kN} \cdot \text{m}^2)}$$

$$= 0.00189 \text{ m} + 0.00094 \text{ m} = +0.00283 \text{ m} = +2.83 \text{ mm} \downarrow$$

To determine the deflection at D, the general deflection formulas of cases 7 and 8 must be used. Thus

$$y_D(+\downarrow) = [y_{x=1.2}]^{\text{due to } P}_{\text{case } 7} + [y_{x=1.2}]^{\text{due to } w}_{\text{case } 8}$$

$$= \frac{Pbx}{6EIL}(L^2 - x^2 - b^2) + \frac{wx}{24EI}(L^3 - 2Lx^2 + x^3)$$

$$= \frac{(8 \text{ kN})(0.8 \text{ m})(1.2 \text{ m})}{6(667 \text{ kN} \cdot \text{m}^2)(2.0)}[(2.0 \text{ m})^2 - (1.2 \text{ m})^2 - (0.8 \text{ m})^2]$$

$$+ \frac{(3 \text{ kN/m})(1.2 \text{ m})}{24(667 \text{ kN} \cdot \text{m}^2)}[(2.0 \text{ m})^3 - 2(2.0 \text{ m})(1.2 \text{ m})^2 + (1.2 \text{ m})^3]$$

$$= 0.00184 \text{ m} + 0.00089 \text{ m} = +0.00273 \text{ m} = +2.73 \text{ mm} \downarrow \quad \blacksquare$$

EXAMPLE 8–9

Use the method of superposition to determine the deflections at the center C and the free end D of the overhanging beam of wide-flange W10 × 22 steel section due to the loads shown.

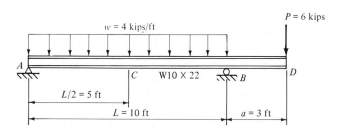

SOLUTION

Note that this is the same problem as Example 8–3, from which the value of EI is

$$EI = 2.46 \times 10^4 \text{ kip-ft}^2$$

To determine the deflection of an overhanging beam, the beam must be divided into two parts: a simple beam between the supports A and B and a cantilever beam for the overhanging part BD, as shown in Figs. (a) and (b).

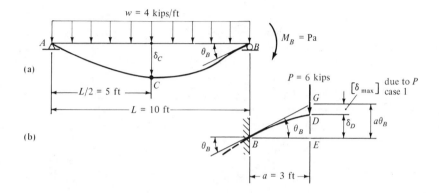

The load on the overhanging part is equivalent to a couple $M_B = Pa$ acting on the simple beam, as shown in Fig. (a). The deflection and slope at any section between A and B can be obtained by superposing cases 8 and 9 in Table 8–1. Thus

$$y_C(+\downarrow) = [y_{max}]_{case\ 8}^{due\ to\ w} - [y_{L/2}]_{case\ 9}^{due\ to\ M_B}$$

$$= \frac{5wL^4}{384EI} - \frac{(Pa)L^2}{16EI}$$

$$= \frac{5(4 \text{ kips/ft})(10 \text{ ft})^4}{384(2.46 \times 10^4 \text{ kip-ft}^2)} - \frac{(6 \text{ kips})(3 \text{ ft})(10 \text{ ft})^2}{16(2.46 \times 10^4 \text{ kip-ft}^2)}$$

$$= 0.0212 \text{ ft} - 0.0046 \text{ ft} = +0.0166 \text{ ft} = +0.199 \text{ in.} \downarrow$$

$$\theta_B(+\diagup) = [\theta_{max}]_{case\ 8}^{due\ to\ w} - [\theta_r]_{case\ 9}^{due\ to\ M_B}$$

$$= \frac{wL^3}{24EI} - \frac{(Pa)L}{3EI}$$

$$= \frac{(4 \text{ kips/ft})(10 \text{ ft})^3}{24(2.46 \times 10^4 \text{ kip-ft}^2)} - \frac{(6 \text{ kips})(3 \text{ ft})(10 \text{ ft})}{3(2.46 \times 10^4 \text{ kip-ft}^2)}$$

$$= 0.00678 - 0.00244 = +0.00434 \text{ rad} \diagup$$

The deflection at the free end D may be obtained by considering the overhanging part as a cantilever beam built in at support B with an initial inclination θ_B. Without the load P, the cantilever deflection curve would be a straight line BG, making an angle θ_B with the horizontal direction, as shown in Fig. (b). Since

θ_B is very small, the distance EG is equal to $a\theta_B$. The force P causes a downward deflection GD equals to

$$GD = [y_{max}]_{case\ 1}^{due\ to\ P} = \frac{Pa^3}{3EI}$$

Therefore, the deflection of end D is

$$y_D(+\uparrow) = ED = EG - GD = a\theta_B - \frac{Pa^3}{3EI}$$

$$= (3\ ft)(0.00434\ rad) - \frac{(6\ kips)(3\ ft)^3}{3(2.46 \times 10^4\ kip\text{-}ft^2)}$$

$$= 0.01302\ ft - 0.00220\ ft = +0.0108\ ft = +0.130\ in.\ \uparrow$$

Note that these answers check with those in Example 8–4.   ■

PROBLEMS

*In Problems **8–19** to **8–22**, use the formulas in Table 8–1 and the method of superposition to determine the maximum deflection of the cantilever beam shown. Use the EI value indicated in the figure.*

8–19

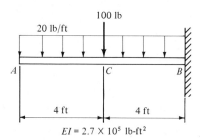

$EI = 2.7 \times 10^5$ lb-ft²

FIGURE P8–19

8–20

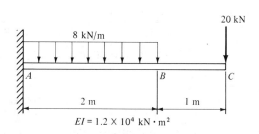

$EI = 1.2 \times 10^4$ kN · m²

FIGURE P8–20

8–21

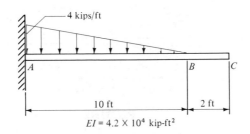

$EI = 4.2 \times 10^4$ kip-ft^2

FIGURE P8–21

8–22   Use the figure for Problem 8–8 (on p. 238).

*In Problems **8–23** to **8–27**, use the formulas in Table 8–1 and the method of superposition to determine the deflection at the center C of the simple beam due to the loading shown. Use the EI value indicated in the figure.*

8–23

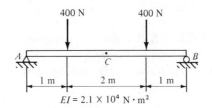

$EI = 2.1 \times 10^4$ N · m^2

FIGURE P8–23

8–24

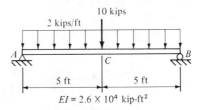

$EI = 2.6 \times 10^4$ kip-ft^2

FIGURE P8–24

8–25

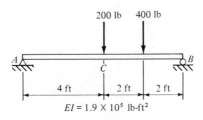

$EI = 1.9 \times 10^5$ lb-ft^2

FIGURE P8–25

8–26

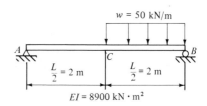

FIGURE P8–26

(**HINT:** $y_C = \frac{1}{2}[y_{max}]_{case\ 8}.$)

8–27

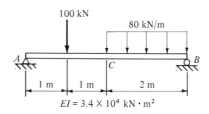

FIGURE P8–27

(See hint in Problem 8–26)

*In Problems **8–28** to **8–32**, use the formulas in Table 8–1 and the method of superposition to determine the deflections at the center C and the free end D of the overhanging beam due to the loading shown. Use the EI value indicated in the figure.*

8–28 Use the figure for Problem 8–13 (on p. 239).

8–29 Use the figure for Problem 8–14 (on p. 239).

8–30 Use the figure for Problem 8–15 (on p. 240).

8–31

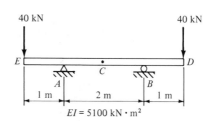

FIGURE P8–31

8–32

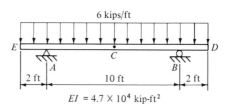

FIGURE P8–32

8–7
DEVELOPMENT OF THE MOMENT-AREA THEOREMS

Consider a simple beam subjected to an arbitrary load, as shown in Fig. 8–3(a). The moment diagram and the deflection curve of the beam are sketched as shown in Fig. 8–3(b) and (c). Let p and q be two points on the elastic curve at a

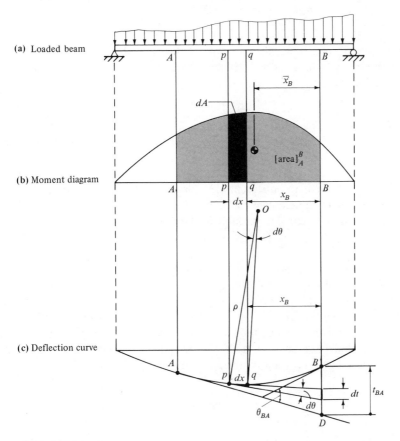

(a) Loaded beam

(b) Moment diagram

(c) Deflection curve

FIGURE 8–3

differential distance dx apart, and let the radius of curvature of the elastic curve at these points be ρ. The differential angle $d\theta$ (measured in radians) between the radii of curvature at points p and q is equal to

$$d\theta = \frac{dx}{\rho} \tag{a}$$

From Eq. (8–1),

$$\frac{1}{\rho} = \frac{M}{EI}$$

Substitution in Eq. (a) gives

$$d\theta = \frac{1}{EI} M \, dx = \frac{1}{EI} dA \tag{b}$$

where dA is the differential area of the element in the moment diagram between p and q. In Fig. 8–3(c), tangents drawn to the elastic curve at p and q make an angle $d\theta$. The angle in radians measured between tangents at any two points A and B on the deflection curve, designated by θ_{BA} in Fig. 8–3(c), can be obtained by integrating $d\theta$ from A to B. That is,

$$\theta_{BA} = \int_A^B d\theta = \int_A^B \frac{1}{EI} dA = \frac{1}{EI} \int_A^B dA = \frac{1}{EI} [\text{area}]_A^B \qquad (8–6)$$

where $[\text{area}]_A^B$ represents the area of the moment diagram between the vertical lines through points A and B [the shaded area in Fig. 8–3(b)]. Equation (8–6) can be stated as follows:

> **Theorem I.** *The angle between the tangents at any two points A and B of the deflection curve is equal to the area of the bending moment diagram between the vertical lines through A and B, divided by EI of the beam.*

In Fig. 8–3(c), if the differential angle $d\theta$ for the element dx is multiplied by the distance x_B of the element to the point B, it gives a distance very nearly equal to dt along the vertical line through B. That is,

$$dt = x_B \, d\theta$$

This approximation is valid, since the slopes are usually very small for beams used in engineering structure.

By integrating this expression from A to B, the vertical distance BD is obtained. Geometrically, this distance represents the deviation of the point B from the tangent to the deflection curve at A. It will be called the *tangential deviation* of point B from the tangent at A and will be designated t_{BA}. Thus

$$t_{BA} = \int_A^B dt = \int_A^B x_B \, d\theta$$

When we substitute the expression for $d\theta$ from Eq. (b), we obtain

$$t_{BA} = \frac{1}{EI} \int_A^B x_B \, dA$$

Using the definition of the centroid of an area we can write this equation in a simpler form:

$$t_{BA} = \frac{1}{EI} [\text{area}]_A^B \, \bar{x}_B \qquad (8–7)$$

where $[\text{area}]_A^B$ is the area of the moment diagram between points A and B and $\bar{x}_B$ is the horizontal distance from the centroid of this area to point B. Equation (8–7) can be stated as follows:

> **Theorem II.** *If A and B are points on the deflection curve, the vertical distance of B from the tangent to the curve at A, t_{BA}, called the tangential deviation, is*

equal to the area of the bending moment diagram between the vertical lines through A and B multiplied by the horizontal distance from the centroid of the area to point B divided by EI of the beam.

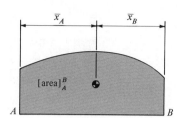

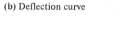

(a) Moment diagram

(b) Deflection curve

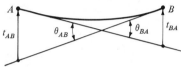

FIGURE 8–4

Although θ_{AB} is equal to θ_{BA}, t_{AB} is usually not equal to t_{BA}. By analogous reasoning the deviation t_{AB} of point A from the tangent to the elastic curve at B, as shown in Fig. 8–4, is

$$t_{AB} = \frac{1}{EI} [\text{area}]_A^B \, \bar{x}_A \qquad (8\text{–}8)$$

where $[\text{area}]_A^B$ is the area of the moment diagram between A and B and $\bar{x}_A$ is the horizontal distance from the centroid of the area to point A.

8–8
MOMENT DIAGRAM BY PARTS

As shown in the preceding section, when the moment-area method is used for calculating beam deflections, the area of the moment diagram and the location of its centroid are needed. To aid in this sort of computations, moment diagrams are drawn by parts. That is, draw the moment diagram of each load and each reaction separately. Then, by the method of superposition, the algebraic sum of all the moment diagrams drawn separately will be equivalent to the moment diagram drawn in the usual manner.

Four fundamental cantilever loadings and their respective moment diagrams are shown in Fig. 8–5. The areas of the moment diagrams are called rectangle, triangle, parabolic spandrel, and cubic parabolic spandrel, respectively. Properties of these areas are listed in Table 8–2.

The method of plotting moment diagrams by parts can be applied to beams of any support conditions. After the reactions of a beam are determined, the reactions are treated as applied loads and an imaginary fixed support can be placed at any section along the beam, the moment diagram can be drawn by parts toward that section using the formulas in Fig. 8–5. The following examples illustrate the method.

TABLE 8–2 Properties of Areas

Shape	Centroid	Area
Rectangle		bh
Triangle		$\dfrac{bh}{2}$
Semiparabolic area		$\dfrac{2bh}{3}$
Parabolic spandrel		$\dfrac{bh}{3}$
Cubic parabolic spandrel		$\dfrac{bh}{4}$

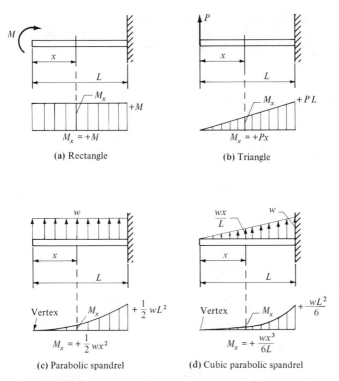

(a) Rectangle

(b) Triangle

(c) Parabolic spandrel

(d) Cubic parabolic spandrel

FIGURE 8–5

━━━ **EXAMPLE 8–10** ━━━

Draw the moment diagram by parts for the cantilever beam shown.

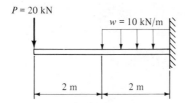

SOLUTION

Since the given beam is a cantilever beam, the formulas in Fig. 8–5 can be applied directly by considering each load separately, as shown in the following figures.

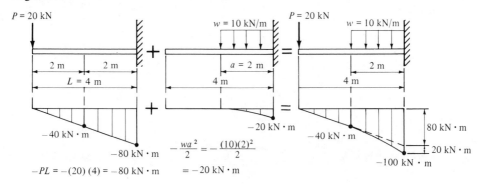

The combined moment diagram is the superposition of the two moment diagrams drawn separately, as shown at the right-hand side of the figure. ∎

EXAMPLE 8–11

Draw the moment diagram by parts for the uniformly loaded simple beam shown.

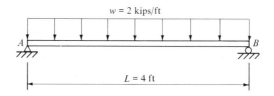

SOLUTION

Due to symmetry, the reaction at each support is equal to one-half of the total load; thus

$$R_A = R_B = \tfrac{1}{2}wL = \tfrac{1}{2}(2 \text{ kips/ft})(4 \text{ ft}) = 4 \text{ kips}$$

Now the moment diagram can be drawn by parts toward any section desired. If we choose to draw the moment diagram toward end B, the following diagrams are obtained.

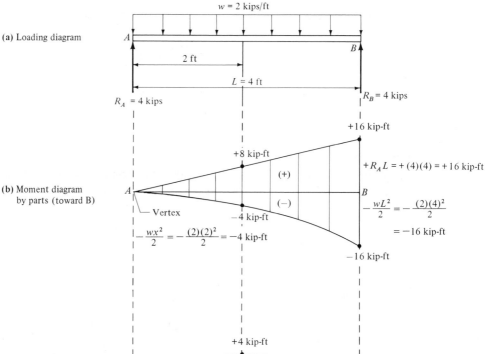

Now we see why it is preferable to draw the moment diagram by parts. If the area and the centroid location of the entire moment diagram are desired, then the moment diagram either by the usual method or by parts can be used. But if the areas and the locations of the centroids of part of the moment diagram between A and some intermediate section are needed, then the moment diagram drawn by parts must be used.

EXAMPLE 8–12

Draw the moment diagram by parts for the overhanging beam subjected to the loading shown in Fig. (a). The reactions are indicated in the figure.

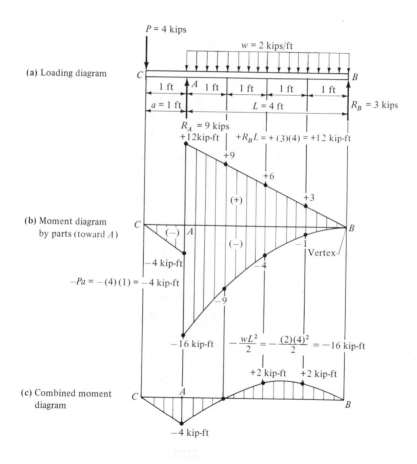

(a) Loading diagram

(b) Moment diagram by parts (toward A)

(c) Combined moment diagram

SOLUTION

The moment diagram is drawn by parts toward section A, as shown in Fig. (b). The parts in Fig. (b) can be superposed into a combined moment diagram shown in Fig. (c), which is the same as the moment diagram obtained in the usual method. We see that the areas in Fig. (c) are very difficult to compute. The areas and their centroids in Fig. (b) can be computed readily by using the formulas in Table 8–2.

PROBLEMS

In Problems **8–33** *to* **8–40***, draw the moment diagram by parts toward the section indicated for each beam and loading shown.*

8–33

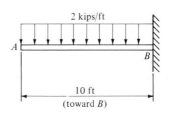

FIGURE P8–33

8–34

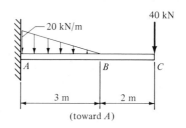

FIGURE P8–34

8–35

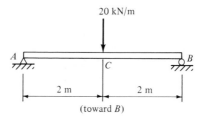

FIGURE P8–35

8–36

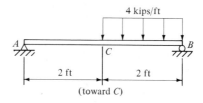

FIGURE P8–36

8–37

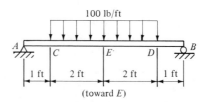

FIGURE P8–37

8–38

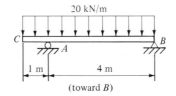

20 kN/m

C *A* *B*

1 m   4 m

(toward *B*)

FIGURE P8–38

8–39

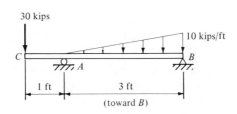

10 kN

4 kN/m

A *B* *C*

5 m   1 m

(toward *B*)

FIGURE P8–39

8–40

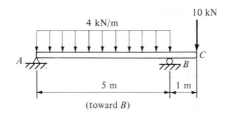

30 kips

10 kips/ft

C *A* *B*

1 ft   3 ft

(toward *B*)

FIGURE P8–40

8–9
BEAM DEFLECTION BY THE MOMENT-AREA METHOD

The two theorems discussed in Section 8–5 will now be used to solve beam deflection problems. The theorems are applicable to beams loaded in any general way.

To simplify the calculations, moment diagrams are usually drawn by parts, as discussed in Section 8–8. The properties of areas in Table 8–2 will be very useful in the computation. The method of superposition will be used in summing up the areas of the moment diagram and the first moments of the areas about a point. In performing the summation, areas corresponding to positive moments are treated as positive, and areas corresponding to the negative moments are treated as negative.

When applying the moment-area method, a carefully sketched deflection curve is always necessary. The angles subtended between the tangents and the tangential deviations can be obtained by direct application of the moment-area theorems. Deflections of points along the beam can be obtained by further consideration of the geometry of the deflection curve, as illustrated in the following examples.

━━━━━ **EXAMPLE 8–13** ━━━━━━━━━━━━━━━━━━━━━━━━━━━━━━━━━━━━━━━

Find the expressions for the maximum slope and the maximum deflection of the cantilever beam due to a concentrated load P, as shown in Fig. (a).

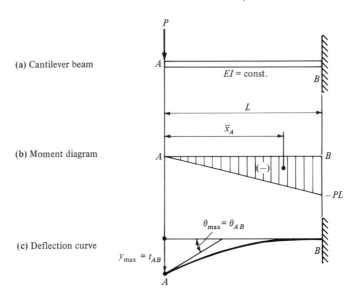

(a) Cantilever beam

(b) Moment diagram

(c) Deflection curve

SOLUTION

The maximum slope and the maximum deflection both occur at the free end A. Since the tangent at the fixed end B is in the horizontal direction, the maximum slope θ_{max} is equal to θ_{AB}, and the maximum deflection y_{max} is equal to the tangential deviation t_{AB}, as shown in Fig. (c).

The moment diagram in Fig. (b) is a triangle; thus from Table 8–2, the area of the moment diagram between AB is

$$[\text{area}]_A^B = \frac{bh}{2} = \frac{(L)(-PL)}{2} = -\frac{PL^2}{2}$$

and the centroid of the area from A is

$$\bar{x}_A = \frac{2b}{3} = \frac{2L}{3}$$

Therefore, from Eq. (8–6),

$$\theta_{max} = \theta_{AB} = \frac{1}{EI}[\text{area}]_A^B = \frac{1}{EI}\left[-\frac{PL^2}{2}\right] = -\frac{PL^2}{2EI}$$

and from Eq. (8–7),

$$y_{max} = t_{AB} = \frac{1}{EI}[\text{area}]_A^B \bar{x}_A = \frac{1}{EI}\left[-\frac{PL^2}{2}\right]\frac{2L}{3} = -\frac{PL^3}{3EI}$$

The negative sign of t_{AB} means that the point A is below the tangent at B.   ∎

EXAMPLE 8–14

Find the maximum deflection of the cantilever beam of wide-flange W12 × 40 steel section due to the uniform load shown.

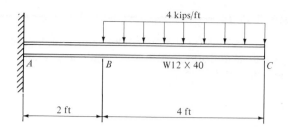

SOLUTION

The given load may be considered to be the superposition of two uniform loads, as shown in Fig. (a). The moment diagram is drawn by parts toward A as shown in Fig. (b), and the deflection curve is shown in Fig. (c).

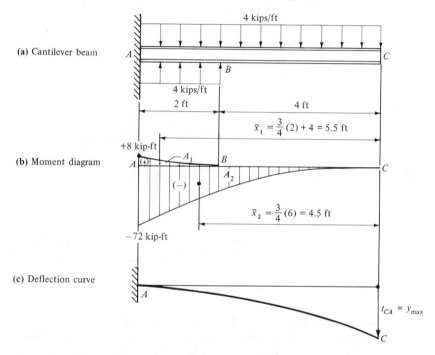

(a) Cantilever beam

(b) Moment diagram

(c) Deflection curve

Note that this is the same problem as Example 8–7, from which the constant EI value is

$$EI = 6.46 \times 10^4 \text{ kip-ft}^2$$

From Eq. (8–7), using units of kip and ft for each term, we have

$$y_{max} = t_{CA} = \frac{1}{EI} [\text{area}]_A^C \, \bar{x}_C = \frac{1}{EI}[A_1 \, \bar{x}_1 + A_2 \, \bar{x}_2]$$

$$= \frac{1}{6.46 \times 10^4} \left\{ \left[\frac{1}{3}(2)(+8)\right](5.5) + \left[\frac{1}{3}(6)(-72)\right](4.5) \right\}$$

$$= -9.58 \times 10^{-3} \text{ ft} = -0.115 \text{ in.} \downarrow$$

Note that this answer checks exactly with that of Example 8–7.

—————— **EXAMPLE 8–15** ——————

Find the expressions for the maximum slope and maximum deflection of a simple beam due to a uniform load w over the entire span length L. The beam has a constant value of EI.

SOLUTION

The simple beam with a uniform load is shown in Fig. (a). Its moment diagram and elastic curve are shown in Figs. (b) and (c), respectively. The moment diagram is drawn in the usual manner.

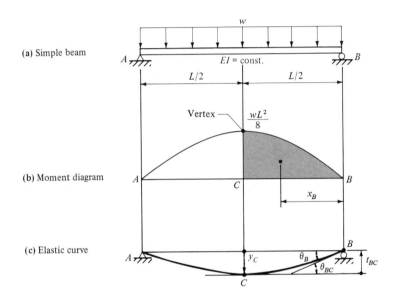

(a) Simple beam

(b) Moment diagram

(c) Elastic curve

Since the loading is symmetrical with respect to the vertical axis passing through the center C, the elastic curve must be also symmetrical with respect to the same axis. Therefore, the tangent at C is horizontal, and thus

$$\theta_B = \theta_{BC}$$

$$y_C = t_{BC}$$

as shown in Fig. (c).

The moment diagram between C and B is a semiparabolic area. Thus, from Table 8–2,

$$[\text{area}]_C^B = \frac{2bh}{3} = \frac{2}{3}\left(\frac{L}{2}\right)\frac{wL^2}{8} = \frac{wL^3}{24}$$

$$\bar{x}_B = \frac{5}{8}b = \frac{5}{8}\left(\frac{L}{2}\right) = \frac{5L}{16}$$

From Eqs. (8–6) and (8–7),

$$\theta_{\max} = \theta_B = \theta_{BC} = \frac{1}{EI}[\text{area}]_C^B = +\frac{wL^3}{24EI}$$

$$y_{\max} = y_C = t_{BC} = \frac{1}{EI}[\text{area}]_C^B \bar{x}_B$$

$$= \frac{1}{EI}\left(\frac{wL^3}{24}\right)\frac{5L}{16} = +\frac{5wL^4}{384EI}$$

where the positive signs for $\theta_{\max}$ and $y_{\max}$ indicate that their assumed directions are correct.

◼

EXAMPLE 8–16

Determine the deflections at points C, D, and E of the beam due to a uniform load over part of the span, as shown in Fig. (a).

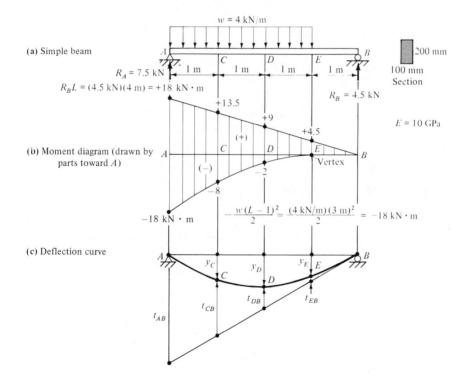

(a) Simple beam

$R_A = 7.5$ kN
$R_B L = (4.5 \text{ kN})(4 \text{ m}) = +18$ kN · m

$w = 4$ kN/m

$R_B = 4.5$ kN

200 mm

100 mm
Section

$E = 10$ GPa

(b) Moment diagram (drawn by parts toward A)

$+13.5$
$+9$
$+4.5$
Vertex
$(+)$
$(-)$
-2
-8
-18 kN · m

$-\dfrac{w(L-1)^2}{2} = \dfrac{(4 \text{ kN/m})(3 \text{ m})^2}{2} = -18$ kN · m

(c) Deflection curve

y_C y_D y_E
t_{AB} t_{CB} t_{DB} t_{EB}

SOLUTION

The reactions of the beam are determined first; the results are shown in Fig. (a). Then the moment diagram of the beam is drawn by parts toward A as shown in Fig. (b). The deflection curve of the beam is sketched in Fig. (c). The tangential deviations at points A, C, D, and E from the tangent to the elastic curve at B are calculated in the following, using units of kN and m.

$$EI = (10 \times 10^6 \text{ kN/m}^2)\left(\frac{0.1 \times 0.2^3}{12} \text{ m}^4\right) = 667 \text{ kN·m}^2$$

$$t_{AB} = \frac{1}{EI} [\text{area}]_A^B \bar{x}_A = \frac{1}{667}\left[\frac{1}{2}(4)(+18)\left(\frac{4}{3}\right) + \frac{1}{3}(3)(-18)\left(\frac{3}{4}\right)\right] = 0.0517 \text{ m}$$

$$t_{CB} = \frac{1}{EI} [\text{area}]_C^B \bar{x}_C = \frac{1}{667}\left[\frac{1}{2}(3)(+13.5)\left(\frac{3}{3}\right) + \frac{1}{3}(2)(-8)\left(\frac{2}{4}\right)\right] = 0.0264 \text{ m}$$

$$t_{DB} = \frac{1}{EI} [\text{area}]_D^B \bar{x}_D = \frac{1}{667}\left[\frac{1}{2}(2)(+9)\left(\frac{2}{3}\right) + \frac{1}{3}(1)(-2)\left(\frac{1}{4}\right)\right] = 0.0087 \text{ m}$$

$$t_{EB} = \frac{1}{EI} [\text{area}]_E^B \bar{x}_E = \frac{1}{667}\left[\frac{1}{2}(1)(+4.5)\left(\frac{1}{3}\right)\right] = 0.0011 \text{ m}$$

From the elastic curve in Fig. (c), by similar triangles, we have

$$\frac{y_C + t_{CB}}{3} = \frac{t_{AB}}{4}$$

from which

$$y_C = \tfrac{3}{4}t_{AB} - t_{CB} = \tfrac{3}{4}(0.0517) - 0.0264 = +0.0124 \text{ m} = +12.4 \text{ mm}$$

Similarly,

$$y_D = \tfrac{1}{2}t_{AB} - t_{DB} = \tfrac{1}{2}(0.0519) - 0.0087 = +0.0172 \text{ m} = +17.2 \text{ mm}$$

$$y_E = \tfrac{1}{4}t_{AB} - t_{EB} = \tfrac{1}{4}(0.0517) - 0.0011 = +0.0118 \text{ m} = +11.8 \text{ mm}$$

The positive signs indicate that the deflections of all the points are downward as assumed.

The deflection diagram of the beam is plotted in the following figure. We see that the maximum deflection of the beam occurs near the midpoint D.

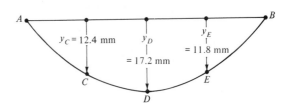

──────── **EXAMPLE 8–17** ────────

Determine the deflection at points C and D of the overhanging beam of wide-flange W10 × 22 steel section due to the loads shown in Fig. (a).

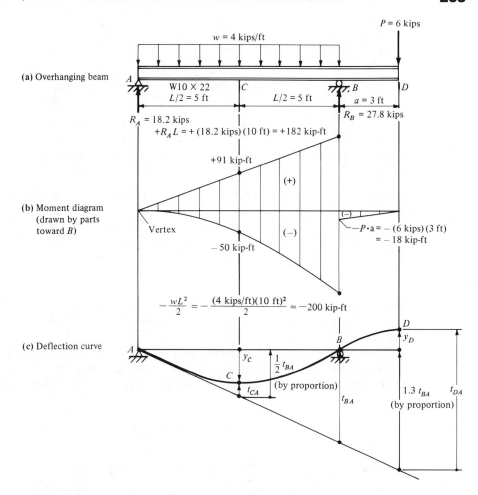

(a) Overhanging beam

(b) Moment diagram (drawn by parts toward B)

(c) Deflection curve

SOLUTION

The reactions of the beam are first determined; the results are shown in Fig. (a). Then the moment diagram of the beam is drawn by parts toward B as shown in Fig. (b). The deflection curve of the beam is sketched in Fig. (c).

Note that this is the same problem as Examples 8–3 and 8–9, from which the EI value is

$$EI = 2.46 \times 10^4 \text{ kip-ft}^2$$

Using the units kips and feet, we obtain the following tangential deviations of points C, B, and D from the tangent at A to the deflection curve:

$$t_{CA} = \frac{1}{EI} [\text{area}]_A^C \bar{x}_C = \frac{1}{2.46 \times 10^4} \left[\frac{1}{2}(5)(91)\frac{5}{3} + \frac{1}{3}(5)(-50)\left(\frac{5}{4}\right) \right]$$

$$= +0.0112 \text{ ft}$$

$$t_{BA} = \frac{1}{EI} [\text{area}]_A^B \bar{x}_B = \frac{1}{2.46 \times 10^4} \left[\frac{1}{2}(10)(182)\frac{10}{3} + \frac{1}{3}(10)(-200)\left(\frac{10}{4}\right) \right]$$

$$= +0.05556 \text{ ft}$$

$$t_{DA} = \frac{1}{EI}[\text{area}]_A^D \bar{x}_D = \frac{1}{2.46 \times 10^4}\left[\frac{1}{2}(10)(182)\left(3 + \frac{10}{3}\right)\right.$$

$$\left. + \frac{1}{3}(10)(-200)\left(3 + \frac{10}{4}\right) + \frac{1}{2}(3)(-18)\left(\frac{2}{3} \times 3\right)\right]$$

$$= +0.0830 \text{ ft}$$

From the geometry of the elastic curve in Fig. (c), the deflections of points C and D are

$$y_C = \tfrac{1}{2}t_{BA} - t_{CA} = \tfrac{1}{2}(+0.05556 \text{ ft}) - (+0.0112 \text{ ft}) \quad = +0.0166 \text{ ft}$$

$$= +0.199 \text{ in.} \downarrow$$

$$y_D = t_{DA} - 1.3t_{BA} = (+0.0830 \text{ ft}) - 1.3(+0.05556 \text{ ft}) = +0.0108 \text{ ft}$$

$$= +0.130 \text{ in.} \uparrow$$

Note that these answers are the same as those obtained in Examples 8–3 and 8–9. The positive signs for y_C and y_D indicate that points C and D are deflected in the assumed direction.

Before the value of y_D is calculated, however, it is difficult to tell whether the deflection of point D is upward or downward. The deflection curve could have been sketched with point D deflected downward, as shown in the following figure.

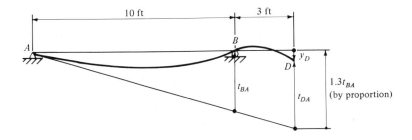

From the figure, the deflection of point D is

$$y_D = 1.3t_{BA} - t_{DA} = 1.3(+0.05556 \text{ ft}) - (+0.0830 \text{ ft}) = -0.0108 \text{ ft}$$

The negative value for y_D indicates that instead of the assumed downward deflection of point D, as shown in the figure, the deflection of point D is actually upward (i.e., above the undeformed axis of the beam). ∎

PROBLEMS

*In Problems **8–41** to **8–43**, use the moment-area method to find expressions for the maximum slope and the maximum deflection of the cantilever beam due to the load shown in the figure.*

8–41

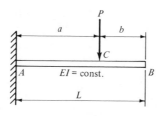

FIGURE P8–41

8–42

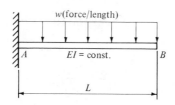

FIGURE P8–42

8–43

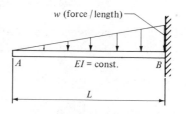

FIGURE P8–43

In Problems **8–44** *to* **8–47**, *use the moment-area method to find the maximum deflection of the cantilever beam due to the loading shown.*

8–44   Use the figure for Problem 8–19 (on p. 248).

8–45   Use the figure for Problem 8–20 (on p. 248).

8–46   Use the figure for Problem 8–21 (on p. 249).

8–47   Use the figure for Problem 8–8 (on p. 238).

8–48   Rework Example 8–15 (on p. 262) using the moment diagram plotted by parts toward the midsection *C*.

8–49   Use the moment-area method to find the expressions for the maximum slope and the maximum deflection of a simple beam of span length *L* due to a concentrated load *P* applied at the center. The beam has a constant value of *EI*.

In Problems **8–50** *to* **8–53**, *use the moment-area method to find the maximum deflection at the center C of the simple beam due to the given symmetrical loading shown.* (**HINT:** *The tangent to the beam at the center C is in the horizontal direction.*)

8–50   Use the figure for Problem 8–23 (on p. 249).

8–51   Use the figure for Problem 8–24 (on p. 249).

8–52

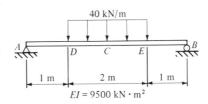

FIGURE P8–52

8–53

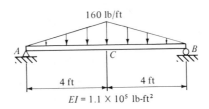

FIGURE P8–53

In Problems **8–54** *to* **8–58,** *use the moment-area method to find the deflection at the center C of the simple beam due to the loading shown.*

8–54   Use the figure for Problem 8–25 (on p. 249).

8–55   Use the figure for Problem 8–26 (on p. 250).

8–56   Use the figure for Problem 8–27 (on p. 250).

8–57   Use the figure for Problem 8–10 (on p. 239).

8–58   Use the figure for Problem 8–12 (on p. 239).

In Problems **8–59** *and* **8–60,** *use the moment-area method to find the deflections at the center C and the free end D of the symmetrical overhanging beam due to the symmetrical loading shown.* (**HINT:** *Due to the symmetrical condition, the tangent to the beam at the center C is in the horizontal direction.*)

8–59   Use the figure for Problem 8–31 (on p. 250).

8–60   Use the figure for Problem 8–32 (on p. 250).

In Problems **8–61** *to* **8–63,** *use the moment-area method to find the deflections at the center C and the free end D of the overhanging beam due to the loading shown.*

8–61   Use the figure for Problem 8–13 (on p. 239).

8–62   Use the figure for Problem 8–14 (on p. 239).

8–63   Use the figure for Problem 8–15 (on p. 240).

*8–10
COMPUTER PROGRAM ASSIGNMENTS

In the following assignments, write the computer programs either in FORTRAN language or in BASIC language to compute beam deflections. Do not use a specific unit system in the program. When consistent units are used in the input data, the computed results are in the same system of units.

C8-1 Develop a computer program that would compute the slopes at the ends and the deflections at every tenth point of the simple beam subjected to the loading shown in Fig. C8-1. Use the method of superposition and the formulas in Table 8–1 for

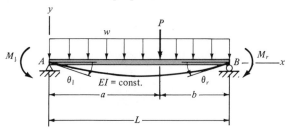

FIGURE C8–1

formulations. Input data for EI, L, w, P, a, M_l, and M_r. Run the program using data in Examples 8–3 (on p. 232), 8–8 (on p. 245) and Problems 8–24 (on p. 249), 8–15 (on p. 240) for the part of beam between the supports.

C8-2 Develop a computer program that would compute the deflections at every fifth point of the cantilever beam loaded as shown in Fig. C8-2. The beam has an

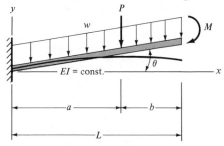

FIGURE C8–2

initial slope θ at the fixed end. Use the method of superposition and the formulas in Table 8–1 for formulations. Input data for EI, L, w, P, a, M, and θ. Run the program using data from Example 8–6 (on p. 241) and Problems 8–19 (on p. 248) and 8–15 (on p. 240) for the overhanging part of the beam.

C8-3 Apply the programs developed in C8-1 and C8-2 to the beam shown in Fig. C8-3. Plot the deflection curve of the beam using exaggerated deflection scale.

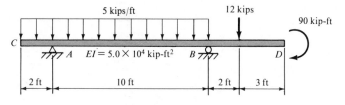

FIGURE C8–3

CHAPTER 9

Statically Indeterminate Problems

9–1
INTRODUCTION

Structural members considered in previous chapters are statically determinate because the support reactions can be determined from equilibrium equations without considering deformations. There are many cases in which the reactions at the supports of a structure cannot be determined by the equilibrium equations alone. In these cases, deformations of the structure must be taken into consideration. These structures are said to be *statically indeterminate*.

Procedures for the solution of statically indeterminate problems are discussed in this chapter. Statically indeterminate axially loaded members are discussed first. Then stresses caused by temperature changes are discussed. Statically indeterminate shafts are analyzed next. Finally, statically indeterminate beams are studied.

Results should be examined to make certain that the maximum stress does not exceed the proportional limit of the material. Since the solutions are based on the elastic behavior of the material when the maximum stress is within the proportional limit of the material.

9–2
STATICALLY INDETERMINATE AXIALLY LOADED MEMBERS

The following examples show how the conditions of the deformation geometries are used, in addition to the equilibrium equations, to solve problems involving statically indeterminate axially loaded members.

From Eq. (3–5), the axial deformation δ of a member with cross-sectional area A and length L subjected to axial load P is

$$\delta = \frac{PL}{AE}$$

where E is the modulus of elasticity of the material of the member.

—— **EXAMPLE 9–1** ——

A bar is built in at both ends to fixed supports. Determine the reactions at the supports A and B caused by the axial force P, which acts at an intermediate point C.

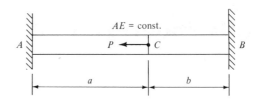

SOLUTION

If the fixed support at A had been removed and been replaced by the reaction R_A as shown in Fig. (a), the axial force diagram of the rod would be as shown in Fig. (b).

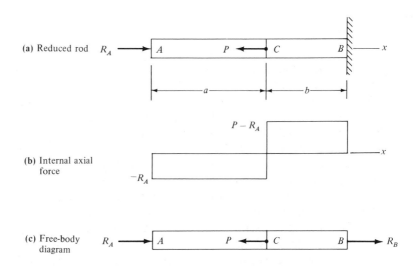

(a) Reduced rod

(b) Internal axial force

(c) Free-body diagram

Since the supports at A and B are fixed, the total axial deformation δ_{AB} between the supports must be equal to zero. Thus

$$\delta_{AB} = \delta_{AC} + \delta_{CB} = \frac{(-R_A)(a)}{AE} + \frac{(P - R_A)(b)}{AE} = 0$$

The constant AE is canceled by multiplying each term in the equation by AE. The equation is reduced to

$$R_A(a + b) = Pb$$

from which

$$R_A = +\frac{b}{a + b}P$$

The equilibrium condition of the free-body diagram of the bar shown in Fig. (c) requires that

$$\xrightarrow{+}\Sigma F_x = R_A + R_B - P = 0$$

from which

$$R_B = P - R_A = P - \frac{b}{a + b}P = +\frac{a}{a + b}P$$

■

EXAMPLE 9–2

A steel cylinder fits loosely in a copper tube as shown. The length of the steel cylinder is 0.001 in. longer than the copper tube. Determine the stresses in the solid steel cylinder and in the copper tube caused by an axial force P of (a) 15 kips, and (b) 55 kips, applied via a rigid cap. The moduli of elasticity are: for steel, $E_{st} = 30 \times 10^3$ ksi, for copper, $E_{cu} = 17 \times 10^3$ ksi.

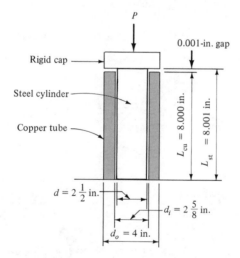

SOLUTION

The cross-sectional areas of the solid steel cylinder and the copper tube are

$$A_{st} = \tfrac{1}{4}\pi d^2 = \tfrac{1}{4}\pi(2.5)^2 = 4.91 \text{ in.}^2$$

$$A_{cu} = \tfrac{1}{4}\pi(d_o^2 - d_i^2) = \tfrac{1}{4}\pi(4^2 - 2.625^2) = 7.15 \text{ in.}^2$$

The force F necessary to close the gap can be determined from the following equation:

$$\delta = 0.001 \text{ in.} = \frac{FL_{st}}{A_{st}E_{st}} = \frac{(F \text{ kips})(8.001 \text{ in.})}{(4.91 \text{ in.}^2)(30\ 000 \text{ kips/in.}^2)}$$

from which

$$F = 18.4 \text{ kips}$$

(a) Since the force $P = 15$ kips is less than the force F necessary to close the gap, the force P will be carried by the steel cylinder alone. Thus the stress in the steel cylinder is

$$\sigma_{st} = \frac{P}{A_{st}} = \frac{15 \text{ kips}}{4.91 \text{ in.}^2} = 3.05 \text{ ksi (C)}$$

and the copper tube is not stressed.

(b) The force $P = 55$ kips is large enough to close the gap. After the gap is closed the copper tube will also be compressed and carry part of the load. Let the forces carried by the steel cylinder and the copper tube be P_{st} and P_{cu}, respectively; then

$$P_{st} + P_{cu} = 55 \text{ kips} \tag{a}$$

as required by the equilibrium condition, $\Sigma F_y = 0$.

The amount of axial deformation in the steel cylinder is 0.001 in. more than that of the copper tube. Thus

$$\delta_{st} = \delta_{cu} + 0.001 \text{ in.}$$

or

$$\frac{P_{st}L_{st}}{A_{st}E_{st}} = \frac{P_{cu}L_{cu}}{A_{cu}E_{cu}} + 0.001 \text{ in.}$$

or

$$\frac{(P_{st} \text{ kips})(8.001 \text{ in.})}{(4.91 \text{ in.}^2)(30\ 000 \text{ kips/in.}^2)} = \frac{(P_{cu} \text{ kips})(8.000 \text{ in.})}{(7.15 \text{ in.}^2)(17\ 000 \text{ kips/in.}^2)} + 0.001 \text{ in.}$$

Multiplying each term by $(4.91 \text{ in.}^2)(30\ 000 \text{ kips/in.}^2)/(8.001 \text{ in.})$, we have

$$P_{st} - 1.21P_{cu} = 18.4 \text{ kips} \tag{b}$$

Solving Eqs. (a) and (b) simultaneously gives

$$P_{st} = 38.4 \text{ kips} \qquad P_{cu} = 16.6 \text{ kips}$$

The stresses in the steel cylinder and in the copper tube are

$$\sigma_{st} = \frac{P_{st}}{A_{st}} = \frac{38.4 \text{ kips}}{4.91 \text{ in.}^2} = 7.82 \text{ ksi (C)}$$

$$\sigma_{cu} = \frac{P_{cu}}{A_{cu}} = \frac{16.6 \text{ kips}}{7.15 \text{ in.}^2} = 2.32 \text{ ksi (C)}$$

PROBLEMS

9–1 A stepped-bar of the same material is supported between two fixed supports, as shown in Fig. P9–1. Determine the stress in each segment of the bar caused by the applied load of 100 kips.

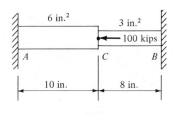

FIGURE P9–1

9–2 In Fig. P9–2, determine the reactions at the fixed supports, and plot the axial force diagram of the rod subjected to the axial loads shown.

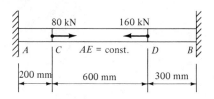

FIGURE P9–2

9–3 A bronze cylinder of 0.0065 m² cross-sectional area fits loosely inside an aluminum tube having 0.0045 m² cross-sectional area (Fig. P9–3). The aluminum tube is 0.3 mm longer than the bronze cylinder before the load is applied. If the modulus of elasticity of bronze is 83 GPa and that of aluminum is 70 GPa, determine the stresses in the bronze cylinder and in the aluminum tube caused by an axial load P of 1000 kN.

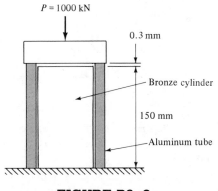

FIGURE P9–3

9–4 A timber post 10 in. by 10 in. is strengthened by four steel angles L2 × 2 × $\frac{3}{16}$, as shown in Fig. P9–4. If the moduli of elasticity are $E_{st} = 30 \times 10^3$ ksi, $E_{wd} = 1.5 \times 10^3$ ksi, and the allowble stresses are $(\sigma_{st})_{allow} = 23$ ksi and $(\sigma_{wd})_{allow} = 1.7$ ksi, determine the allowable load P.

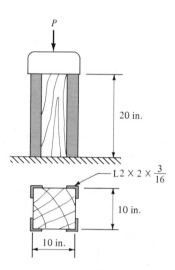

FIGURE P9–4

9–5 As shown in Fig. P9–5, a 5-kip weight is lifted by two steel wires; one wire is initially 10.000 ft long and the other wire is initially 10.002 ft long. The cross-sectional area and modulus of elasticity of each wire are $A = 0.25$ in.^{2} and $E = 30 \times 10^3$ ksi. Determine the stress in each wire.

5 kips

FIGURE P9–5

9–6 A load $P = 50$ kN is applied to a rigid bar suspended by three wires of the same cross-sectional area, as shown in Fig. P9–6. The outside wires are copper ($E_{cu} = 120$ GPa) and the inside wire is steel ($E_{st} = 210$ GPa). There is no slack or tension in the wires before the load is applied. Determine the load carried by each wire. (**HINT:** Due to symmetry, the axial deformation of each wire is identical.)

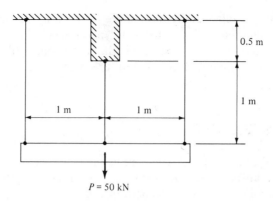

FIGURE P9–6

9–7 A rigid beam is supported by a hinge at A and two identical steel rods at C and E, as shown in Fig. P9–7. Determine the forces in the rods due to the load $W = 40$ kN. Before the load is applied, the beam hangs in the horizontal position.

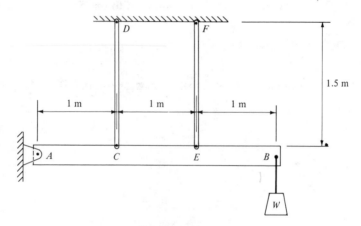

FIGURE P9–7

9–8 The steel bolt is loosely fitted in an aluminum tube as shown. The steel bolt has a pitch of 20 threads per inch (one turn of the nut moves it a distance of $\frac{1}{20}$ in. along the bolt). Determine the stresses in the steel bolt and in the aluminum tube

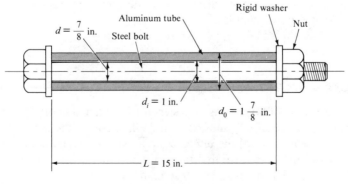

FIGURE P9–8

if the nut is tightened by turning the nut $\frac{1}{4}$ turn. For steel, $E_{st} = 30 \times 10^6$ psi, and for aluminum, $E_{al} = 10 \times 10^6$ psi. (**HINT:** The elongation of the bolt plus the contraction of the tube equals the movement of the nut along the bolt. The tensile force in the bolt is equal to the compressive force in the tube.)

9–3
THERMAL STRESSES

Most materials expand when the temperature increases and contract when the temperature decreases. In a statically determinate system having no restrictions on dimensional changes, the deformations caused by temperature change may be disregarded, since the members are free to expand or contract. In statically in-determinate systems, however, deformations are partially or fully restrained in certain directions, and significant stresses may be caused by temperature change.

If a member is free to deform, it expands or contracts due to a temperature change. Let the deformation due to a temperature change of ΔT over a length L be δ_T. The ratio δ_T/L is called the *thermal strain*, ε_T. It has been observed that the thermal strain is directly proportional to the change of temperature ΔT. That is,

$$\varepsilon_T \propto \Delta T$$

Let the constant of proportionality be α; we write

$$\varepsilon_T = \frac{\delta_T}{L} = \alpha \, \Delta T$$

or

$$\delta_T = \alpha L \, \Delta T \qquad\qquad (9\text{–}1)$$

where α is the *coefficient of thermal expansion* and has units of length per unit length per degree temperature change. For the unit of temperature, degrees Fahr-enheit is used in U.S. customary units, and degrees Celsius is used in SI units. Average values of α for some common materials are given in Table A–7 in the Appendix Tables.

The solutions of indeterminate problems involving temperature changes follow the general procedures discussed in the preceding section, as illustrated by the following two examples.

——— **EXAMPLE 9–3** ———————————————————————————

A steel rod 4.000 m long is fastened to a fixed support A as shown in Fig. (a). There is a gap of 0.4 mm between the free end of the rod and the fixed wall B at 10°C. Determine (a) the temperature at which the gap is just closed but no stress develops in the rod, (b) the stress in the rod at the temperature of 38°C. For steel, $E = 210$ GPa, $\alpha = 12 \times 10^{-6}$ m/m/°C.

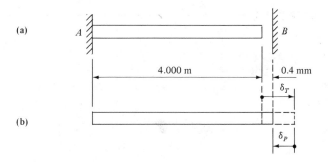

SOLUTION

(a) The temperature at which the gap is just closed can be determined from Eq. (9–1). Thus

$$\delta_T = \alpha L\ \Delta T$$

$$0.0004\ \text{m} = (12 \times 10^{-6}\ \text{m/m/°C})(4\ \text{m})(T°C - 10°C)$$

Solving for T gives

$$T = \frac{0.0004\ \text{m}}{(12 \times 10^{-6}\ \text{m/m/°C})(4\ \text{m})} + 10°C = 18.3°C$$

(b) If the rod were free to expand, the free expansion of the rod δ_T due to a temperature rise of $\Delta T = 38° - 10° = 28°C$ is

$$\delta_T = \alpha L\ \Delta T = (12 \times 10^{-6}\ \text{m/m/°C})(4\ \text{m})(28°C) = 0.00134\ \text{m} = 1.34\ \text{mm}$$

This is more than the dimension of the gap. Since the support and the wall are both fixed, the bar can expand only 0.4 mm to close the gap. The supports must exert a compressive force P large enough to deform the bar an amount δ_P to prevent the additional expansion, as shown in Fig. (b). From the geometry of the figure

$$\delta_P = \delta_T - 0.0004\ \text{m}$$

or

$$\frac{PL}{AE} = \frac{\sigma L}{E} = 0.001\ 34\ \text{m} - 0.0004\ \text{m} = 0.000\ 94\ \text{m}$$

from which

$$\sigma = \frac{(0.000\ 94)E}{L} = \frac{(0.000\ 94\ \text{m})(210 \times 10^{3}\ \text{MN/m}^2)}{4\ \text{m}} = 49.4\ \text{MPa (C)}$$

■

──────── **EXAMPLE 9-4** ────────────────────────────────────

A steel bolt of $\frac{3}{4}$ in. diameter is closely fitted in an aluminum sleeve of $1\frac{1}{4}$ in. outside diameter, as shown in Fig. (a). At 60°F, the nut is hand-tightened until it fits snugly against the end of the sleeve. Determine the stresses in the bolt and sleeve if the temperature rises to 170°F. For steel, $E_{st} = 30 \times 10^6$ psi and $\alpha_{st} = 6.5 \times 10^{-6}$ in./in./°F. For aluminum, $E_{al} = 10 \times 10^6$ psi and $\alpha_{al} = 13.0 \times 10^{-6}$ in./in./°F.

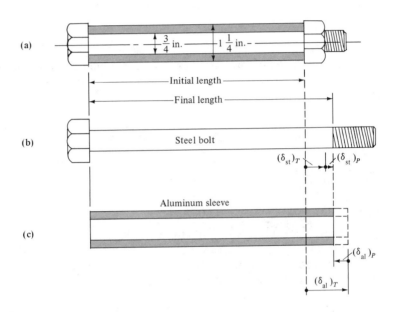

SOLUTION

Since α_{al} is greater than α_{st}, the thermal expansion of the aluminum sleeve is greater than that of the steel bolt. But the final length of the bolt and the sleeve must be the same; therefore, the bolt is subjected to tension and the sleeve is subjected to compression.

From the static equilibrium condition, the tensile force in the bolt must be equal to the compressive force in the sleeve. That is,

$$P_{st} = |P_{al}| = P$$

From the deformation geometry shown in Figs. (b) and (c), it is seen that the thermal expansion $(\delta_{st})_T$ of the steel bolt plus the stretch $(\delta_{st})_P$ of the bolt is equal to the thermal expansion $(\delta_{al})_T$ of the aluminum sleeve minus the absolute value of the contraction $(\delta_{al})_P$ of the sleeve. That is,

$$(\delta_{st})_T + (\delta_{st})_P = (\delta_{al})_T - |(\delta_{al})_P|$$

or

$$\alpha_{st} L \, \Delta T + \frac{PL}{A_{st}E_{st}} = \alpha_{al} L \, \Delta T - \frac{PL}{A_{al}E_{al}}$$

After the length L is canceled from each term, the equation becomes

$$\alpha_{st} \, \Delta T + \frac{P}{A_{st}E_{st}} = \alpha_{al} \, \Delta T - \frac{P}{A_{al}E_{al}} \qquad \text{(a)}$$

in which the temperature change ΔT is

$$\Delta T = 170°F - 60°F = 110°F$$

and the cross-sectional areas of the steel bolt and the aluminum sleeve are, respectively,

$$A_{st} = \tfrac{1}{4}\pi(0.75)^2 = 0.442 \text{ in.}^2$$

$$A_{al} = \tfrac{1}{4}\pi(1.25^2 - 0.75^2) = 0.785 \text{ in.}^2$$

When we substitute the numerical values, Eq. (a) becomes

$$(6.5 \times 10^{-6} \text{ ft/ft/°F})(110°F) + \frac{P, \text{ lb}}{(0.442 \text{ in.}^2)(30 \times 10^6 \text{ lb/in.}^2)}$$

$$= (13.0 \times 10^{-6} \text{ ft/ft/°F})(110°F) - \frac{P, \text{ lb}}{(0.785 \text{ in.}^2)(10 \times 10^6 \text{ lb/in.}^2)}$$

If we multiply both sides by 10^6, the equation becomes

$$715 + 0.0754P = 1430 - 0.1274P$$

from which

$$P = 3526 \text{ lb}$$

The stresses for the steel bolt and the aluminum sleeve are, respectively,

$$\sigma_{st} = \frac{P}{A_{st}} = \frac{3526 \text{ lb}}{0.442 \text{ in.}^2} = 7980 \text{ psi (T)}$$

$$\sigma_{al} = \frac{P}{A_{al}} = \frac{3526 \text{ lb}}{0.785 \text{ in.}^2} = 4490 \text{ psi (C)}$$

PROBLEMS

9–9 A steel structural member is supported between two fixed supports so that it cannot expand. At 60°F there is no stress in the member. Determine the stresses in the member at 100°F. For steel, $E = 30 \times 10^6$ psi and $\alpha = 6.5 \times 10^{-6}$ in./in./°F.

9–10 If the member in Problem 9–9 is 40 ft long and is supported between two fixed supports with a clearance of $\tfrac{1}{16}$ in. at 60°F, determine the stress in the member at 100°F.

9–11   A steel wire is held taut between two unyielding supports. At 10°C the wire is tightened so that it has a tensile stress of 100 MPa. Determine the temperature at which the wire would become slack. For steel, $E = 210$ GPa and $\alpha = 12 \times 10^{-6}$ m/m/°C.

9–12   The composite steel and brass rod shown in Fig. P9–12 is attached to unyielding supports with no initial stress at 25°C. Determine the stress in each portion when the temperature is reduced to -5°C. For steel, $E_{st} = 210$ GPa and $\alpha_{st} = 12 \times 10^{-6}$ m/m/°C, and for brass, $E_{br} = 100$ GPa and $\alpha_{br} = 19 \times 10^{-6}$ m/m/°C.

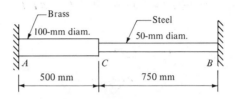

FIGURE P9–12

9–13   Three short posts of equal length support the 100-kN rigid concrete block shown in Fig. P9–13. Determine **(a)** the stress in each post, **(b)** the temperature decrease that will relieve the aluminum post of any stress. For steel, $E_{st} = 210$ GPa and $\alpha_{st} = 12 \times 10^{-6}$ m/m/°C, and for aluminum, $E_{al} = 70$ GPa and $\alpha_{al} = 23.4 \times 10^{-6}$ m/m/°C.

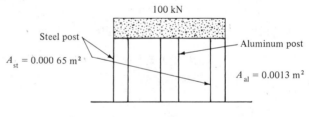

FIGURE P9–13

9–14   Solve Example 9–4 if the steel bolt is 1 in. in diameter, the aluminum sleeve has a $1\frac{1}{2}$-in. outside diameter, and the temperature rises to 150°F.

9–15   In Problem 9–8 (on p. 277), the bolt is put in tension by turning the nut $\frac{1}{4}$ turn. Determine the drop in temperature that would loosen the bolt. For steel $\alpha_{st} = 6.5 \times 10^{-6}$ in./in./°F, and for aluminum, $\alpha_{al} = 13.0 \times 10^{-6}$ in./in./°F.

*9–4
STATICALLY INDETERMINATE TORSIONAL SHAFTS

Torsional shafts may also be statically indeterminate. For example, if a shaft is fixed at both ends and a torque is applied at some intermediate point, the torsional reactions at the supports cannot be determined immediately from the static equilibrium condition alone. The geometry of the angle of twist of the shaft must be considered.

From Eq. (5–8), the angle of twist ϕ (in radians) along a shaft of length L subjected to torque T is

$$\phi = \frac{TL}{JG}$$

where J is the polar moment of inertia of the shaft section and G is the shear modulus of the shaft material.

EXAMPLE 9–5

Determine the required diameter of a shaft built in to fixed supports and subjected to forces $F = 60$ kips, as shown in Fig. (a). The forces produce a torque at section C. The allowable shear stress of the shaft is $\tau_{\text{allow}} = 8$ ksi.

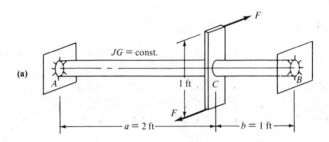

(a)

SOLUTION

If the fixed support at A had been removed and been replaced by the torsional reaction T_A as shown in Fig. (b), the internal torque diagram of the shaft would be as shown in Fig. (c), where the "right-hand rule" sign convention (see Section

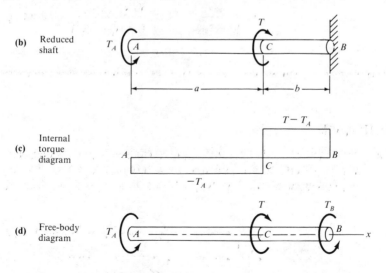

(b) Reduced shaft

(c) Internal torque diagram

(d) Free-body diagram

5–3) is used. Since the supports at A and B are fixed, the total angle of twist ϕ_{AB} between the supports must be equal to zero. Thus

$$\phi_{AB} = \phi_{AC} + \phi_{CB} = \frac{(-T_A)(a)}{JG} + \frac{(T - T_A)(b)}{JG} = 0 \qquad \text{(a)}$$

The constant JG is canceled from each term when the equation is multiplied by JG. The equation is reduced to

$$T_A(a + b) = Tb$$

from which

$$T_A = + \frac{b}{a + b} T = + \frac{1 \text{ ft}}{(2 + 1) \text{ ft}} (60 \text{ kip-in.}) = +20 \text{ kip-in.}$$

The equilibrium condition of the free-body diagram of the shaft shown in Fig. (d) requires that

$$+ \underset{\curvearrowright}{\longrightarrow} \Sigma M_x = T_A + T_B - T = 0$$

from which

$$T_B = T - T_A = 60 - 20 = +40 \text{ kip-in.}$$

Hence the shaft is subjected to a torque of 20 kip-in. in part AC and a torque of 40 kip-in. in part CB. The shaft should be designed to carry the larger torque without exceeding the allowable shear stress of 8 ksi. Thus

$$\tau_{max} = \frac{T_{max}c}{J} = \frac{T_{max}(d/2)}{\pi d^4/32} = \frac{16T_{max}}{\pi d^3} \leq \tau_{allow}$$

from which

$$d \geq \sqrt[3]{\frac{16T_{max}}{\pi(\tau_{allow})}} = \sqrt[3]{\frac{16(40 \text{ kip-in.})}{\pi(8 \text{ kips/in.}^2)}} = 2.94 \text{ in.}$$

$$d_{req} = 2.94 \text{ in.} \qquad \blacksquare$$

PROBLEMS

9–16 Both ends of a solid circular shaft are built in to fixed supports A and B, as shown in Fig. P9–16. If $d = 50$ mm, $a = 200$ mm, and $b = 300$ mm, determine the maximum shear stress in the shaft due to a torque $T = 4$ kN · m.

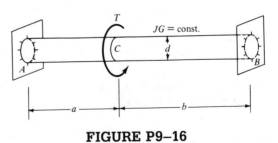

FIGURE P9–16

9–17 The stepped shaft of homogeneous material is held between fixed supports at A and B, as shown in Fig. P9–17. Determine the torsional reactions at the supports and the maximum shear stress in the shaft due to a torque $T = 10$ kip-in. applied at the section C.

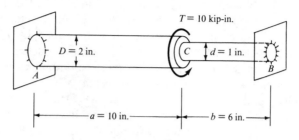

$T = 10$ kip-in.

$D = 2$ in.

C $d = 1$ in.

A

B

$a = 10$ in.

$b = 6$ in.

FIGURE P9–17

9–18 The composite shaft built in between two fixed supports is subjected to a torque $T = 40$ kip-in. at section C, as shown in Fig. 9–18. If the allowable shear stress in steel is 8000 psi and the allowable sheer stress in aluminum is 4000 psi, determine **(a)** the ratio of length a/b for which each material will be stressed to its allowable limit, **(b)** the required diameter of the shaft. For steel, $G_{st} = 11.6 \times 10^6$ psi, and for aluminum, $G_{al} = 3.8 \times 10^6$ psi.

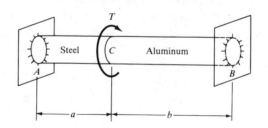

T

Steel C Aluminum

A

B

a

b

FIGURE P9–18

9–19 A shaft with both ends built in to fixed supports is twisted by two torques at sections C and D, as shown in Fig. P9–19. If $M_C = 3000$ N · m, $M_D = 1500$ N · m, $a = 500$ mm, $b = 700$ mm, and $c = 300$ mm, determine the torque in each portion of the shaft.

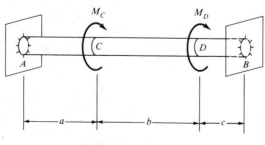

M_C M_D

A C D B

a b c

FIGURE P9–19

9–5

STATICALLY INDETERMINATE BEAMS

Three types of beams have been considered previously: (1) a simple beam, (2) an overhanging beam, and (3) a cantilever beam. These are all statically determinate beams in which the reactions at the supports can be determined from equilibrium conditions. Beams in which the equilibrium equations alone are insufficient to solve for reactions at the supports are statically indeterminate. Additional equations based on deflection conditions must be introduced to solve for the external reactions on a statically indeterminate beam.

Three equilbrium equations are available for the determination of beam reactions. In statically indeterminate beams we have more than three reaction components. Hence there are more constraints than needed to maintain the equilibrium of the beam. The superfluous constraints, which are not necessary for equilibrium, are called *redundant constraints*. The number of the redundant constraints of a beam is referred to as the *degree of indeterminacy* of the beam. Figure 9–1 shows three examples of statically indeterminate beams. Part (a) shows a propped beam with one redundant constraint, and it is statically indeterminate to the first degree. Part (b) shows a continuous beam with two redundant constraints, and it is statically indeterminate to the second degree. Part (c) shows a fixed beam with three redundant constraints. It is statically indeterminate to the third degree.

The principle of superposition is applicable to beams made of linearly elastic material undergoing small elastic deformation. The first step in solving a statically indeterminate beam problem, using the method of superposition, is to remove the redundant constraints so that the beam becomes statically determinate. The reactions at redundant constraints being removed are considered as externally applied loads. The magnitudes of these reactions are obtained by using proper deflection conditions at the redundant constraints of the original beam. Which constraints should be regarded as redundant is primarily a matter of convenience.

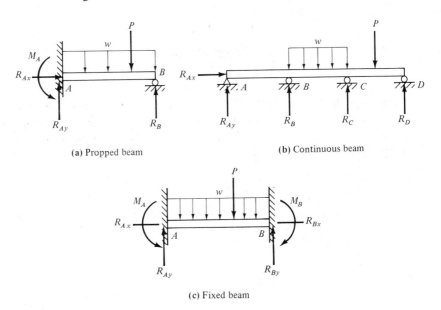

(a) Propped beam

(b) Continuous beam

(c) Fixed beam

FIGURE 9–1

Once the redundant reactions are determined, the beam becomes statically de-terminant; the other support reactions can be determined from statics. Stresses and deflections can be determined by the methods previously introduced.

------- **EXAMPLE 9–6** ---

Determine the reactions and plot the shear force and bending moment diagrams for a uniformly loaded propped beam shown. The value of EI for the beam is a constant.

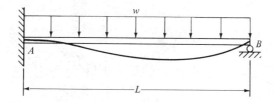

SOLUTION

The beam is statically indeterminate to the first degree. Either one of the fol-lowing two methods may be used, depending on which reaction component is treated as the redundant constraint.

[Method I]   Treating M_A as the Redundant Constraint

By removing the rotational constraint at the fixed support A, the member is reduced to a simple beam. The reaction M_A of the redundant constraint is treated as an external load applied to the simple beam, as shown in Fig. (a).

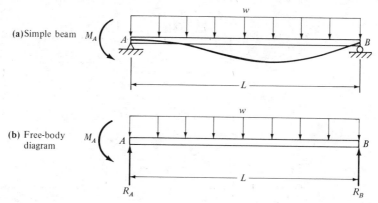

(a) Simple beam

(b) Free-body diagram

The magnitude of M_A should be such that it will make the slope at end A equal to zero and thus satisfy the fixed-ended condition of the original beam. By the method of superposition the slope at A of the simple beam due to M_A and w can be obtained by the algebraic sum of cases 10 and 8 from Table 8–1. Thus

$$\theta_A(\measuredangle+) = [\theta_l]_{\text{case }10}^{\text{due to } M_A} - [\theta_l]_{\text{case }8}^{\text{due to } w} = 0$$

or

$$\frac{M_A L}{3EI} - \frac{wL^3}{24EI} = 0$$

from which

$$M_A = +\frac{1}{8}\,wL^2 \qquad M_A = \frac{1}{8}\,wL^2\ \circlearrowright$$

With M_A determined, the other reaction components can be calculated by the equilibrium equations applied to the free-body diagram shown in Fig. (b). Thus

$$\oplus \Sigma M_B = R_A(L) - M_A - wL\left(\frac{L}{2}\right) = 0$$

$$R_A = \frac{1}{L}\left(\frac{wL^2}{8}\right) + \frac{wL}{2} = +\frac{5}{8}\,wL \qquad R_A = \frac{5}{8}\,wL\ \uparrow$$

$$\oplus \Sigma M_A = R_B(L) + M_A - wL\left(\frac{L}{2}\right) = 0$$

$$R_B = -\frac{1}{L}\left(\frac{wL^2}{8}\right) + \frac{wL}{2} = +\frac{3}{8}\,wL \qquad R_B = \frac{3}{8}\,wL\ \uparrow$$

$$+\uparrow\Sigma F_y = \frac{5}{8}wL + \frac{3}{8}wL - wL = 0 \qquad\qquad \text{(checks)}$$

After all the reactions are determined, the loading diagram of the beam is drawn in Fig. (c) and the shear and moment diagrams are plotted by using the summation method as shown in Figs. (d) and (e).

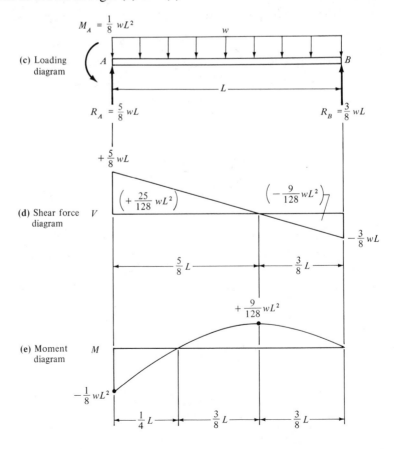

(c) Loading diagram

(d) Shear force diagram

(e) Moment diagram

[Method II] Treating R_B as the Redundant Constraint

By removing the roller support at the end B, the member is reduced to a cantilever beam. The reaction R_B is treated as external load applied on the cantilever beam, as shown in Fig. (f).

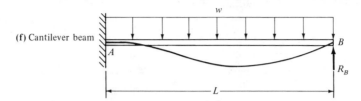

(f) Cantilever beam

The magnitude of R_B should be such that the deflection at the end B is zero to satisfy the support condition of the original beam. By the method of superposition, the deflection at B of the cantilever beam due to R_B and w can be obtained by algebraic sum of cases 1 and 3 from Table 8–1. Thus

$$y_B(+\uparrow) = [y_{max}]_{case\ 1}^{due\ to\ R_B} - [y_{max}]_{case\ 3}^{due\ to\ w} = 0$$

or

$$\frac{R_B L^3}{3EI} - \frac{wL^4}{8EI} = 0$$

from which

$$R_B = +\frac{3}{8}wL \qquad R_B = \frac{3}{8}wL \uparrow$$

The remainder of the problem can be solved by considering the static equilibrium conditions of the cantilever beam in Fig. (f). The results will be identical to those found in method I. ∎

EXAMPLE 9–7

Determine the reactions for the continuous beam subjected to a concentrated load as shown in Fig. (a).

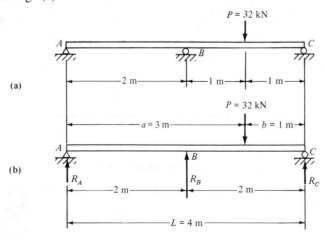

(a)

(b)

SOLUTION

Treat the reaction R_B at the roller support B as redundant; then the member is reduced to the simple beam shown in Fig. (b). The magnitude of R_B should be such that it will make the deflection at B equal to zero to satisfy the support condition of the original beam. Thus

$$y_B(+\uparrow) = [y_{max}]^{due\ to\ R_B}_{case\ 6} - [y_{L/2}]^{due\ to\ P}_{case\ 7} = 0$$

From Table 8-1 we have

$$\frac{R_B L^3}{48EI} - \frac{Pb}{48EI}(3L^2 - 4b^2) = 0$$

$$\frac{R_B(4\ m)^3}{48EI} - \frac{(32\ kN)(1\ m)}{48EI}[3(4\ m)^2 - 4(1\ m)^2] = 0$$

from which

$$R_B = +22\ kN \qquad R_B = 22\ kN \uparrow$$

The equilibrium conditions of the simple beam in Fig. (b) require that

$$\circlearrowleft\Sigma M_C = R_A(4) - P(1) + R_B(2) = 0$$

from which

$$R_A = \frac{(32 \times 1) - (22 \times 2)}{4} = -3\ kN \qquad R_A = 3\ kN \downarrow$$

and

$$\circlearrowright\Sigma M_A = R_C(4) - P(3) + R_B(2) = 0$$

from which

$$R_C = \frac{32(3) - 22(2)}{4} = +13\ kN, \qquad R_C = 13\ kN \uparrow$$

Check

$$+\uparrow\Sigma F y = -3 + 22 - 32 + 13 = 0 \qquad\qquad (checks) \qquad \blacksquare$$

────── **EXAMPLE 9–8** ──────────────────────────────────────

Determine the moments M_A and M_B at the fixed supports (called the fixed-end moments) for the fixed beam due to a concentrated load shown in Fig. (a).

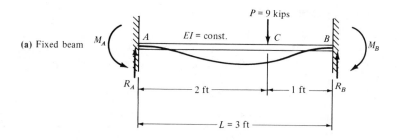

(a) Fixed beam

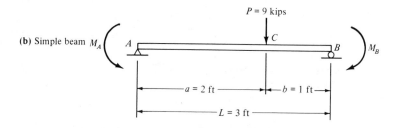

(b) Simple beam

SOLUTION

Since there is no horizontal force applied to the beam, the horizontal components of reactions are zero. There are four unknown reaction components, as shown in Fig. (a), and only two equilibrium equations are available ($\Sigma F_x = 0$ is a trivial equation). The beam is therefore statically indeterminate to the second degree. Treating M_A and M_B as redundant, the beam is reduced to a simple beam as shown in Fig. (b). The magnitudes of M_A and M_B should be such that the slopes at A and B are equal to zero to satisfy the support conditions of the original beam. Thus

$$\theta_A(\measuredangle +) = [\theta_l]_{\text{case } 10}^{\text{due to } M_A} + [\theta_l]_{\text{case } 9}^{\text{due to } M_B} - [\theta_l]_{\text{case } 7}^{\text{due to } P} = 0$$

From Table 8–1 we have

$$\frac{M_A L}{3EI} + \frac{M_B L}{6EI} - \frac{Pb(L^2 - b^2)}{6EIL} = 0$$

or

$$\frac{M_A(3 \text{ ft})}{3EI} + \frac{M_B(3 \text{ ft})}{6EI} - \frac{(9 \text{ kips})(1 \text{ ft})[(3 \text{ ft})^2 - (1 \text{ ft})^2]}{6EI(3 \text{ ft})} = 0$$

or

$$M_A + 0.5M_B = 4 \text{ kip-ft} \tag{a}$$

and

$$\theta_B(+\measuredangle) = [\theta_r]_{\text{case } 10}^{\text{due to } M_A} + [\theta_r]_{\text{case } 9}^{\text{due to } M_B} - [\theta_r]_{\text{case } 7}^{\text{due to } P} = 0$$

From Table 8–1 we have

$$\frac{M_A L}{6EI} + \frac{M_B L}{3EI} - \frac{Pab(2L - b)}{6EIL} = 0$$

or

$$\frac{M_A(3 \text{ ft})}{6EI} + \frac{M_B(3 \text{ ft})}{3EI} - \frac{(4 \text{ kips})(2 \text{ ft})(1 \text{ ft})(2 \times 3 \text{ ft} - 1 \text{ ft})}{6EI(3 \text{ ft})} = 0$$

or

$$0.5M_A + M_B = 5 \text{ kip-ft} \qquad\qquad (b)$$

Solving Eqs. (a) and (b) simultaneously gives

$$M_A = +2 \text{ kip-ft} \qquad M_A = 2 \text{ kip-ft} \circlearrowleft$$

$$M_B = +4 \text{ kip-ft} \qquad M_B = 4 \text{ kip-ft} \circlearrowright$$

■

PROBLEMS

*For the beams loaded as shown in the figures for Problems **9–20** to **9–23**, determine the magnitude of the redundant reactions indicated in the figures. In each problem, the value of EI of the beam is a constant.*

9–20

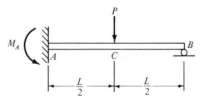

FIGURE P9–20

9–21

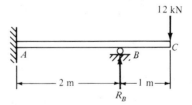

FIGURE P9–21

9–22

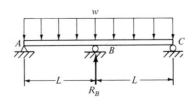

FIGURE P9–22

9–23

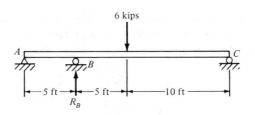

FIGURE P9–23

*For the beams loaded as shown in the figures for Problems **9–24** to **9–27**, determine the reactions in all the supports. In each problem, the value of EI of the beam is a constant.*

9–24

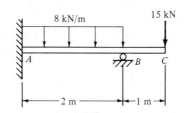

FIGURE P9–24

9–25

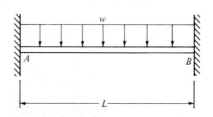

FIGURE P9–25

9–26

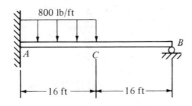

FIGURE P9–26

9–27

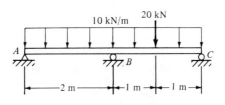

FIGURE P9–27

*For the beams loaded as shown in the figures for Problems **9–28** to **9–31**, determine the reactions and plot shear and moment diagrams. In each problem, the value of EI of the beam is a constant.*

9–28

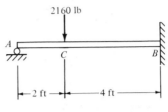

FIGURE P9–28

9–29

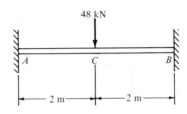

FIGURE P9–29

9–30

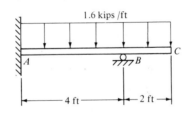

FIGURE P9–30

9–31

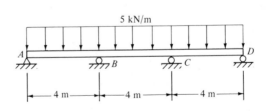

FIGURE P9–31

CHAPTER 10

Combined Stresses

10–1
INTRODUCTION

The fundamental formulas for calculating the stresses in a member subjected to only one type of loading have been developed in previous chapters. These formulas are listed for reference in Table 10–1. In deriving the formulas, it was assumed that (1) only a single internal force or moment was acting on a cross section of a member, and (2) the maximum stress in the member was within the elastic limit of the material so that stress was proportional to strain.

In many engineering applications, more than one internal force may exist at a section of a member. Therefore, a technique is needed for finding the combined stress in a member subjected to several internal forces acting simultaneously.

To determine the combined stresses caused by two or more internal forces, the method of superposition is used. That is, stresses caused by each internal force are determined separately, using the fundamental formulas in Table 10–1. The algebraic sum of these stresses gives the combined stresses caused by all the internal forces acting simultaneously. The method of superposition is valid only if the maximum stress is within the elastic limit of the material and if the deformations are very small.

To begin with, the determination of normal stress caused by the simultaneous action of axial force and bending moment is discussed. This is followed by the discussions of bending about two perpendicular axes, as well as the effects of eccentric loading. Then the superposition of shear stresses in circular members is studied. Finally, we develop the formulas for calculating stresses in thin-walled pressure vessels, which provide a good example of the biaxial stress condition (normal stresses occur in two perpendicular directions).

10–2

STRESSES DUE TO BENDING MOMENT AND AXIAL FORCE

Consider the cantilever beam subjected to a force F applied at the centroid of the section at the free end, as shown in Fig. 10–1(a). The force F is resolved into

TABLE 10–1   List of the Fundamental Formulas

Internal Force or Moment	Stress	Formula	Equation Number
Axial force	Axial stress	$\sigma = \dfrac{P}{A}$	(2–3)
Bending moment	Flexural stress	$\sigma = \dfrac{My}{I}$	(7–3)
		$\sigma_{max} = \dfrac{Mc}{I} = \dfrac{M}{S}$	(7–2) (7–7)
Torque (torsional moment in circular shaft)	Torsional shear stress	$\tau = \dfrac{T\rho}{J}$	(5–2)
		$\tau_{max} = \dfrac{Tc}{J}$	(5–1)
Shear force in beam	Beam shear stress	$\tau = \dfrac{VQ}{It}$	(7–10)
		$\tau_{max} = 1.5\dfrac{V}{A}$	(7–11)
		(for rectangular section)	
		$\tau_{max} = \dfrac{4V}{3A}$	(Problem 7–24)
		(for circular section)	

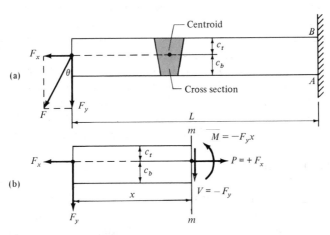

FIGURE 10–1

horizontal and vertical components F_x and F_y. From the free-body diagram in Fig. 10–1(b), the internal forces at an arbitrary section m–m are

$$\text{axial force:} \qquad P = +F_x$$

$$\text{bending moment:} \quad M = -F_y x$$

$$\text{shear force:} \qquad V = -F_y$$

The axial force and the bending moment produce normal stresses on the section while the shear force produces shear stresses. The shear stresses can be calculated by using the beam shear formula.

Using the method of superposition, consider the axial force and the bending moment acting separately. The normal stresses are

Axial stress due to axial force:

$$\sigma' = +\frac{P}{A}$$

Flexure stress due to bending moment:

$$\sigma'' = \pm\frac{My}{I}$$

Both stresses are normal to the cross section and hence they can be added algebraically. Thus the normal stresses at points on cross section *m–m* are

$$\sigma = \sigma' + \sigma'' = +\frac{P}{A} \pm \frac{My}{I} \tag{10–1}$$

where the positive sign indicates tension and the negative sign indicates compression. The normal stresses at points A and B are

$$\sigma_A = +\frac{P}{A} - \frac{Mc_b}{I} = +\frac{F_x}{A} - \frac{F_y Lc_b}{I} \tag{a}$$

$$\sigma_B = +\frac{P}{A} + \frac{Mc_t}{I} = +\frac{F_x}{A} + \frac{F_y Lc_t}{I} \tag{b}$$

In Eq. (10–1), σ is a linear function of y; therefore, the normal stress in the section at the fixed end varies linearly from σ_A at the bottom to σ_B at the top.

EXAMPLE 10–1

A cast-iron frame for a punch press has the dimensions shown in the figure. The cross section and section properties are given as shown. Determine the normal stress distribution in section 1–1 if $F = 45$ kN.

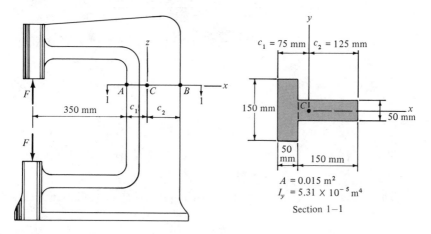

$A = 0.015 \text{ m}^2$
$I_y = 5.31 \times 10^{-5} \text{ m}^4$

Section 1–1

SOLUTION

Use the method of sections cutting the frame at section 1–1. The free-body diagram of the upper part of the frame and its equilibrium equations are shown as follows:

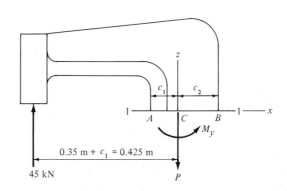

$$+\uparrow \Sigma F_z = 45 \text{ kN} - P = 0 \qquad P = +45 \text{ kN (T)}$$

$$\oplus \Sigma M_C = M_y - (45 \text{ kN})(0.425 \text{ m}) = 0$$

$$M_y = +19.1 \text{ kN} \cdot \text{m}$$

The axial stress due to axial force P is constant across the cross section and is equal to

$$\sigma' = +\frac{P}{A} = +\frac{45 \text{ kN}}{0.015 \text{ m}^2} = +3000 \text{ kN/m}^2 = +3.0 \text{ MPa}$$

The flexural stresses at A and B due to the moment M are, respectively,

$$\sigma''_A = +\frac{M_y c_1}{I} = +\frac{(19.1 \text{ kN} \cdot \text{m})(0.075 \text{ m})}{5.31 \times 10^{-5} \text{ m}^4}$$

$$= +27.0 \times 10^3 \text{ kN/m}^2 = +27.0 \text{ MPa}$$

$$\sigma''_B = -\frac{M_y c_2}{I} = -\frac{(19.1 \text{ kN} \cdot \text{m})(0.125 \text{ m})}{5.31 \times 10^{-5} \text{ m}^4}$$

$$= -45.0 \times 10^3 \text{ kN/m}^2 = -45.0 \text{ MPa}$$

Since the axial force and bending moment act simultaneously, the normal stresses at A and B are, respectively,

$$\sigma_A = \sigma' + \sigma''_A = +3.0 \text{ MPa} + 27.0 \text{ MPa} = +30.0 \text{ MPa (T)}$$

$$\sigma_B = \sigma' + \sigma''_B = +3.0 \text{ MPa} - 45.0 \text{ MPa} = -42.0 \text{ MPa (C)}$$

The stress distribution in section 1–1 is plotted in the following figures.

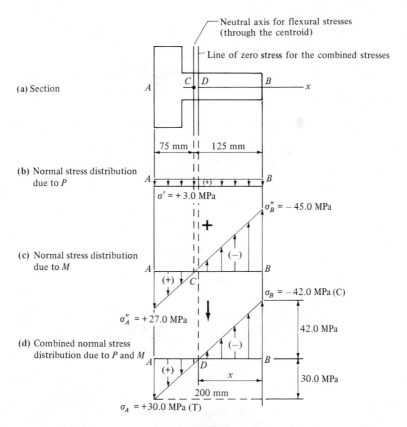

(a) Section

(b) Normal stress distribution due to P

(c) Normal stress distribution due to M

(d) Combined normal stress distribution due to P and M

Neutral axis for flexural stresses (through the centroid)

Line of zero stress for the combined stresses

75 mm   125 mm

$\sigma' = +3.0$ MPa

$\sigma''_B = -45.0$ MPa

$\sigma''_A = +27.0$ MPa

$\sigma_B = -42.0$ MPa (C)

42.0 MPa

30.0 MPa

200 mm

$\sigma_A = +30.0$ MPa (T)

Note that for flexural stresses due to bending moment M only, the neutral axis passes through the centroid of the section, as shown in Fig. (c), but for the combined stresses due to both axial force and bending moment, the line of zero stress shifts to the right. The distance BD, denoted by x in Fig. (d), can be determined by proportion:

$$\frac{x}{42.0 \text{ MPa}} = \frac{200 \text{ mm}}{(42 + 30) \text{ MPa}} \qquad x = 116.7 \text{ mm}$$

EXAMPLE 10–2

A crane with swinging arm as shown in Fig. (a) is designed to hoist a maximum weight of 2 kips. If the allowable compressive stress is 13 ksi, select a wide-flange section for the arm AB.

SOLUTION

The free-body diagram of the arm AB is constructed as shown in Fig. (b), where $\mathbf{T}$ is the tension in the rod CD and T_x and T_y are the components of $\mathbf{T}$. The equilibrium equations of the free body give

$$\circlearrowleft \Sigma M_A = \frac{5}{13} T(12 \text{ ft}) - (2 \text{ kips})(12 \text{ ft} + 8 \text{ ft}) = 0 \qquad T = 8.67 \text{ kips}$$

$$\xrightarrow{+} \Sigma F_x = R_{Ax} - \frac{12}{13}(8.67 \text{ kips}) = 0 \qquad R_{Ax} = 8.00 \text{ kips}$$

$$+\uparrow \Sigma F_y = -R_{Ay} + \frac{5}{13}(8.67 \text{ kips}) - 2 \text{ kips} = 0 \qquad R_{Ay} = 1.33 \text{ kips}$$

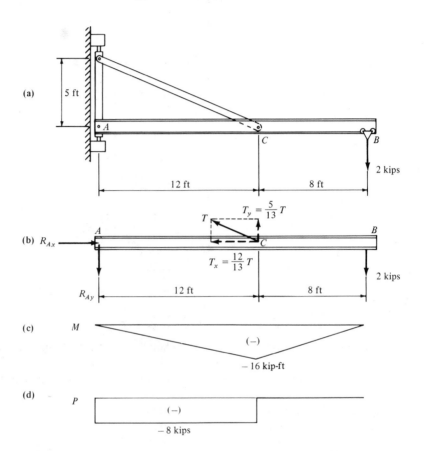

The bending moment and the axial force diagrams of the arm AB are shown in Figs. (c) and (d), respectively. From these diagrams we see that the critical section occurs just to the left of C, where the maximum negative moment is 16 kip-ft (or 192 kip-in.) and the compressive axial force is 8 kips.

For a tentative selection of a W shape, consider the bending moment only, which requires

$$S_{\text{req}} = \frac{M}{\sigma_{\text{allow}}^{(C)}} = \frac{192 \text{ kip-in.}}{13 \text{ kips/in.}^2} = 14.8 \text{ in.}^3$$

From Table A–1, try W8 × 18 ($A = 5.26$ in.^{2}, $S = 15.2$ in.^{3}). The maximum compressive stress at the critical section is

$$|\sigma_{\text{max}}^{(C)}| = \frac{P}{A} + \frac{M}{S} = \frac{8 \text{ kips}}{5.26 \text{ in.}^2} + \frac{192 \text{ kip-in.}}{15.2 \text{ in.}^3}$$

$$= 14.2 \text{ ksi} > \sigma_{\text{allow}}^{(C)} = 13 \text{ ksi} \quad \text{N.G.}$$

Try the next larger size W8 × 21 ($A = 6.16$ in.², $S = 18.2$ in.³).

$$|\sigma_{max}^{(C)}| = \frac{P}{A} + \frac{M}{S} = \frac{8 \text{ kips}}{6.16 \text{ in.}^2} + \frac{192 \text{ kip-in.}}{18.2 \text{ in.}^3}$$

$$= 11.8 \text{ ksi} < \sigma_{allow}^{(C)} = 13 \text{ ksi} \quad \text{O.K.}$$

Hence a W8 × 21 section is satisfactory. ∎

PROBLEMS

10–1 A timber beam of rectangular section supports a load applied at the free end, as shown in Fig. P10–1. Determine the normal stresses at points A and B.

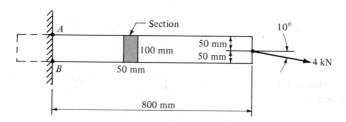

FIGURE P10–1

10–2 A steel bracket is loaded as shown in Fig. P10–2. **(a)** Determine the normal stresses at points A and B, **(b)** Plot the normal stress distribution along AB. **(c)** Determine the location of the line of zero stress at section AB.

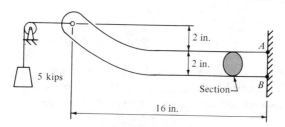

FIGURE P10–2

10–3 A concrete block of rectangular section is subjected to the loads shown in Fig. P10–3. Determine **(a)** the magnitude of the load F such that the normal stress at A is equal to zero, **(b)** the normal stress at B when the normal stress at A is zero.

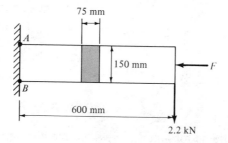

FIGURE P10–3

10–4 The short steel post of wide-flange section W14 × 34 is subjected to the load shown in Fig. P10–4. **(a)** Determine the normal stress at A and B. **(b)** Plot the normal stress distribution along A–B. **(c)** Determine the location of the line of zero stress at section AB.

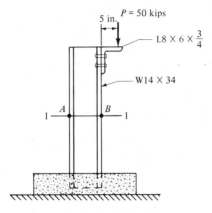

FIGURE P10–4

10–5 A machine part for transmitting a pull of 100 kN is offset as shown in Fig. P10–5. Determine the normal stress at A and B.

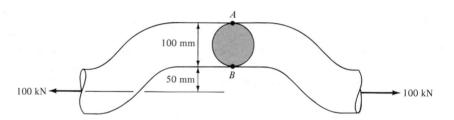

FIGURE P10–5

10–6 The frame of a hydraulic press has the dimensions shown in Fig. P10–6. If $P = 1600$ kN, determine the maximum tensile and compressive stresses at section 1–1. The cross section and the section properties are given.

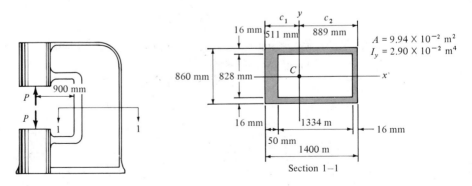

FIGURE P10–6

10–7 A cast-iron frame for a punch press has the dimensions and the properties of the cross section 1–1, as shown in Fig. P10–7. If the allowable stresses are 4000 psi in tension and 12 000 psi in compression, determine the maximum allowable force P that can be applied. Consider section 1–1 as the controlling section.

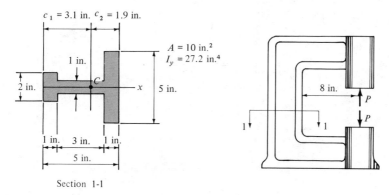

Section 1-1

FIGURE P10–7

10–8 A beam consisting of two standard C9 × 15 steel channels arranged back to back is loaded and supported as shown in Fig. P10–8. Determine the maximum tensile and compressive stresses along the beam.

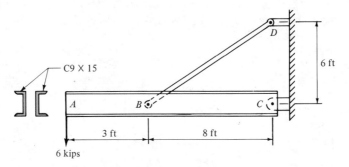

FIGURE P10–8

10–9 The horizontal beam of the jib crane shown in Fig. P10–9 is made of two standard steel channels. The maximum load including the weight of the moving cart that the crane is designed to carry is 8 kips. If the allowable compressive stress is 15 ksi, select a proper size for the pair of channels.

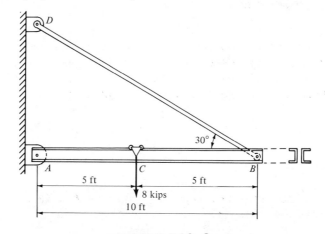

FIGURE P10–9

10–10 A wide-flange steel beam W10 × 100 is lifted by a crane, as shown in Fig. P10–10. Determine the maximum tensile and compressive stresses in the beam.

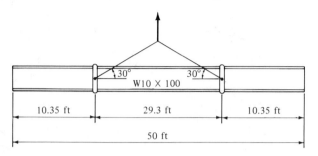

FIGURE P10–10

10–3
STRESSES DUE TO BENDING ABOUT TWO AXES

Consider a simple example of a cantilever, whose cross section has two axes of symmetry, as shown in Fig. 10–2(a). The load P acts in the cross section at the

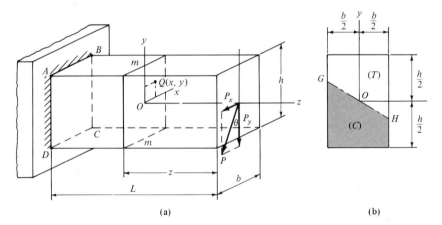

FIGURE 10–2

free end through the centroid and makes an angle θ with the longitudinal vertical plane. In calculating the normal stresses in the beam, the method of superposition will be used. The load P is resolved into two components P_x and P_y in the directions of the two axes of symmetry of the cross section. Thus

$$P_x = P \sin \theta$$
$$P_y = P \cos \theta$$

At an arbitrary section m–m, the component P_y produces a bending moment M_x about the x-axis equal to

$$M_x = P_y z$$

and the component P_x produces a bending moment M_y about the y-axis equal to

$$M_y = P_x z$$

From the directions of the two components and their location with respect to the x- and y-axes, we see that the moment M_x produces tension at points with a positive y, and the moment M_y produces tension at points with a positive x. The normal stress at a point $Q(x, y)$ in section m–m is obtained by taking the algebraic sum of the bending stresses produced by M_x and M_y separately. Thus

$$\sigma_Q = \frac{M_x y}{I_x} + \frac{M_y x}{I_y} \tag{10–2}$$

The equation of the line of zero stress can be found by setting σ_Q to zero; then Eq. (10–2) becomes

$$\frac{M_x}{I_x} y + \frac{M_y}{I_y} x = 0 \tag{10–3}$$

In this equation, when x is equal to zero, y must also be equal to zero; therefore, the line of zero stress passes through the centroid of the cross section, as shown in Fig. 10–2(b). The tension zone and compression zone are located on each side of the line of zero stress GH, as labeled in the figure.

The maximum tension occurs at point B of the built-in section where moments M_x and M_y and the x- and y-coordinates are maximum. The maximum value is obtained by substituting $(M_x)_{max} = P_y L$, $(M_y)_{max} = P_x L$, $x = b/2$, and $y = h/2$ in Eq. (10–2); thus

$$\sigma_{max}^{(T)} = \sigma_B = \frac{P_y L(h/2)}{I_x} + \frac{P_x L(b/2)}{I_y}$$

The maximum compression occurs at D, and it has the same magnitude as the maximum tension at B.

─────── **EXAMPLE 10–3** ───────────────────────────

A timber beam of rectangular cross section 150 mm × 200 mm carries a uniform load w of 4 kN/m over a simple span of 3 m and is supported at the ends in the tilted position as shown in Fig. (a). Determine the location of the line of zero stress and the maximum normal stress in the beam.

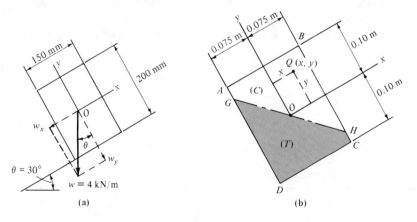

(a)   (b)

SOLUTION

The uniform load w is first resolved into x- and y-components as

$$w_x = w \sin \theta = (4 \text{ kN/m}) \sin 30° = 2.00 \text{ kN/m}$$

$$w_y = w \cos \theta = (4 \text{ kN/m}) \cos 30° = 3.46 \text{ kN/m}$$

The maximum bending moments at the midspan about the x- and y-axes are, respectively,

$$M_x = \frac{w_y L^2}{8} = \frac{(3.46 \text{ kN/m})(3 \text{ m})^2}{8} = 3.89 \text{ kN·m}$$

$$M_y = \frac{w_x L^2}{8} = \frac{(2.00 \text{ kN/m})(3 \text{ m})^2}{8} = 2.25 \text{ kN·m}$$

The moments of inertia of the section about the x- and y-axes are, respectively,

$$I_x = \frac{(0.15 \text{ m})(0.20 \text{ m})^3}{12} = 1.0 \times 10^{-4} \text{ m}^4$$

$$I_y = \frac{(0.20 \text{ m})(0.15 \text{ m})^3}{12} = 5.63 \times 10^{-5} \text{ m}^4$$

The normal stresses at point $Q(x, y)$ due to M_x and M_y are both compression; thus

$$\sigma_Q = -\frac{M_x y}{I_x} - \frac{M_y x}{I_y} = -\frac{(3.89 \text{ kN·m})y(\text{m})}{1.0 \times 10^{-4} \text{ m}^4} - \frac{(2.25 \text{ kN·m})x(\text{m})}{5.63 \times 10^{-5} \text{ m}^4}$$

$$= -(38.9y + 40.0x) \times 10^3 \text{ kN/m}^2 = -(38.9y + 40.0x) \text{ MPa} \quad \text{(a)}$$

The line of zero stress is determined by setting σ_Q to zero:

$$\sigma_Q = -(38.9y + 40.0x) = 0$$

or

$$38.9y + 40.0x = 0$$

From this equation it follows that when $x = -0.075$ m, $y = +0.0773$ m; and when $x = +0.075$ m, $y = -0.0773$ m. These results locate points G and H through which the line of zero stress passes, as shown in Fig. (b). This line also passes through the original O. The tension zone and compression zone are labeled in the figure.

The maximum tensile stress occurs at point D of the midspan; its value can be determined by substituting the coordinates of D, $x = -0.075$ m and $y = 0.10$ m in Eq. (a). Thus

$$\sigma_{max}^{(T)} = \sigma_D = -38.9(-0.10) + 40.0(-0.075) = +6.89 \text{ MPa}$$

The magnitude of maximum compression at B is found to be 6.89 MPa, the same value as the maximum tension at D.

━━━ **EXAMPLE 10–4** ━━━

A bridge-type crane consists of a movable beam on rails and a cart that can be moved along the beam, as shown in Fig. (a). The beam of W12 × 26 wide-flange steel section has a 12-ft span and carries a maximum weight, including the movable cart of $P = 6$ kips. When the beam moves on the rail, the load P makes an angle θ of 15° with the longitudinal vertical plane, as shown in Fig. (b). Determine the maximum flexural stress in the beam.

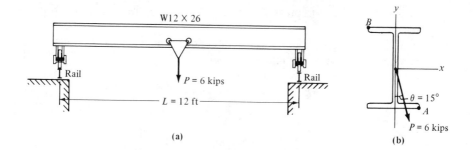

(a)   (b)

SOLUTION

The loading condition that causes maximum moment occurs when the movable cart is located at the midspan of the beam and the critical section is at the midspan. Resolve P into components along the x- and y-axes as

$$P_x = P \sin \theta = 6 \sin 15° = 1.55 \text{ kips}$$

$$P_y = P \cos \theta = 6 \cos 15° = 5.80 \text{ kips}$$

The weight of the beam acts along the y-axis; thus

$$w_y = 26 \text{ lb/ft} = 0.026 \text{ kip/ft}$$

The maximum moments at the midspan about the x- and y-axes are, respectively,

$$(M_x)_{max} = \frac{P_y L}{4} + \frac{w_y L^2}{8} = \frac{(5.80 \text{ kips})(12 \text{ ft})}{4} + \frac{(0.026 \text{ kip/ft})(12 \text{ ft})^2}{8}$$

$$= 17.9 \text{ kip-ft} = (17.9 \text{ kip-ft})\left(\frac{12 \text{ in.}}{1 \text{ ft}}\right) = 214 \text{ kip-in.}$$

$$(M_y)_{max} = \frac{P_x L}{4} = \frac{(1.55 \text{ kips})(12 \text{ ft})}{4}$$

$$= 4.65 \text{ kip-ft} = (4.65 \text{ kip-ft})\left(\frac{12 \text{ in.}}{1 \text{ ft}}\right) = 55.8 \text{ kip-in.}$$

From Table A–1, the section moduli of a W12 × 26 section are

$$S_x = 33.4 \text{ in.}^3 \qquad S_y = 5.34 \text{ in.}^3$$

The maximum tensile stress occurs at point A of the midsection; it is equal to

$$\sigma_{max}^{(T)} = \sigma_A = \frac{(M_x)_{max}}{S_x} + \frac{(M_y)_{max}}{S_y}$$

$$= \frac{214 \text{ kip-in.}}{33.4 \text{ in.}^3} + \frac{55.8 \text{ kip-in.}}{5.34 \text{ in.}^3} = 16.9 \text{ ksi}$$

The maximum compressive stress occurs at point B of the midsection and the magnitude of the compressive stress is the same as the magnitude of the maximum tensile stress.

If load P were in the vertical direction, that is, $\theta = 0°$, then the maximum bending moment in the midspan would be

$$M_{max} = \frac{PL}{4} + \frac{wL^2}{8} = \frac{(6 \text{ kips})(12 \text{ ft})}{4} + \frac{(0.026 \text{ kip/ft})(12 \text{ ft})^2}{8}$$

$$= 18.5 \text{ kip-ft} = (18.5 \text{ kip-ft})\left(\frac{12 \text{ in.}}{1 \text{ ft}}\right) = 222 \text{ kip-in.}$$

and the maximum bending stress in the beam would be

$$\sigma_{max} = \frac{M_{max}}{S_x} = \frac{222 \text{ kip-in.}}{33.4 \text{ in.}^3} = 6.65 \text{ ksi}$$

Hence it is seen that when the load is tilted from the vertical plane, the maximum flexural stress increases from 6.65 ksi to 16.9 ksi, an increase of 2.5 times. This is because S_y is much smaller than S_x for a wide-flange section. Therefore, in beam design care must be taken to investigate the adverse effect of bending about the y-axis of the section. ∎

PROBLEMS

10–11 A 10-ft-long simply supported timber beam is supported in such a way that the vertical concentrated load $P = 2$ kips applied at the centroid of the midspan passes through diagonal AC, as shown in Fig. P10–11. Neglecting the weight of the beam, locate the line of zero stress and find the normal stresses at points A, B, C, and D in the midspan.

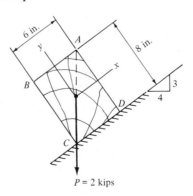

$P = 2$ kips

FIGURE P10–11

10–12 A cantilever beam has a 2-m horizontal span and is built into a concrete pier on one end at a tilted position, as shown in Fig. P10–12. This beam has a full-sized timber section of 50 mm by 100 mm. A vertical load $P = 270$ N is applied at the free end through the centroid. Neglecting the weight of the beam, locate the line of zero stress and find the maximum flexural stress in the beam at the built-in end.

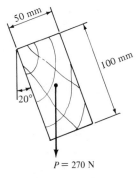

$P = 270$ N

FIGURE P10–12

10–13 Suppose that the cantilever timber beam shown in Fig. P10–13 has a span length $L = 4$ ft and a full-sized rectangular section with a $= 3$ in. It carries a horizontal load $P_x = 120$ lb and a vertical load $P_y = 300$ lb, as shown. Neglecting the weight of the beam, determine **(a)** the normal stress at point Q in terms of its coordinates x and y, **(b)** the location of the zero stress line, **(c)** the normal stresses at points A, B, C, and D.

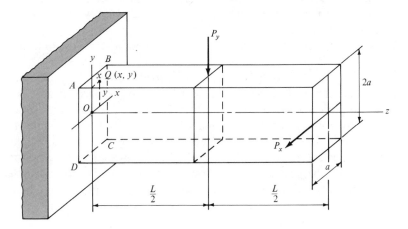

FIGURE P10–13

10–14 Suppose that the cantilever timber beam shown in Fig. P10–13 has a span length $L = 3$ m and carries a horizontal load P_x of 600 N and a vertical load P_y of 1000 N. If the allowable bending stress for timber is 10 MPa, determine the required dimension a of the section. Neglect the weight of the beam.

10–15 Suppose that the simply supported standard steel I-beam S6 × 17.25 shown in Fig. P10–15 has a span length $L = 12$ ft and carries a load $P = 1.5$ kips applied at the midspan. The load passes through the centroid of the section and makes an angle $\theta = 20°$ with the longitudinal vertical plane as shown. Determine the maximum bending stress in the beam.

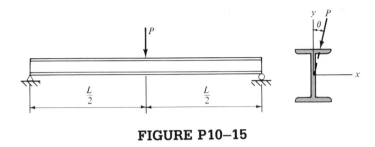

FIGURE P10–15

10–16 Suppose that the simply supported steel I-beam shown in Fig. P10–15 has a span length $L = 16$ ft and carries a load $P = 6$ kips applied at the midspan. The load passes through the centroid of the section and makes an angle $\theta = 10°$ with the longitudinal vertical plane as shown. If the allowable bending stress is 20 ksi, select a proper size for the standard steel I-beam.

10–4
STRESSES IN ECCENTRICALLY LOADED MEMBERS

Eccentric loading is a special case of the combination of axial and flexural stresses. The method of superposition applies to short posts that have lengths less than 10 times their minimum lateral dimension. In these cases the deflections are so small that the effect of the deflections can be neglected when compared to the effect produced by the eccentricity of the load.

Consider a short block of rectangular section subjected to a force P applied at E with eccentricities y_0 from the x-axis and x_0 from the y-axis, as shown in Fig. 10–3(a). To determine the equivalent load system, let us consider the step-by-step development shown in Figure 10–3(b) to (e). Two equal and opposite forces P are placed at D, as shown in Fig. 10–3(b). The downward force at E and the upward force at D, being equal and opposite, form a couple $M_x = Py_0$. The system is thus reduced to a downward force P at D and a couple M_x about the x-axis, as shown in Fig. 10–3(c). Similarly, two equal and opposite forces P are placed at the centroid C, as shown in Fig. 10–3(d). The downward force at D and the upward force at C, being equal and opposite, form a couple $M_y = Px_0$. Thus the system is finally reduced to an equivalent loading, as shown in Fig. 10–3(e); it consists of the following:

The axial compressive force through the centroid of the section: P

The bending moment about the x-axis: $M_x = Py_0$

The bending moment about the y-axis: $M_y = Px_0$

The same loading acts unchanged at any cross section in the block. Using the method of superposition, we find that the normal stress at point $Q(x, y)$ in section m–m is

$$\sigma_Q = -\frac{P}{A} + \frac{M_x y}{I_x} - \frac{M_y x}{I_y}$$

or

$$\sigma_Q = -\frac{P}{A} + \frac{(Py_0)y}{I_x} - \frac{(Px_0)x}{I_y} \qquad (10\text{–}4)$$

where A is the cross-sectional area of the member and I_x and I_y are the moments of inertia of the cross-sectional area with respect to the x- and y-axes, respectively. The positive signs correspond to tensile stresses and the negative signs correspond to compressive stresses.

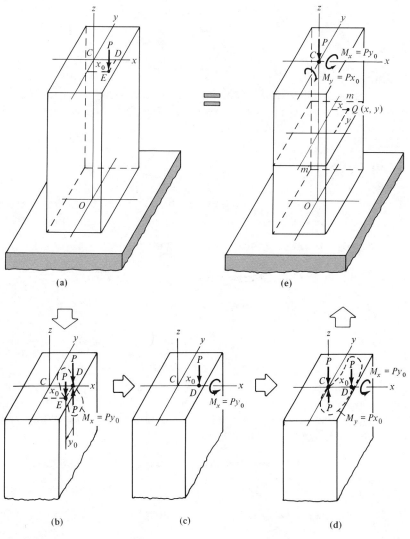

FIGURE 10–3

━━━━━ **EXAMPLE 10–5** ━━━━━

The rectangular block is subjected to the axial force $P = 20$ kips applied at E shown in Fig. (a). Neglecting the weight of the block, determine (a) the normal stress at point Q in terms of its coordinates x and y, (b) the location of the line of zero stress, (c) the normal stresses at points A, B, C, and D.

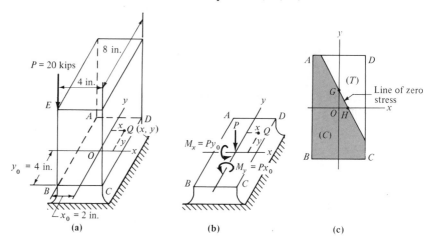

(a)   (b)   (c)

SOLUTION

(a) The equivalent loading for the section $ABCD$ is shown in Fig. (b). The loading consists of axial compressive force P and bending moments M_x and M_y. The normal stress at point Q is

$$\sigma_Q = -\frac{P}{A} + \frac{(Py_0)y}{I_x} + \frac{(Px_0)x}{I_y}$$

$$= -\frac{20\ \text{kips}}{4 \times 8\ \text{in.}^2} + \frac{(20\ \text{kips})(4\ \text{in.})y(\text{in.})}{(4\ \text{in.})(8\ \text{in.})^3/12} + \frac{(20\ \text{kips})(2\ \text{in.})x(\text{in.})}{(8\ \text{in.})(4\ \text{in.})^3/12}$$

$$= [-0.625 + 0.469y(\text{in.}) + 0.938x(\text{in.})]\ \text{ksi} \qquad\qquad (a)$$

(b) To locate the line of zero stress, set σ_Q to zero. Thus

$$\sigma_Q = -0.625 + 0.469y + 0.938x = 0$$

From this equation, when $x = 0$, $y = +1.33$ in.; and when $y = 0$, $x = +0.666$ in. These results locate two points $G(0, +1.33)$ and $H(+0.666, 0)$ through which the line of zero stress passes, as shown in Fig. (c). The tension and compression zones are indicated in the figure, from which we see that points A, B, and C are in compression, and point D is in tension.

(c) The normal stresses at points A, B, C, and D can be obtained by substituting the coordinates of each point in Eq. (a). Thus

$$\sigma_A = -0.625 + 0.469(+4) + 0.938(-2) = -0.625\ \text{ksi (C)}$$

$$\sigma_B = -0.625 + 0.469(-4) + 0.938(-2) = -4.38\ \ \text{ksi (C)}$$

$$\sigma_C = -0.625 + 0.469(-4) + 0.938(+2) = -0.625\ \text{ksi (C)}$$

$$\sigma_D = -0.625 + 0.469(+4) + 0.938(+2) = +3.13\ \ \text{ksi (T)}$$

────── **EXAMPLE 10–6** ──────────────────────────────────────

Find the zone over which a vertical downward force may be applied without causing tensile stresses at any point in the rectangular block shown. Neglect the weight of the block.

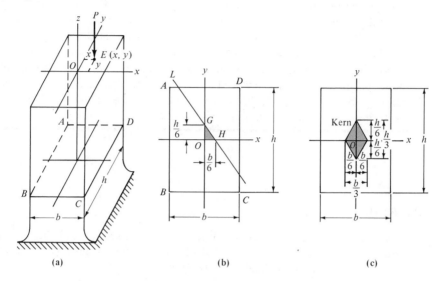

(a)   (b)   (c)

SOLUTION

Let the force P be placed at point $E(x, y)$ in the first quadrant of the x-y coordinate system shown. If tensile stresses exist at any point in section $ABCD$, the maximum tensile stress must occur at point B and be equal to

$$\sigma_B = -\frac{P}{A} + \frac{M_x}{S_x} + \frac{M_y}{S_y} = -\frac{P}{A} + \frac{Py}{bh^2/6} + \frac{Px}{hb^2/6}$$

Since $A = bh$, we have

$$\sigma_B = \frac{P}{A}\left(-1 + \frac{6y}{h} + \frac{6x}{b}\right)$$

Setting the stress at B equal to zero fulfills the limiting condition of the problem.

$$\sigma_B = \frac{P}{A}\left(-1 + \frac{6y}{2h} + \frac{6x}{b}\right) = 0$$

or

$$\frac{6x}{b} + \frac{6y}{h} = 1$$

which is an equation of a straight line. From this equation, it follows that when $x = 0$, $y = h/6$; and when $y = 0$, $x = b/6$. These results locate two points, $G(0, h/6)$ and $H(b/6, 0)$, which define the line L, as shown in Fig. (b). A vertical force applied to the block at a point along the line L between GH will cause the stress at B to be equal to zero and compressive stresses will occur at all the other points in the section. If the force P acts anywhere to the upper right of this line, tensile stresses will occur in the member. As far as the first quadrant is concerned, the point of application of the force P should be within the shaded area shown in Fig. (b) so that no tensile stresses occur anywhere in the section.

A similar situation occurs in other quadrants. A shaded-area zone called the *kern* of the section is shown in Fig. (c). A vertical compressive force applied at a point within the kern will not cause tensile stresses at any point in the member.

From the discussion above, it can be concluded that if a compressive force P (or the resultant force acting at the section) applied along an axis of symmetry is inside the kern or within the middle third of a rectangular cross section, there will be no tensile stress in the material at that section. Since a gravitational dam must not be subjected to tensile stress at any point at its base, the resultant force must be acting within the middle third of the base. ∎

PROBLEMS

10–17 A timber bar has a cross section 3 in. by 4 in. At section 1–1 the 4-in. width is reduced to 2 in., as shown in Fig. P10–17. Determine the normal stresses at points A and B due to an axial load $P = 1800$ lb.

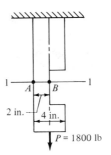

FIGURE P10–17

10–18 In Fig. P10–18, find the normal stresses at points A and B due to an eccentrically applied axial compressive load $P = 600$ kN acting on a circular post as shown.

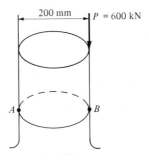

FIGURE P10–18

10–19 Prove that the kern of a circular section is a concentric circular area having a diameter equal to one-fourth of the diameter of the section, as shown in Fig. P10–19.

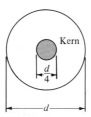

FIGURE P10–19

10–20 In Fig. 10–20, determine the maximum eccentricity e at which the vertical compressive load can be applied to the wide-flange W14 × 90 steel section without causing tensile stress anywhere in the section. Neglect the weight of the section.

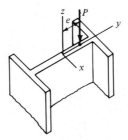

FIGURE P10–20

10–21 The short timber block shown in Fig. 10–21 supports an eccentric load $P = 80$ kN, as shown. Determine **(a)** an equation expressing the normal stress at point Q in terms of its coordinates x and y, **(b)** the location of the line of zero stress, **(c)** the normal stresses at points A, B, C, and D. Neglect the weight of the block.

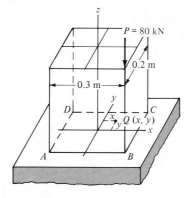

FIGURE P10–21

10–22 An aluminum-alloy block is subjected to an eccentric axial compressive load P, as shown in Fig. P10–22. The linear strain produced by the load at point A in the vertical direction is measured to be 7.2×10^{-4} in./in. If the modulus of elasticity of the aluminum alloy is 10×10^6 psi, determine the magnitude of the load P.

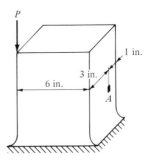

FIGURE P10–22

10–23   A rectangular short post is subjected to the loads $P = 30$ kN and $Q = 4$ kN, as shown in Fig. P10–23. Determine the normal stresses at points A, B, C, and D and locate the line of zero stress. Neglect the weight of the post.

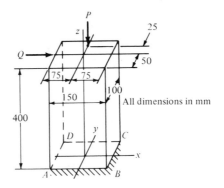

FIGURE P10–23

10–24   A gravity dam has a cross section 3 ft by 6 ft, as shown in Fig. P10–24. The water pressure varies linearly from zero at the free surface to γh at the bottom, where γ is the weight of water per unit volume, equal to 62.4 lb/ft^3. If the weight of concrete per unit volume is 150 lb/ft^3, determine the height of the water level h at which the foundation pressure at A is just equal to zero. (**HINT:** For the purpose of calculation, consider one linear foot of length along the longitudinal direction of the dam.)

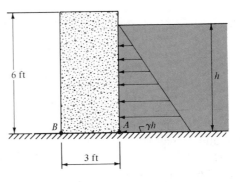

FIGURE P10–24

*10-5
SUPERPOSITION OF SHEAR STRESSES IN CIRCULAR MEMBERS

A member may be subjected both to torsion and to transverse shear forces. These loads cause shear stresses. The shear stress caused by each of these loads was discussed earlier. In the chapter on torsion (Chapter 5), only circular shafts were considered. This limits the type of problems that can be solved here to circular members. The combined shear stresses due to torsion and shear force acting simultaneously can be determined by the method of superposition.

The maximum shear stress occurs at the critical section with the most serious combination of internal torque and internal shear force. The maximum shear stress on the outside surface due to torsion is

$$\tau'_{max} = \frac{Tc}{J} = \frac{TR}{\frac{1}{2}\pi R^4} = \frac{2T}{\pi R^3} \tag{10-5}$$

The maximum shear stress at the neutral axis of a circular shaft due to the transverse shear force is (see Problem 7-24)

$$\tau''_{max} = \frac{4V}{3A} = \frac{4V}{3\pi R^2} \tag{10-6}$$

At the point where the maximum shear stresses calculated from Eqs. (10-5) and (10-6) act in the same direction, the two stresses can be added to obtain the maximum shear stress due to the combined loading. Thus

$$\tau_{max} = \tau'_{max} + \tau''_{max} \tag{10-7}$$

──────── **EXAMPLE 10-7** ──

Find the maximum shear stress at section *m–m* in the 2-in.-diameter circular member due to the applied force *P* acting as shown in Fig. (a).

SOLUTION

The equivalent loading at the section *m–m* due to the applied load *P* is shown in Fig. (b). Torque *T* and shear force *V* cause shear stresses, while the bending moment *M* causes flexural stresses.

Due to torque *T*, the maximum shear stress at *A* is shown in a small element in Fig. (c). The direction of the shear stress agrees with the direction of *T*. From Eq. (10-5), the value of the shear stress is

$$\tau'_{max} = \frac{2T}{\pi R^3} = \frac{2(2 \text{ kips})(6 \text{ in.})}{\pi(1 \text{ in.})^3} = 7.64 \text{ ksi}$$

Due to shear force *V*, the maximum shear stress at *A* is shown in a small element in Fig. (d). The direction of shear stress agrees with the direction of the shear force *V*. From Eq. (10-6), the value of the shear stress is

$$\tau''_{max} = \frac{4V}{3\pi R^2} = \frac{4(2 \text{ kips})}{3\pi(1 \text{ in.})^2} = 0.85 \text{ ksi}$$

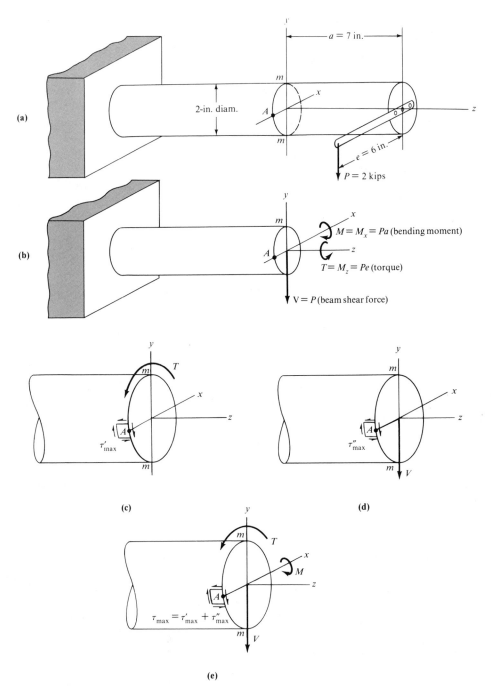

Since the shear stresses at point A in Figs. (c) and (d) act in the same direction, the maximum combined shear stress is the sum of the two shear stresses. Thus

$$\tau_{max} = \tau'_{max} + \tau''_{max} = 7.64 + 0.85 = 8.49 \text{ ksi}$$

The maximum shear stress acting on a rectangular element is shown in Fig. (e). Note that no normal stress acts on this element, since the element is located on the neutral axis.

PROBLEMS

10–25 A solid circular shaft is subjected to a load P as shown in Fig. P10–25. Given $P = 20$ kips, $d = 6$ in., and $a = 10$ in. Determine the normal and shear stresses on the element at point A. Neglect the weight of the shaft.

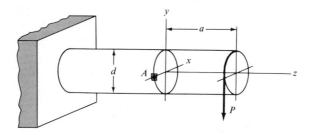

FIGURE P10–25

10–26 Rework Problem 10–25 if $P = 25$ kN, $d = 100$ mm, and $a = 150$ mm.

10–27 Determine the normal and shear stresses acting on the small elements at points A and B of the shaft shown in Fig. P10–27. Neglect the weights of the shaft and the pulley.

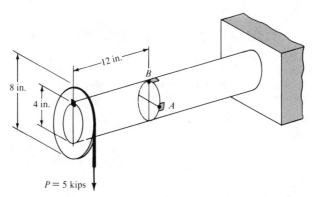

FIGURE P10–27

10–28 A solid circular shaft is loaded as shown in Fig. P10–28. Determine the normal and shear stresses at points A and B. Neglect the weights of the shaft and the pulley.

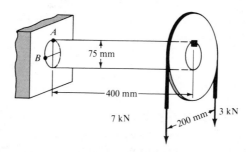

FIGURE P10–28

10–29 A rectangular sign board is rigidly attached to a solid steel rod $3\frac{1}{2}$ in. in diameter, as shown in Fig. P10–29. The wind force, which has an intensity of 35 lb/ft², is blowing on the sign board along the positive y-direction. Determine the normal and shear stresses acting on the rectangular elements at points A and B. Neglect the weights of the rod and the sign board and also neglect the wind force on the rod.

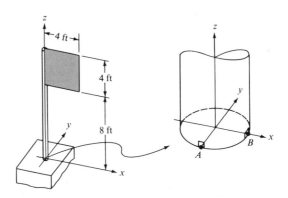

FIGURE P10–29

10–6
STRESSES IN THIN-WALLED PRESSURE VESSELS

When a thin-walled cylindrical vessel such as a boiler, shown in Fig. 10–4(a), is subjected to internal pressure p, normal stresses develop on the wall of the vessel, as shown on an element in Fig. 10–4(b). The normal stress σ_c acting along the circumferential direction is called the *circumferential stress* (or sometimes called the *hoop stress*). The normal stress σ_l acting along the longitudinal direction is called the *longitudinal stress*. Since the internal pressure p acting inside the vessel is much smaller than σ_c and σ_l, the pressure p acting on the element along the z-direction can be neglected, and the element is subjected to normal stresses only in two directions. Such an element is in a state of *biaxial stress*.

The formulas for σ_c and σ_l can be derived based on the assumption that these stresses are uniformly distributed throughout the thickness of the wall. The stresses computed are accurate for a wall thickness $t \le r/10$, where r is the inside radius of the vessel.

The formulas for σ_c can be derived by considering the equilibrium of one-half of the cylindrical wall segment with enclosed fluid under a uniform internal gauge pressure p (pressure above the atmospheric pressure). The free-body diagram and the equilibrium equation $\Sigma F_y = 0$ are shown in Fig. 10–4(c), from which the circumferential stress σ_c is

$$\sigma_c = \frac{pr}{t} \qquad\qquad (10\text{–}8)$$

where r is the inside radius of the cylinder.

The formula for σ_l may be derived by considering the equilibrium of the cylindrical vessel to the left of cross section 1–1. The free-body diagram and the

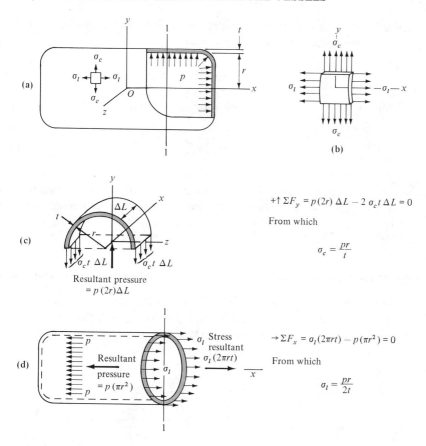

$$+\uparrow \Sigma F_y = p(2r)\,\Delta L - 2\,\sigma_c t\,\Delta L = 0$$

From which

$$\sigma_c = \frac{pr}{t}$$

$$\rightarrow \Sigma F_x = \sigma_l(2\pi rt) - p(\pi r^2) = 0$$

From which

$$\sigma_l = \frac{pr}{2t}$$

FIGURE 10-4

equilibrium equation $\Sigma F_x = 0$ are shown in Fig. 10-4(d), from which the longitudinal stress is

$$\sigma_l = \frac{pr}{2t} \tag{10-9}$$

Note that from Eqs. (10-8) and (10-9), $\sigma_c = 2\sigma_l$.

A similar method can be used to derive an expression for the normal stress in thin-walled spherical pressure vessels as shown in Fig. 10-5(a). By passing a section through the center of the sphere, a hemisphere with enclosed fluid under

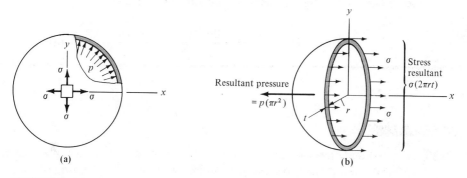

Resultant pressure

$$= p(\pi r^2)$$

(a)   (b)

FIGURE 10-5

a uniform internal pressure p is isolated, as shown in Fig. 10–5(b). The equilibrium condition along the x direction requires

$$\xrightarrow{+}\Sigma F_x = \sigma(2\pi rt) - p(\pi r^2) = 0$$

From which the normal stress in the wall of the spherical pressure vessel is

$$\sigma = \frac{pr}{2t} \tag{10–10}$$

where r is the inside radius of the spherical vessel. Since any section that passes through the center of the sphere yields the same result, the normal stress in the wall of spherical vessel is the same value along any direction, as given by Eq. (10–10). This stress condition is called *all-around tension*.

──────── **EXAMPLE 10–8** ────────────────────────────────

A cylindrical pressure vessel of 10-in. inside radius has a wall thickness of $\frac{1}{2}$ in. The vessel is subjected to an internal gage pressure of 400 psi. If the cylinder is also subjected to an axial pull of 100 kips, determine the state of stress of the element shown.

SOLUTION

The longitudinal stress σ_x is the sum of the stresses due to the axial force and the internal pressure. Thus

$$\sigma_x = +\frac{P}{A} + \sigma_l = +\frac{P}{2\pi rt} + \frac{pr}{2t} = \frac{100\ 000\ \text{lb}}{2\pi(10\ \text{in.})(\frac{1}{2}\ \text{in.})} + \frac{(400\ \text{lb/in.}^2)(10\ \text{in.})}{2(\frac{1}{2}\ \text{in.})}$$

$$= 3183\ \text{psi} + 4000\ \text{psi} = 7180\ \text{psi}$$

The circumferential stress σ_y is due to internal pressure alone, since the axial force does not cause any stress in this direction. Thus

$$\sigma_y = \sigma_c = \frac{pr}{t} = \frac{(400\ \text{lb/in.}^2)(10\ \text{in.})}{\frac{1}{2}\ \text{in.}} = 8000\ \text{psi}$$

■

PROBLEMS

10–30 A cylindrical pressure vessel of 8-in. radius is made of $\frac{1}{8}$-in. steel plate. Determine the maximum permissible pressure within the vessel if the tensile stress must not exceed 8000 psi. (The allowable stress is set at a lower value to provide for the corrosion effects.)

10–31 A stainless steel cylindrical pressure vessel of 300 mm inside radius is subjected to an internal pressure of 3.5 MPa. If the allowable tensile stress is 140 MPa, determine the minimum thickness of the wall.

10–32 A spherical pressure vessel with an inside diameter of 8 in. and a wall thickness of $\frac{1}{4}$ in. is made of a material having an allowable tensile stress of 6000 psi. Determine the maximum allowable gage pressure that the vessel can withstand.

10–33 A 30-kN hydraulic jack has the dimensions shown in Fig. P10–33. Determine the minimum wall thickness t if the cylinder is made of steel having an allowable tensile stress of 140 MPa.

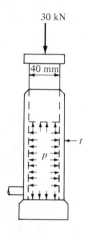

FIGURE P10–33

10–34 A steel pipe has an inside diameter of 15.0 in. and a wall thickness of 0.500 in. If the ultimate strength of steel is 65 ksi, determine the bursting pressure for the pipe.

10–35 A piece of steel pipe of an inside diameter of 480 mm and a wall thickness of 15 mm was sealed at the ends, as shown in Fig. P10–35. The assembly was then mounted on a testing machine and simultaneously subjected to an axial pull P of 200 kN and an internal pressure of 3 MPa. Determine the stresses σ_x and σ_y on the element shown.

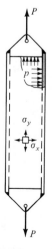

FIGURE P10–35

10–36 A cylindrical pressure vessel has an inside diameter of 12 in. and a wall thickness of $\frac{1}{4}$ in. In addition to subjecting to an internal pressure of 400 psi, the vessel is also subjected to a torque $T = 200$ kip-in., as shown in Fig. P10–36. Determine the state of stress on an element at the surface of the vessel.

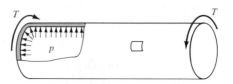

FIGURE P10–36

CHAPTER 11

Transformation of Plane Stress

11–1
INTRODUCTION

The general state of plane stress (stresses exist only along the directions in a plane) at a point, as expressed on a small rectangular element enclosing the point, consists of normal and shear stresses acting simultaneously on the element. In this chapter equations are derived for calculating the normal and shear stresses on an inclined plane in an element where the general state of plane stress in two perpendicular directions is given. The graphical representation of the equations, called the Mohr's circle, is also introduced. From the Mohr's circle, the formulas for calculating the maximum and minimum normal and shear stresses are derived.

11–2
EQUATIONS OF TRANSFORMATION OF PLANE STRESS

When an element is subjected to normal and shear stresses only in two directions, as shown in Fig. 11–1(a), the element is said to be in a state of plane stress. Figure 11–1(b) shows the planar representation of the plane stress condition where

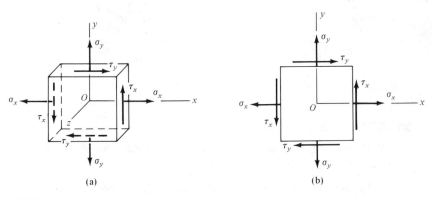

(a)   (b)

FIGURE 11–1

σ_x and τ_x are the normal and shear stresses acting on the vertical planes whose normals are along the x-axis, and σ_y and τ_y are the normal and shear stresses acting on the horizontal planes whose normals are along the y-axis. With σ_x, σ_y, τ_x, and τ_y all acting simultaneously, the element is in a state of general plane stress. The uniaxial stress (normal stress along only one direction), the biaxial stress (normal stresses along two perpendicular directions), and the pure shear (element subjected to shear stresses only) are special cases of plane stress.

The state of general plane stress usually occurs in the combined loading condition. For example, when a thin-walled cylindrical vessel is subjected simultaneously to internal pressure and external torque, a rectangular element on the wall is in a state of general plane stress, as shown in Fig. 11–2.

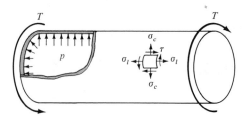

FIGURE 11–2

Consider an element in general plane stress, as shown in Fig. 11–3(a). We want to find the normal and shear stresses σ_θ and τ_θ acting on the plane BC, which is inclined at an angle θ from the vertical plane AB.

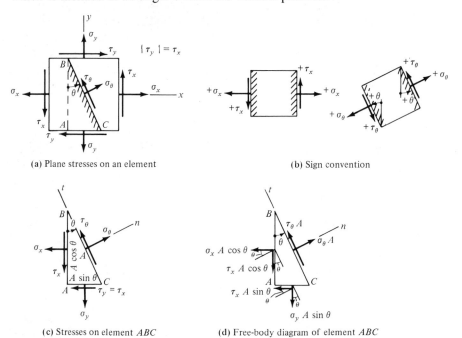

(a) Plane stresses on an element　　　　　(b) Sign convention

(c) Stresses on element ABC　　　　(d) Free-body diagram of element ABC

FIGURE 11–3

Before the general formulas are derived, it is important to establish the sign conventions for the quantities involved.

1. *The sign for the normal stress:* Tensile stresses are considered positive and compressive stresses are considered negative.

2. *The sign for the shear stress:* Shear stresses are considered positive when the pair of shear stresses acting on opposite and parallel sides of an element form a counterclockwise couple, as shown in Fig. 11–3(b).
3. *The sign for the angle of inclination:* The angle of inclination θ is considered positive when measured from the vertical plane toward the inclined plane in a counterclockwise direction, as shown in Fig. 11–3(b).

From these sign conventions, it is seen that all the stresses are shown in the positive sense in the element in Fig. 11–3(a) except τ_y, which is negative. Because the shear stresses on perpendicular planes must be equal (Section 5–5), the absolute value of the shear stress τ_y must be equal to τ_x.

To derive the formula for σ_θ and τ_θ, a triangular element *ABC* is isolated from the element in Fig. 11–3(a). The inclined plane *BC* makes an angle θ with the vertical direction; thus if the area of the inclined plane is A, the area of the horizontal plane *AC* is $A \sin \theta$ and the area of the vertical plane *AB* is $A \cos \theta$, as shown in Fig. 11–3(c). A free-body diagram of the wedge is drawn as shown in Fig. 11–3(d), where the normal and shear forces acting on an area are obtained by multiplying the normal and shear stresses by the area. By writing the equilibrium equations for the triangular element, stresses σ_θ and τ_θ are obtained as follows.

$$+\nearrow \Sigma F_n = \sigma_\theta A - (\sigma_x A \cos \theta)(\cos \theta) - (\sigma_y A \sin \theta)(\sin \theta)$$
$$- (\tau_x A \cos \theta)(\sin \theta) - (\tau_x A \sin \theta)(\cos \theta) = 0$$

from which

$$\sigma_\theta = \sigma_x \cos^2 \theta + \sigma_y \sin^2 \theta + 2\tau_x \sin \theta \cos \theta \qquad \text{(a)}$$

$$+\nwarrow \Sigma F_t = \tau_\theta A + (\sigma_x A \cos \theta)(\sin \theta) - (\sigma_y A \sin \theta)(\cos \theta)$$
$$- (\tau_x A \cos \theta)(\cos \theta) + (\tau_x A \sin \theta)(\sin \theta) = 0$$

from which

$$\tau_\theta = -(\sigma_x - \sigma_y) \sin \theta \cos \theta + \tau_x(\cos^2 \theta - \sin^2 \theta) \qquad \text{(b)}$$

From trigonometry, we have the following well-known trigonometric identities:

$$\cos^2 \theta = \tfrac{1}{2}(1 + \cos 2\theta)$$

$$\sin^2 \theta = \tfrac{1}{2}(1 - \cos 2\theta)$$

$$\sin \theta \cos \theta = \tfrac{1}{2} \sin 2\theta$$

Substituting these identities in Eqs. (a) and (b), we have

$$\sigma_\theta = \frac{\sigma_x + \sigma_y}{2} + \frac{\sigma_x - \sigma_y}{2} \cos 2\theta + \tau_x \sin 2\theta \qquad (11\text{–}1)$$

$$\tau_\theta = -\frac{\sigma_x - \sigma_y}{2} \sin 2\theta + \tau_x \cos 2\theta \qquad (11\text{–}2)$$

These are the formulas for the determination of normal and shear stresses acting on any inclined plane. The equations are referred to as the *transformation formulas for plane stress*.

EXAMPLE 11–1

Determine the normal and shear stresses on the inclined plane m–m of the member in Example 2–9 subjected to uniaxial compression as shown in Fig. (a).

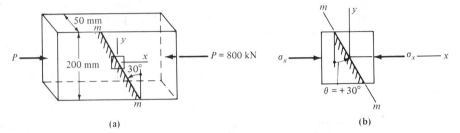

(a)   (b)

SOLUTION

The small rectangular element shown in Fig. (b) is subjected to uniaxial compression; thus it is a special case of plane stress. The state of plane stress of the element is

$$\sigma_x = -\frac{P}{A} = -\frac{800 \text{ kN}}{0.2 \times 0.05 \text{ m}^2} = -80\ 000 \text{ kN/m}^2 = -80 \text{ MPa}$$

$$\sigma_y = 0$$

$$\tau_x = \tau_y = 0$$

The angle θ from the vertical to the inclined plane m–m is 30° measured in the counterclockwise direction. Its sign is positive according to the sign convention established in this section. Therefore,

$$\theta = +30° \quad \text{and} \quad 2\theta = +60°$$

From Eqs. (11–1) and (11–2), the normal and shear stresses on the inclined plane m–m are, respectively,

$$\sigma_\theta = \frac{\sigma_x + \sigma_y}{2} + \frac{\sigma_x - \sigma_y}{2} \cos 2\theta + \tau_x \sin 2\theta$$

$$= \frac{(-80 \text{ MPa}) + 0}{2} + \frac{(-80 \text{ MPa}) - 0}{2} \cos (+60°) + 0 = -60 \text{ MPa}$$

$$\tau_\theta = -\frac{\sigma_x - \sigma_y}{2} \sin 2\theta + \tau_x \cos 2\theta$$

$$= -\frac{(-80 \text{ MPa}) - 0}{2} \sin (+60°) + 0 = +34.6 \text{ MPa}$$

These stresses are shown to act on the inclined plane m–m in the following figure.

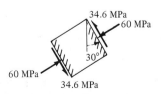

Note that these results are identical to those obtained in Example 2–9. ∎

──────── **EXAMPLE 11–2** ────────

For a small element with the state of plane stress given as shown in Fig. (a), determine the stresses acting on the inclined plane *m–m*.

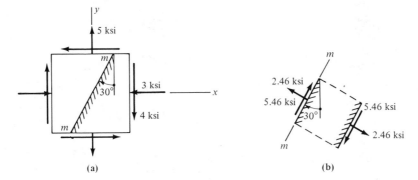

(a)   (b)

SOLUTION

According to the sign convention established in this section, the state of plane stress of the element and the angle of inclination of the inclined plane are

$$\sigma_x = -3 \text{ ksi}$$

$$\sigma_y = +5 \text{ ksi}$$

$$\tau_x = -4 \text{ ksi}$$

$$\theta = -30° \qquad 2\theta = -60°$$

From Eqs. (11–1) and (11–2) the normal and shear stresses on the inclined plane are

$$\sigma_\theta = \frac{-3 \text{ ksi} + 5 \text{ ksi}}{2} + \frac{-3 \text{ ksi} - 5 \text{ ksi}}{2} \cos(-60°) + (-4 \text{ ksi}) \sin(-60°)$$

from which

$$\sigma_\theta = +2.46 \text{ ksi}$$

$$\tau_\theta = -\frac{-3 \text{ ksi} - 5 \text{ ksi}}{2} \sin(-60°) + (-4 \text{ ksi}) \cos(-60°)$$

from which

$$\tau_\theta = -5.46 \text{ ksi}$$

These stresses are shown to act on the inclined plane *m–m* in Fig. (b). ■

PROBLEMS

*In Problems **11–1** to **11–6**, an element and the state of stress on the element are given. Determine the normal and shear stresses on the inclined plane m–m. Show by a sketch the stresses acting on the inclined plane.*

11–1

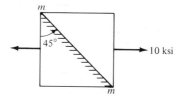

FIGURE P11–1

11–2

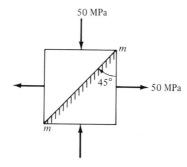

FIGURE P11–2

11–3

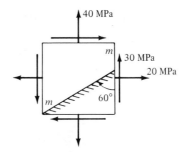

FIGURE P11–3

11–4

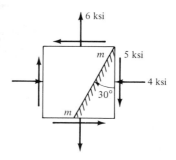

FIGURE P11–4

11–5

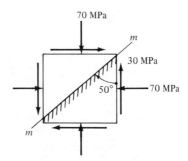

FIGURE P11–5

11–6

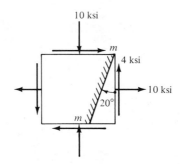

FIGURE P11–6

11–3
MOHR'S CIRCLE OF STRESS

When the parameter θ is eliminated from Eqs. (11–1) and (11–2), a functional relationship between σ_θ and τ_θ is obtained. To do this, Eqs. (11–1) and (11–2) are rewritten as

$$\sigma_\theta - \frac{\sigma_x + \sigma_y}{2} = \frac{\sigma_x - \sigma_y}{2} \cos 2\theta + \tau_x \sin 2\theta$$

$$\tau_\theta = -\frac{\sigma_x - \sigma_y}{2} \sin 2\theta + \tau_x \cos 2\theta$$

Squaring both sides of each equation gives

$$\left(\sigma_\theta - \frac{\sigma_x + \sigma_y}{2}\right)^2 = \left(\frac{\sigma_x - \sigma_y}{2}\right)^2 \cos^2 2\theta$$
$$+ (\sigma_x - \sigma_y)\tau_x \sin 2\theta \cos 2\theta + \tau_x^2 \sin^2 2\theta \qquad (a)$$

$$\tau_\theta^2 = \left(\frac{\sigma_x - \sigma_y}{2}\right)^2 \sin^2 2\theta$$
$$- (\sigma_x - \sigma_y)\tau_x \sin 2\theta \cos 2\theta + \tau_x^2 \cos^2 2\theta \qquad (b)$$

Adding (a) and (b) and using the trigonometric identity $\sin^2 2\theta + \cos^2 2\theta = 1$, we get

$$\left(\sigma_\theta - \frac{\sigma_x + \sigma_y}{2}\right)^2 + \tau_\theta^2 = \left(\frac{\sigma_x - \sigma_y}{2}\right)^2 + \tau_x^2 \qquad \text{(c)}$$

where the values of σ_x, σ_y, and τ_x are known for a given problem. If we let

$$a = \frac{\sigma_x + \sigma_y}{2}$$

and

$$r = \sqrt{\left(\frac{\sigma_x - \sigma_y}{2}\right)^2 + \tau_x^2} \qquad \text{(11–3)}$$

then Eq. (c) becomes

$$(\sigma_\theta - a)^2 + (\tau_\theta - 0)^2 = r^2 \qquad \text{(11–4)}$$

This equation is recognized as the equation of a circle with center located at $(a, 0)$ or $[\frac{1}{2}(\sigma_x + \sigma_y), 0]$, and radius r given by Eq. (11–3). When this circle is plotted on σ–τ coordinate axes, as shown in Fig. 11–4, the coordinates $(\sigma_\theta, \tau_\theta)$ of a point S on the circle represent the normal and shear stress on a certain inclined plane. The circle is called *Mohr's circle of stress*.

Mohr's circle of stress is widely used in practice for the solution of stress transformation problems. The recommended procedure for constructing and using the Mohr's circle is outlined in the following paragraphs.

1. Sketch a rectangular element and indicate on the element the given state of plane stress with proper sense [Fig. 11–5(a)]. The sign conventions established in Section 11–2 must be strictly followed.

2. Set up a rectangular coordinate system with the origin O located at a convenient point. The normal stress axis is in the horizontal direction, to the right of

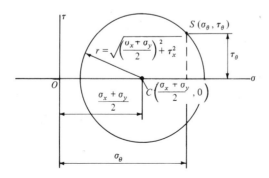

FIGURE 11–4

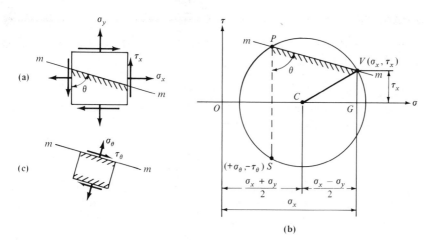

FIGURE 11–5

O as positive; the shear stress axis is in the vertical direction, above O as positive, as shown in Fig. 11–5(b).

3. To draw the Mohr's circle [Fig. 11–5(b)], first locate the center C along the horizontal σ-axis at $\frac{1}{2}(\sigma_x + \sigma_y)$ from the origin O. Then plot point V at (σ_x, τ_x) corresponding to the stress condition on the vertical plane. With C as the center and CV as the radius, draw a circle passing through V. The circle constructed is the Mohr's circle for the element with the given state of plane stress.

From Fig. 11–5(b), we obtain

$$CG = OG - OC = \sigma_x - \frac{\sigma_x + \sigma_y}{2} = \frac{\sigma_x - \sigma_y}{2}$$

and by the Pythagorean theorem, we can prove that CV is indeed the radius of the circle:

$$CV = \sqrt{CG^2 + VG^2} = \sqrt{\left(\frac{\sigma_x - \sigma_y}{2}\right)^2 + \tau_x^2} = r$$

4. To find the stresses acting on an inclined plane, from point V draw line VP parallel to the inclined plane and locate point P on the circle. The coordinates of S, a point on the circle vertically opposite from P, give the stresses acting on the inclined plane. The coordinates of the point S are identified as $(+\sigma_\theta, -\tau_\theta)$. A positive σ_θ indicates a tensile stress, and a negative value of τ_θ indicates that the shear stresses on the opposite sides of an element form a clockwise couple. On this basis the stresses acting on the inclined plane m–m are shown in Fig. 11–5(c).

5. To determine the direction of a plane whose stresses are indicated by the point S, we can proceed by reversing the procedure in step 4. That is, locate the point P on the circle vertically opposite from S; then the stresses given by the coordinates of point S are acting on the inclined plane parallel to the line PV.

───── **EXAMPLE 11–3** ─────

Rework Example 11–2 (on p. 329) by constructing Mohr's circle for the element with the given stresses.

SOLUTION
The element and the given state of plane stress are shown in Fig. (a). The center C of Mohr's circle is located at

$$\tfrac{1}{2}(\sigma_x + \sigma_y) = \tfrac{1}{2}(-3 \text{ ksi} + 5 \text{ ksi}) = +1 \text{ ksi}$$

on the σ-axis. The stresses on the vertical plane of the element are $(-3, -4)$, which are the coordinates of point V on the circle. The Mohr's circle is drawn by using C as the center and CV as the radius, as shown in Fig. (b).

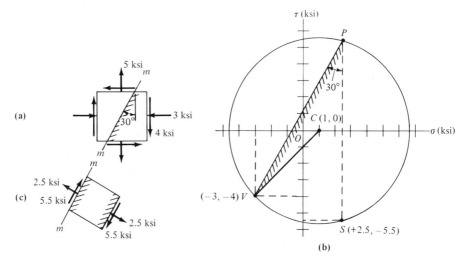

(a)

(c)

(b)

A line VP drawn parallel to the plane $m-m$ locates point P. The coordinates of point S vertically opposite from P give the stresses acting on the inclined plane $m–m$. From Fig. (b) the coordinates of point S are

$$S(+2.5, -5.5)$$

Therefore,

$$\sigma_\theta = +2.5 \text{ ksi} \qquad \tau_\theta = -5.5 \text{ ksi}$$

These stresses are shown acting on the inclined plane $m–m$ in Fig. (c).

■

PROBLEMS

*In Problems **11–7** to **11–12**, solve the problem indicated by drawing the Mohr's circle of stress for each element.*

11–7 Solve Problem 11–1 (on p. 330).

11–8 Solve Problem 11–2 (on p. 330).

11–9 Solve Problem 11–3 (on p. 330).

11–10 Solve Problem 11–4 (on p. 330).

11–11 Solve Problem 11–5 (on p. 331).

11–12 Solve Problem 11–6 (on p. 331).

11–13 A boiler of diameter 1 m and wall thickness 20 mm is subjected to an internal pressure of 6 MN/m². Determine **(a)** the state of stress in the rectangular element shown in Fig. P11–13, **(b)** the normal and shear stress along the inclined plane *m–m*.

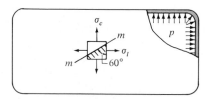

FIGURE P11–13

11–14 A simply supported timber beam with a full-sized cross section 2 in. by 4 in. supports a concentrated load at the midspan, as shown in Fig. P11–14. Determine **(a)** the state of stress of point *C* at section 1–1 just to the left of the concentrated load, **(b)** the normal and shear stresses on the inclined plane *m–m* passing through point *C* as shown.

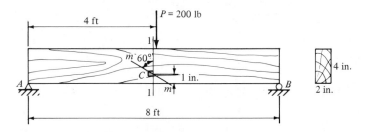

FIGURE P11–14

11–4
PRINCIPAL STRESSES AND MAXIMUM SHEAR STRESSES

The maximum and minimum normal stress in an element are called *principal stresses*. The planes where principal stresses occur are called *principal planes*. Mohr's circle can be used readily to determine the principal stresses and principal planes. Consider a stressed element and its corresponding Mohr's circle shown in Fig. 11–6(a) and (b); we see that the principal stresses are given by the points on the Mohr's circle that have maximum and minimum σ values. Thus point *A* gives maximum normal stress σ_1 or σ_{max} and point *B* gives minimum normal

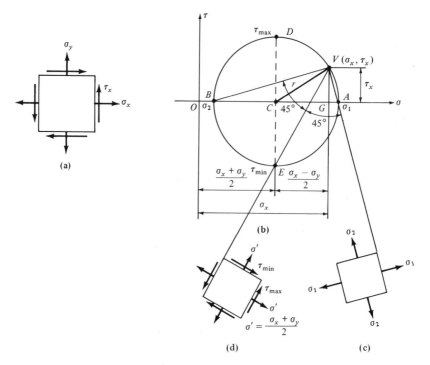

FIGURE 11–6

stress σ_2 or σ_{min}. The corresponding principal planes are parallel to lines VA (for σ_1) and VB (for σ_2), and since angle $\angle AVB$ is 90°, the principal planes are perpendicular to each other. Figure 11–6(c) shows the principal planes and the principal stresses. Note that no shear stress acts on the principal planes.

The radius of the Mohr's circle is

$$BC = CA = r = \sqrt{\left(\frac{\sigma_x - \sigma_y}{2}\right)^2 + \tau_x^2}$$

Therefore, the formulas for the principal stresses are

$$\sigma_1 = \sigma_{max} = OC + CA = \frac{\sigma_x + \sigma_y}{2} + \sqrt{\left(\frac{\sigma_x - \sigma_y}{2}\right)^2 + \tau_x^2} \qquad (11\text{–}5)$$

$$\sigma_2 = \sigma_{min} = OC - BC = \frac{\sigma_x + \sigma_y}{2} - \sqrt{\left(\frac{\sigma_x - \sigma_y}{2}\right)^2 + \tau_x^2} \qquad (11\text{–}6)$$

The maximum and minimum shear stresses (note that the minimum shear stress is not zero, but the negative shear stress with the largest absolute value) are given by the τ values at points D and E on the Mohr's circle, respectively,

$$\tau_{\substack{max \\ min}} = \pm\sqrt{\left(\frac{\sigma_x - \sigma_y}{2}\right)^2 + \tau_x^2} \qquad (11\text{–}7)$$

From step 5 in Section 11–3, we see that the maximum shear stress acts on the plane parallel to line VE (since E is vertically opposite to D) and the minimum

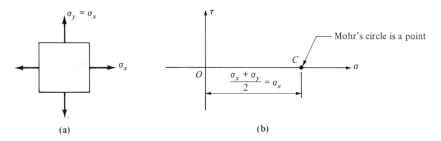

(a)   (b)

FIGURE 11–7

shear stress acts on the plane parallel to line VD; the two planes are 90° apart. Note that the angle between lines VE and VA is 45°; therefore, the plane of maximum shear stress and the principal planes are 45° apart. Figure 11–6(d) shows the maximum and minimum shear stresses and the planes on which they act. It shows also that the associated normal stresses on these planes of maximum and minimum shear stresses are

$$\sigma' = \frac{\sigma_x + \sigma_y}{2} \tag{11–8}$$

Two special cases of plane stress deserve special attention:

1. All-around tension (or compression): In Section 10–6 it was pointed out that the normal stress in the wall of a spherical vessel subjected to internal pressure is the same along any direction. Thus the state of stress on an element isolated from the wall is as shown in Fig. 11–7(a), where σ_x and σ_y are equal. The Mohr's circle of the element is a single point [Fig. 11–7(b)]. Therefore, the normal stress on any inclined plane is equal to σ_x and shear stresses do not exist on any plane. This state of stress is called *all-around tension*.

2. Pure shear: A condition of *pure shear* occurs when only shear stresses exist in two mutually perpendicular directions, as shown in Fig. 11–8(a). The Mohr's circle of the element is shown in Fig. 11–8(b). The principal stresses and principal planes are shown in Fig. 11–8(c).

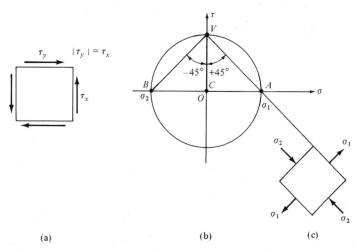

(a)   (b)   (c)

FIGURE 11–8

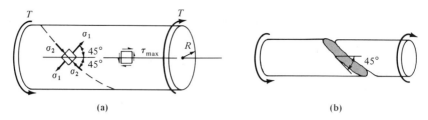

(a)   (b)

FIGURE 11–9

Note that the principal planes form angles of $+45°$ and $-45°$ with the vertical direction. The principal stresses and the maximum shear stress are equal to

$$\sigma_1 = -\sigma_2 = \tau_{max} = \tau_x$$

The condition of pure shear occurs at an element on the surface of a torsion bar, as shown in Fig. 11–9(a). The maximum shear stress and the principal stresses are shown acting on the respective element. By Eq. (5–1),

$$\sigma_1 = -\sigma_2 = \tau_{max} = \frac{Tc}{J} = \frac{TR}{\frac{1}{2}\pi R^4} = \frac{2T}{\pi R^3}$$

Most brittle materials are weak in tension. When subjected to torsion, brittle materials fail by tearing in a line perpendicular to the direction of σ_1. This may be demonstrated in the classroom by twisting a piece of chalk to failure, as shown in Fig. 11–9(b). The failure takes place along a helix at $45°$ from the longitudinal direction. Shafts made from brittle materials such as cast iron, sandstone, or concrete fail in the same manner in a torsion test.

———— **EXAMPLE 11–4** ————————————————————————

The state of stress of an element at a point is given as shown in Fig. (a). Determine (a) the principal stresses, (b) the maximum and minimum stresses and the associated normal stresses. Show the results for both cases on properly oriented elements.

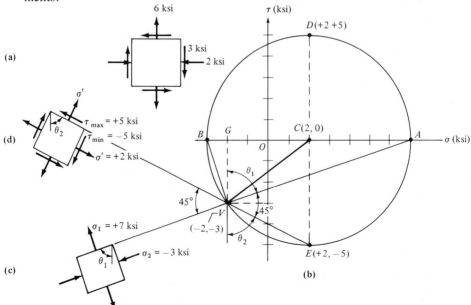

SOLUTION

Mohr's circle of stress is constructed based on the following quantities:

1. Center C along σ axis: $(-2 \text{ ksi} + 6 \text{ ksi})/2 = +2 \text{ ksi}$.
2. From the values of normal and shear stresses on the vertical plane, the coordinates of V are $(-2, -3)$.
3. From triangle CVG, the radius of the circle is $r = CV = \sqrt{4^2 + 3^2} = 5 \text{ ksi}$.

The Mohr's circle is shown in Fig. (b), from which we obtain

$$\sigma_1 = \sigma_{max} = +7 \text{ ksi}$$

$$\sigma_2 = \sigma_{min} = -3 \text{ ksi}$$

$$\tau_{max} = -\tau_{min} = 5 \text{ ksi}$$

(a) The maximum normal stress σ_1 acts on the plane parallel to line VA, and the minimum normal stress σ_2 acts on the plane parallel to line VB. The principal planes and the principal stresses are shown in Fig. (c). The orientation of the principal planes can be determined either by measuring the angle with a protractor or determined by trigonometric relations. By direct measurement, the angle θ_1 is found to be approximately $71\frac{1}{2}°$. Or more accurately, from triangle VAG, the angle θ_1 can be determined analytically as follows:

$$\tan \theta_1 = \frac{AG}{VG} = \frac{9}{3} = 3$$

$$\theta_1 = \tan^{-1} 3 = 71.57°$$

(b) The maximum shear stress τ_{max} and the associated normal stress σ' are given by the coordinates of point D $(+2, +5)$. Thus

$$\tau_{max} = +5 \text{ ksi} \qquad \sigma' = +2 \text{ ksi}$$

These stresses act on the plane parallel to line VE, as shown in Fig. (d). Note that the plane of maximum shear is located at $45°$ from the principal planes. The angle θ_2, which defines the orientation of the plane with the maximum shear stress, is

$$\theta_2 = 180° - \theta_1 - 45° = 63.43°$$

■

──────── **EXAMPLE 11–5** ────────

A solid circular shaft is subjected to a vertical force $P = 10 \text{ kN}$ as shown. Determine the maximum tensile and compressive stresses, and the maximum shear stress in the shaft due to the load P.

SOLUTION

The critical section is the fixed section on the left, where the force P causes the following loading:

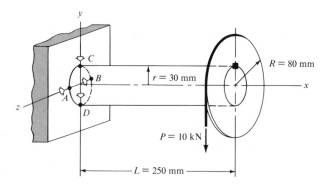

Bending moment: $M_z = PL = (10 \text{ kN})(0.250 \text{ m}) = 2.5 \text{ kN} \cdot \text{m}$

Torque: $T = M_x = PR = (10 \text{ kN})(0.080 \text{ m}) = 0.8 \text{ kN} \cdot \text{m}$

Direct shear force: $V = P = 10 \text{ kN}$

Stresses at points A, B, C, and D are to be investigated. The arrow next to each point in the figure shows the direction in which a rectangular element at the point is viewed.

At point A:

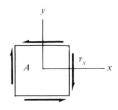

$\sigma_x = 0$ (A is at the neutral axis)

$\sigma_y = 0$

$$\tau_x = \frac{Tc}{J} + \frac{4V}{3A} = \frac{(0.8 \text{ kN} \cdot \text{m})(0.030 \text{ m})}{\frac{1}{2}\pi(0.030 \text{ m})^4} + \frac{4(10 \text{ kN})}{3\pi(0.030 \text{ m})^2}$$

$$= 18\ 860 \text{ kN/m}^2 + 4720 \text{ kN/m}^2 = 23\ 580 \text{ kN/m}^2 = 23.6 \text{ MPa}$$

The element is in pure shear; therefore,

$$\sigma_1 = -\sigma_2 = \tau_x = 23.6 \text{ MPa}$$

$$\tau_{\max} = \tau_x = 23.6 \text{ MPa}$$

At point B:

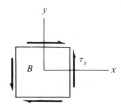

$$\sigma_x = 0 \ (B \text{ is at the neutral axis})$$

$$\sigma_y = 0$$

$$\tau_x = \frac{Tc}{J} - \frac{4V}{3A} = 18\ 860 \text{ kN/m}^2 - 4720 \text{ kN/m}^2$$

$$= 14\ 140 \text{ kN/m}^2 = 14.14 \text{ MPa}$$

The stresses at this point are not critical, since the stresses at point A are greater.

At point C:

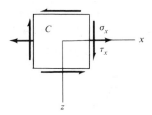

$$\sigma_x = \frac{M_z c}{I_z} = \frac{(2.5)(0.030)}{\frac{1}{4}\pi(0.030)^4} = 118\ 000 \text{ kN/m}^2 = 118 \text{ MPa}$$

$$\sigma_z = 0$$

$$\tau_x = \frac{Tc}{J} = 18\ 900 \text{ kN/m}^2 = 18.9 \text{ MPa}$$

From Eqs. (11–5), (11–6), and (11–7), we have

$$\sigma_1 = \frac{\sigma_x + \sigma_z}{2} + \sqrt{\left(\frac{\sigma_x - \sigma_z}{2}\right)^2 + \tau_x^2}$$

$$= \frac{118 + 0}{2} + \sqrt{\left(\frac{118 - 0}{2}\right)^2 + (-18.9)^2}$$

$$= 59.0 + 62.0 = +121 \text{ MPa}$$

$$\sigma_2 = 59.0 - 62.0 = -3.0 \text{ MPa}$$

$$\tau_{\max} = \sqrt{\left(\frac{\sigma_x - \sigma_z}{2}\right)^2 + \tau_x^2} = \sqrt{\left(\frac{118 - 0}{2}\right)^2 + (-18.9)^2} = 62.0 \text{ MPa}$$

At point D:

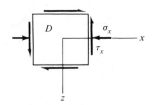

$$\sigma_x = -\frac{M_z c}{I_z} = -118 \text{ MPa}$$

$$\sigma_z = 0$$

$$\tau_x = \frac{Tc}{J} = 18.9 \text{ MPa}$$

$$\sigma_1 = \frac{\sigma_x + \sigma_z}{2} + \sqrt{\left(\frac{\sigma_x - \sigma_z}{2}\right)^2 + \tau_x^2}$$

$$= \frac{-118 + 0}{2} + \sqrt{\left(\frac{-118 - 0}{2}\right)^2 + (18.9)^2}$$

$$= -59.0 + 62.0 = +3.0 \text{ MPa}$$

$$\sigma_2 = -59.0 - 62.0 = -121 \text{ MPa}$$

$$\tau_{max} = \sqrt{\left(\frac{\sigma_x - \sigma_z}{2}\right)^2 + \tau_x^2}$$

$$= \sqrt{\left(\frac{-118 - 0}{2}\right)^2 + (18.9)^2} = 62.0 \text{ MPa}$$

Therefore, the stresses are critical at points C and D. The maximum tensile stress at C is 121 MPa and the maximum compressive stress at D is -121 MPa. The maximum shear stress is 62 MPa, which occurs at both points C and D.

PROBLEMS

*In Problems **11–15** to **11–18**, draw Mohr's circle of stress for the state of stress given in each figure. (a) Show the principal stresses in a properly oriented element. (b) Repeat (a) for the maximum and minimum shear stresses and the associated normal stresses.*

11–15

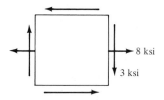

8 ksi

3 ksi

FIGURE P11–15

11–16

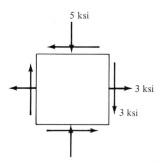

5 ksi

3 ksi

3 ksi

FIGURE P11–16

11–17

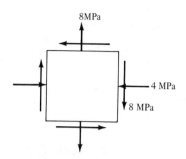

FIGURE P11–17

11–18

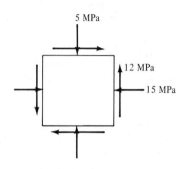

FIGURE P11–18

In Problems **11–19** *to* **11–22**, *show the data given in each problem on a rectangular element, following the sign conventions established in Section 11–2. Draw the Mohr's circle of stress and find (a) the principal stresses, (b) the maximum shear stresses and the associated normal stresses. In each case, show the results on a properly oriented element.*

11–19 $\sigma_x = +4$ ksi
$\sigma_y = +16$ ksi
$\tau_x = +8$ ksi

11–20 $\sigma_x = -20$ MPa
$\sigma_y = +40$ MPa
$\tau_x = -30$ MPa

11–21 $\sigma_x = +400$ psi
$\sigma_y = -800$ psi
$\tau_x = -800$ psi

11–22 $\sigma_x = +18$ ksi
$\sigma_y = 0$
$\tau_x = -12$ ksi

11–23 A timber beam 4 in. wide by 6 in. high has a simple span of 6 ft and supports a uniform load of 500 lb/ft, including its own weight. Determine the principal stresses and the directions of the principal planes at points *A*, *B*, and *C*, at the section shown in Fig. P11–23. Show the results on a properly oriented element.

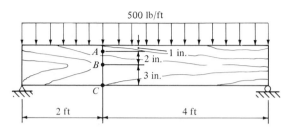

FIGURE P11–23

11–24 A cantilever cast-iron beam supports a uniform load of 6 kN/m, including its own weight. Determine the principal stresses at points A, B, and C at the fixed support shown in Fig. P11–24. The cross section and the section properties are given.

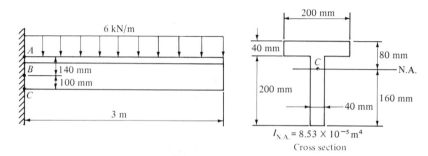

FIGURE P11–24

11–25 The maximum shear stress at point A in the simple beam of Fig. P11–25 is 100 psi. Determine the magnitude of the force P. Neglect the weight of the beam.

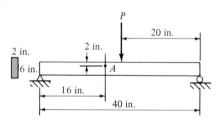

FIGURE P11–25

11–26 A clevis transmits a force $P = 10$ kN to a bracket, as shown in Fig. P11–26. Determine **(a)** the state of stress of the element at point A, **(b)** its principal stresses, **(c)** its maximum and minimum shear stresses and the associated normal stresses. Show the results on properly oriented elements.

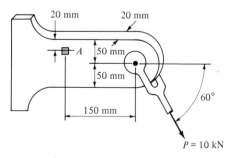

FIGURE P11–26

11–27 A cylindrical pressure vessel of 500 mm inside diameter and 15 mm wall thickness is subjected simultaneously to internal pressure $p = 6$ MPa and an external torque T. If the maximum shear stress in the wall is limited to 100 MPa, determine the maximum permissible torque T.

11–28 Determine the principal stresses and the directions of the principal planes at points A and B of Problem 10–28 (on p. 319).

11–29 Determine **(a)** the principal stresses, **(b)** the maximum shear stresses and the associated normal stress at point A of the beam in Problem 10–29 (on p. 320). Show the results in properly oriented elements.

11–30 A solid circular shaft 100 mm in diameter is subjected to an axial tensile force of 500 kN and a torque of 6 kN · m. Determine the maximum tensile stress in the shaft.

11–31 If the maximum tensile stress at point A is limited to 20 ksi, determine the maximum force P that can be applied to the shaft shown in Fig. P11–31. Neglect the weight of the shaft.

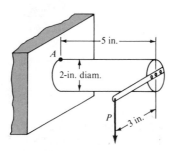

FIGURE P11–31

*11–5
COMPUTER PROGRAM ASSIGNMENTS

In the following assignments, write computer programs either in FORTRAN language or in BASIC language to compute stresses on inclined planes. Do not use a specific unit system in the program. When consistent units are used in the input data, the computed results are in the same system of units.

C11–1 Develop a computer program that would compute the values of σ_θ and τ_θ at the inclined plane m–m for the element shown in Fig. C11–1. Use the sign conventions established in Section 11–2 and Eqs. (11–1) and (11–2). These equations are shown in the following for convenience:

$$\sigma_\theta = B + C \cos A + D \sin A$$

$$\tau_\theta = -C \sin A + D \cos A$$

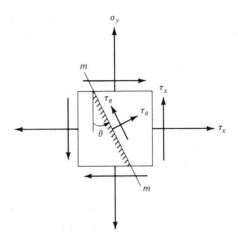

FIGURE C11–1

where

$$A = 2\theta$$

$$B = \frac{\sigma_x + \sigma_y}{2}$$

$$C = \frac{\sigma_x - \sigma_y}{2}$$

$$D = \tau_x$$

Input data for σ_x, σ_y, τ_x, and θ (in degrees). Output results for σ_θ and τ_θ. Run the program using data in Problems 11–5 and 11–6 (on p. 331).

C11–2 Develop a computer program that would compute the values of **(a)** the principal stresses and the orientation of the principal planes, **(b)** the maximum and the minimum shear stresses, the associated normal stresses, and the orientation of the plane for the maximum shear stress, for the element shown in Fig. C11–1. Use the sign conventions established in Section 11–2 and Eqs. (11–5) to (11–8). These equations are shown in the following for convenience:

$$\sigma_{max} = \sigma_1 = B + R \qquad \sigma_{min} = \sigma_2 = B - R$$

$$\tau_{max} = R \qquad \tau_{min} = -R \qquad \sigma' = B$$

where

$$B = \frac{\sigma_x + \sigma_y}{2}$$

$$C = \frac{\sigma_x - \sigma_y}{2}$$

$$D = \tau_x$$

$$R = \sqrt{C^2 + D^2}$$

The following equations can be used to determine the orientations of the plane of maximum normal stress and the plane of maximum shear stress.

$$\theta_1 = \tan^{-1} \frac{R - C}{D}$$

which defines the plane of maximum normal stress.

$$\theta_2 = \tan^{-1} \frac{-C}{R + D}$$

which defines the plane of maximum shear stress. Input data for σ_x, σ_y, and τ_x. Output results for σ_{max}, σ_{min}, θ_1, τ_{max}, τ_{min}, σ', and θ_2. Run the program using the data in Problems 11–16 (on p. 342) and 11–17 (on p. 343).

CHAPTER 12

Columns

12–1
INTRODUCTION

Short bars compressed by axial forces have been discussed in Chapter 2. Short compression members subjected to eccentric axial loads were considered in Chapter 10. In both cases, the members were short enough so that the load-carrying capacity of the member depended solely on the strength of the materials. Failure of these members occurs only when the normal or shear stresses become excessive. When the length of a compression member is large compared to the transverse dimensions, however, the member tends to buckle before high stress levels are reached. When buckling occurs, the member tends to deflect laterally and lose load-carrying stability. A small additional axial load will cause the member to collapse suddenly, without warning. Such long compression members are called *columns*. Stability considerations of columns are the primary concern of this chapter. Formulas will be established for computing the load-carrying capacity of columns of different lengths and end conditions.

12–2
EULER FORMULA FOR PIN-ENDED COLUMNS

The long column shown in Fig. 12–1 has pin supports at both ends and is subjected to an axial compressive load. When the load P is small, as shown in Fig. 12–1(a), the column remains straight. A small lateral load causes the column to deflect laterally, but the column will spring back to its original straight position once the lateral load is removed. Thus the column is in stable equilibrium. When the compressive load is gradually increased to a critical value P_{cr}, as shown in Fig. 12–1(b), the column will remain in the slightly deflected position after a small lateral load is applied and then removed. The column can be in equilibrium in an infinite number of slightly deflected positions. This condition is called neutral equilibrium. When the axial force exceeds the critical value P_{cr}, as shown in Fig. 12–1(c), the column becomes highly unstable and any small disturbance or imperfection in the column material or geometry could trigger the buckling of the

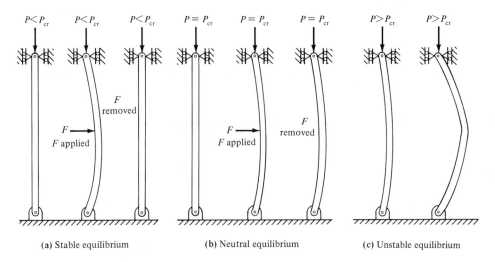

(a) Stable equilibrium   (b) Neutral equilibrium   (c) Unstable equilibrium

FIGURE 12–1

column, followed by a sudden collapse. This condition is referred to as *unstable equilibrium*.

In 1757, Leonard Euler, a famous Swiss mathematician, developed the following formula, now called the *Euler formula* for the *critical (buckling) load* on a long column with pinned ends made of homogeneous material:

$$P_{cr} = \frac{\pi^2 EI}{L^2} \qquad (12\text{--}1)$$

where

P_{cr} = critical (buckling) load, or the largest axial compressive load that a long column can carry before failure due to buckling

E = modulus of elasticity of the column material

I = least moment of inertia of the cross-sectional area of the column (buckling usually occurs about the axis with respect to which the moment of inertia is the smallest)

L = length of the column between the pins

Note that the critical load is independent of the strength of the column material. The only material property involved is the elastic modulus E, which represents the stiffness characteristic of the material. Hence, according to the Euler formula, a column made of high-strength alloy steel will have the same buckling load as a column made of ordinary structural steel, since the elastic moduli of the two kinds of steel are the same.

12–3
EULER FORMULA FOR COLUMNS
WITH OTHER END RESTRAINTS

The Euler formula, Eq. (12–1), applies to columns with pinned ends. This condition is referred to as the *fundamental case*. For other types of end restraints, the shape of the deflection curve of the column can be used to modify the Euler

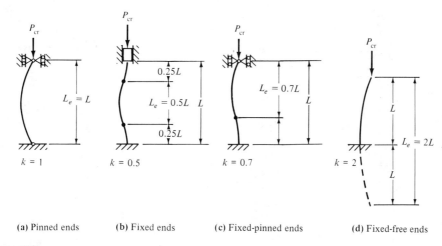

(a) Pinned ends   (b) Fixed ends   (c) Fixed-pinned ends   (d) Fixed-free ends

FIGURE 12–2

column formula. Figure 12–2 shows the deflection curves for several different end restraints. In each case, the length used in the Euler formula is the distance between either the inflection points (points where the curve concavity changes) or the pinned ends. This distance is called the *effective length* L_e. For the fundamental case [Fig. 12–2(a)], L_e is equal to L, but for the cases shown in Fig. 12–2(b), (c), and (d), L_e is equal to $0.5L$, $0.7L$, and $2L$, respectively. For a general case, $L_e = kL$, where k is the effective length factor depending on the end restraints.

The generalized Euler formula for columns of any supporting conditions is

$$P_{cr} = \frac{\pi^2 EI}{L_e^2} = \frac{\pi^2 EI}{(kL)^2} \tag{12–2}$$

where the effective length factor k depends on the end restraints. For example, with both ends fixed as shown in Fig. 12–2(b), $k = 0.5$, from Eq. (12–2), the critical load is

$$P_{cr} = \frac{\pi^2 EI}{(0.5L)^2} = \frac{4\pi^2 EI}{L^2}$$

which is four times larger than the value of the fundamental pin-ended case.

When a pin-ended column is braced at the midpoint, as shown in Fig. 12–3(a), the effective length of the column is $L/2$. The braced column, in effect, can be regarded as two columns, one on top of the other. Similarly, when a pin-ended column is braced at the third point, as shown in Fig. 12–3(b), the effective length of the column is $L/3$. To increase the load-carrying capacity, a column is sometimes braced at its intermediate points in the weaker direction. In this case, buckling about both axes must be investigated.

In some columns, the end restraints of two perpendicular directions may be different. For example, the connecting rod of an engine can be considered to be pin-ended in the plane of rotation, as shown in Fig. 12–4(a). In the perpendicular plane, the ends are considered fixed, as shown in Fig. 12–4(b).

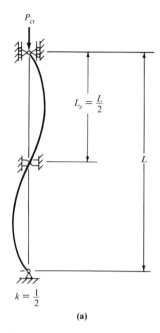

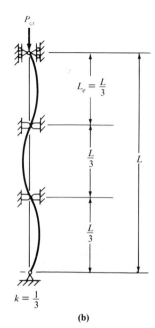

(a)

(b)

FIGURE 12–3

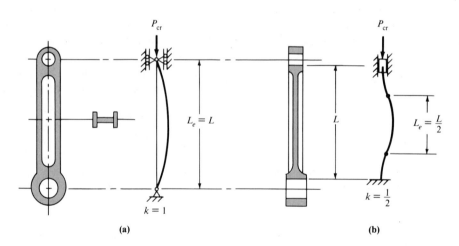

(a)

(b)

FIGURE 12–4

12–4
LIMITATION OF THE EULER FORMULA

The Euler formula was derived based on the elastic behavior of the material. Therefore, for the Euler formula to be applicable, the stress in the column must be within the proportional limit of the material. To see the significance of this limitation, Eq. (12–2) will be written in a different form.

By definition, $I = Ar^2$, where A is the cross-sectional area and r is the radius of gyration of the column cross section. Substituting this in Eq. (12–2) gives

$$P_{cr} = \frac{\pi^2 EI}{(kL)^2} = \frac{\pi^2 EA r^2}{(kL)^2}$$

or

$$\sigma_{cr} = \frac{P_{cr}}{A} = \frac{\pi^2 E}{(kL/r)^2} \qquad (12\text{--}3)$$

where the *critical stress* σ_{cr} is the average compressive stress in the column at the critical load P_{cr}. The quantity r is the *least* radius of gyration of the cross section of the column corresponding to the minimum value of I. The ratio kL/r, defined as the ratio of the effective length of the column to the least radius of gyration of the column section, is called the *slenderness ratio*. A column will always buckle in the direction of its least strength, which means in the direction with greatest slenderness ratio.

The Euler formula applies when σ_{cr} is less than the stress at the proportional limit σ_p, that is, when

$$\sigma_{cr} = \frac{\pi^2 E}{(kL/r)^2} \le \sigma_p$$

Therefore, for the elastic Euler formula to apply, the minimum slenderness ratio must be

$$\left(\frac{kL}{r}\right)_{min} = \sqrt{\frac{\pi^2 E}{\sigma_p}} \qquad (12\text{--}4)$$

For example, for structural steel with $\sigma_p = 30$ ksi and $E = 30 \times 10^3$ ksi, the minimum slenderness ratio is

$$\left(\frac{kL}{r}\right)_{min} = \sqrt{\frac{\pi^2 \times 30 \times 10^3 \text{ ksi}}{30 \text{ ksi}}} \approx 100$$

A graphical representation of Eq. (12–3) for structural steel is shown in Fig. 12–5 by plotting kL/r as the abscissa and σ_{cr} as the ordinate. The point A is the upper limit of applicability of the Euler formula. The Euler curve is valid for long columns with $kL/r > 100$ (the part of curve to the right of point A). The Euler formula is not valid for the AC part of the curve where $kL/r < 100$, since in this region the compressive stress is greater than σ_p, the material no longer behaving elastically.

Many columns encountered in machine and building design involve slenderness ratios that are too small for the Euler formula to apply. For these cases, many semiempirical formulas have been developed. The most frequently used formula in machine design and structural steel design is the J. B. Johnson formula, to be discussed in the next section.

When the slenderness ratio is less than 30, a compression member is in stable equilibrium under any load and the load-carrying capacity of the member will depend only on its strength.

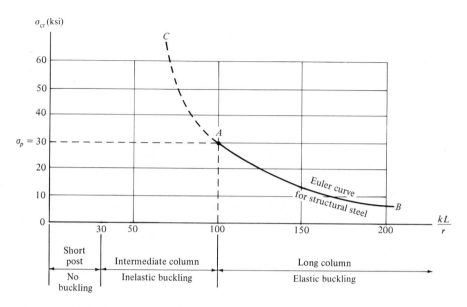

FIGURE 12–5

A column is designed to carry an allowable load P_{allow} equal to a fraction of the critical buckling load P_{cr}. The ratio of the critical load to the allowable load is the factor of safety, F. S. That is,

$$\text{F.S.} = \frac{P_{\text{cr}}}{P_{\text{allow}}} \tag{12–5}$$

The factor of safety to be used depends on many factors and is usually specified by the design code. For structural steel design, a factor of safety of 1.92 is used for long columns. For conditions of greater uncertainty, a factor of safety of 3 or more is recommended.

─── **EXAMPLE 12–1** ───────────────

A 1.5-m-long pin-ended Douglas fir column has a rectangular cross section 50 mm $\times$ 100 mm. Determine the maximum compressive axial load that the column can carry before buckling. For Douglas fir, $E = 12$ GPa and $\sigma_p = 28$ MPa.

SOLUTION

The minimum moment of inertia of the rectangular cross section is

$$I_{\text{min}} = \frac{(0.100 \text{ m})(0.050 \text{ m})^3}{12} = 1.04 \times 10^{-6} \text{ m}^4$$

Hence, by definition,

$$r = r_{\text{min}} = \sqrt{\frac{I_{\text{min}}}{A}} = \sqrt{\frac{1.04 \times 10^{-6} \text{ m}^4}{0.100 \times 0.050 \text{ m}^2}} = 0.0144 \text{ m}$$

Or from the expression in Problem 4–9 (on p. 99),

$$r = r_{min} = 0.289b = 0.289(0.050 \text{ m}) = 0.0144 \text{ m}$$

The slenderness ratio is

$$\frac{kL}{r} = \frac{1(1.5 \text{ m})}{0.0144 \text{ m}} = 104$$

From Eq. (12–4), the minimum slenderness ratio for which the Euler formula applies is

$$\left(\frac{kL}{r}\right)_{min} = \sqrt{\frac{\pi^2 E}{\sigma_p}} = \sqrt{\frac{\pi^2(12 \times 10^3 \text{ MPa})}{28 \text{ MPa}}} = 65$$

Since $kL/r = 104$ is greater than $(kL/r)_{min} = 65$, the Euler formula applies. From Eq. (12–3),

$$\sigma_{cr} = \frac{\pi^2 E}{(kL/r)^2} = \frac{\pi^2(12 \times 10^6 \text{ kN/m}^2)}{(104)^2} = 11\ 000 \text{ kN/m}^2$$

Thus the critical buckling load is

$$P_{cr} = \sigma_{cr}A = (11\ 000 \text{ kN/m}^2)(0.100 \times 0.050 \text{ m}^2) = 55 \text{ kN}$$ ∎

EXAMPLE 12–2

Determine the critical load and the allowable load of a 36-ft-long W10 × 36 steel column with pinned ends. The column is braced at the midpoint in the weaker direction. Use $E = 29 \times 10^3$ ksi, $\sigma_p = 34$ ksi, and F.S. = 1.92.

SOLUTION
From Eq. (12–4), the minimum slenderness ratio for which the Euler equation applies is

$$\left(\frac{kL}{r}\right)_{min} = \sqrt{\frac{\pi^2 E}{\sigma_p}} = \sqrt{\frac{\pi^2(29\ 000 \text{ ksi})}{34 \text{ ksi}}} = 92$$

From Table A–1 the radii of gyration for the W10 × 33 steel section are $r_x = 4.19$ in. and $r_y = 1.94$ in., and the cross-sectional area of the section is $A = 9.31$ in.^{2}.
The slenderness ratio about the two axes are

$$\frac{kL_x}{r_x} = \frac{1(36 \times 12 \text{ in.})}{4.19 \text{ in.}} = 103$$

$$\frac{kL_y}{r_y} = \frac{1(18 \times 12 \text{ in.})}{1.94 \text{ in.}} = 111$$

Buckling will occur about the y-axis, since the slenderness ratio about this axis is greater. Since $kL_y/r_y = 111$ is greater than $(kL/r)_{min} = 92$, the column is a long column and the Euler formula applies. From Eq. (12–3),

$$\sigma_{cr} = \frac{\pi^2 E}{(kL/r)^2} = \frac{\pi^2(29\ 000\ \text{ksi})}{(111)^2} = 23.2\ \text{ksi}$$

which is less than σ_p as is expected for a long column. Thus

$$P_{cr} = \sigma_{cr} A = (23.2\ \text{kips/in.}^2)(9.31\ \text{in.}^2) = 216\ \text{kips}$$

$$P_{allow} = \frac{P_{cr}}{\text{F.S.}} = \frac{216\ \text{kips}}{1.92} = 112.5\ \text{kips}$$

$\blacksquare$

PROBLEMS

*In Problems **12–1** to **12–4**, determine the slenderness ratio of the columns specified in the following table.*

	Section	Length	End Restraints
12–1	Rectangular: 2 in. × 4 in.	8 ft	Pinned ends
12–2	Circular: 100 mm diameter	6 m	Fixed ends
13–3	Tubular: $d_o = 200$ mm $d_i = 150$ mm	3 m	One end fixed, one end free
12–4	Wide-flange section: W12 × 87	40 ft	One end fixed, one end pinned

12–5 If a square section and a circular section have the same cross-sectional area A, which one is a better column section? (**HINT:** Express their radii of gyration in terms of the cross-sectional area A and compare the values.)

*For materials whose compression stress–strain diagrams are as shown in Problems **12–6** to **12–9**, determine the lowest limit of slenderness ratio for which the elastic Euler formula applies.*

12–6

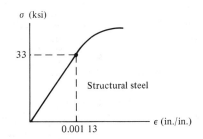

FIGURE P12–6

12–7

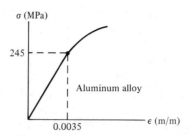

FIGURE P12–7

12–8

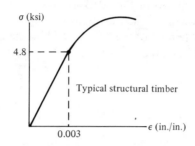

FIGURE P12–8

12–9

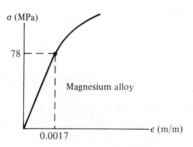

FIGURE P12–9

In Problems **12–10** *to* **12–13**, *determine the critical buckling loads of the columns specified in the following table. Assume that elastic buckling occurs.*

	Materials	*E*	*Section*	*Length*	*End Restraints*
12–10	Timber	12 GPa	100 mm × 100 mm square	2.5 m	Pinned ends
12–11	Steel	29 × 10³ ksi	Steel angle L4 × 4 × ½	18 ft	One end fixed, one end pinned
12–12	Aluminum	70 GPa	80-mm-diameter circle	4 m	Fixed ends
12–13	Steel	29 × 10³ ksi	2-in. standard steel pipe	4 ft	One end fixed, one end free

12–14 Find the minimum required dimension b of a 1.2-m-long pin-ended steel strut of square section that must support an axial compressive load of 20 kN. Use the Euler formula with F.S. $= 2$ and $E = 200$ GPa. [**HINT:** Express I in terms of the required dimension b and solve Eq. (18–1) for b.]

12–15 A 200-lb worker climbs a flagpole made of $\frac{3}{4}$-in. standard steel pipe. If the pole is 10 ft tall, can he get to the top before the pole buckles? If not, how high can he climb? The pole is fixed at the bottom and is free at the top. Neglect the weight of the pole and assume that the man's center of gravity is always along the axis of the pole. Use the Euler formula and $E = 29 \times 10^6$ psi.

12–16 The jib crane shown in Fig. P12–16 has a steel boom of square section 60 mm $\times$ 60 mm. Determine the maximum weight W in kN that the crane can lift based on the allowable load of the boom AB. Use the Euler formula with F.S. $= 3$ and $E = 200$ GPa. Neglect the weight of the structure.

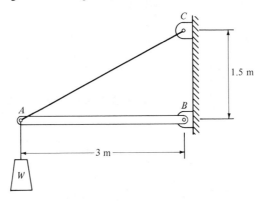

FIGURE P12–16

12–17 For the jib crane shown in Fig. P12–17, determine the smallest size of the standard steel pipe that can be used for member AB if the crane has a capacity of $2\frac{1}{2}$ tons. Use the Euler formula with F.S. $= 3$ and $E = 29 \times 10^3$ ksi. Neglect the weight of the structure.

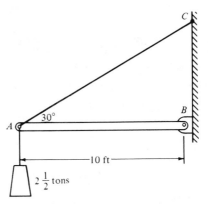

FIGURE P12–17

12–18 A pin-ended timber member AB 2 m long has a rectangular cross section of 50 mm $\times$ 100 mm. Referring to Fig. P12–18, determine the maximum weight W that can be supported by the structure based on the allowable compressive load of member AB. Use the Euler formula with F.S. $= 2$ and $E = 12$ GPa.

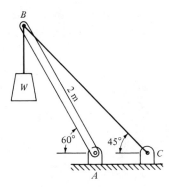

FIGURE P12–18

12–19 A toggle press is a mechanism that causes a large compression force to be exerted on the block D, as shown in Fig. P12–19. If the weight $W = 1\frac{1}{2}$ tons, determine the minimum required diameter of the circular steel rod for the two arms AB and AC. Use the Euler formula with F.S. $= 3$ and $E = 29 \times 10^3$ ksi.

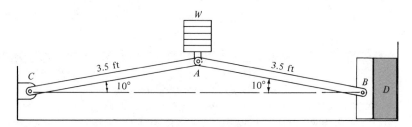

FIGURE P12–19

12–20 Determine the allowable force F that can be applied at D based on the allowable compressive load of column AB shown in Fig. P12–20. The steel column with 1 in. $\times$ 2 in. rectangular section has rounded ends and is braced at the midpoint in the weak direction. Use $E = 30 \times 10^3$ ksi, $\sigma_p = 30$ ksi, and F.S. $= 2$.

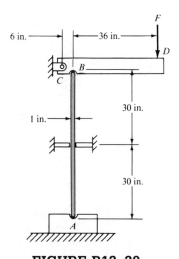

FIGURE P12–20

12–21 A diesel locomotive engine has a steel connecting rod with an I-beam section as shown in Fig. P12–21. In the xy-plane the rod is considered to be pin-ended. In the *yz*-plane, the rod is considered to be fix-ended. If $E = 200$ GPa, $\sigma_p = 240$ MPa, and F.S. = 3, determine the allowable compressive load that can be applied to the rod.

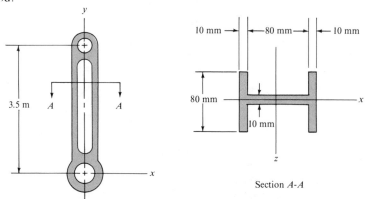

FIGURE P12–21

12–5
J. B. JOHNSON FORMULA FOR INTERMEDIATE COLUMNS

Since the Euler formula does not apply for the intermediate columns, many semi-empirical formulas have been developed. One of these formulas, the J. B. Johnson formula, is used extensively by engineers to determine the critical load for ductile steel columns. As indicated in Fig. 12–6, the J. B. Johnson formula is the equation of a parabola having its vertex at the point on the vertical axis with ordinate equal to σ_y, and is tangent to the Euler curve at the transition slenderness ratio $kL/r = C_c$, corresponding to one-half of the yield stress σ_y of the steel. To fulfill this

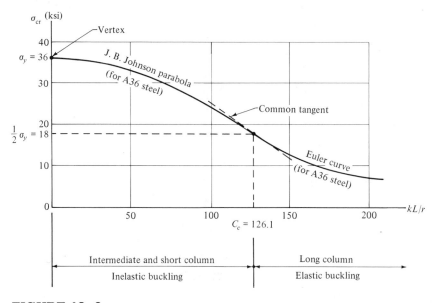

FIGURE 12–6

condition, the value of the transition slenderness ratio can be determined from Eq. (12–3). Thus

$$\sigma_{cr} = \frac{1}{2}\sigma_y = \frac{\pi^2 E}{(kL/r)^2} = \frac{\pi^2 E}{C_c^2}$$

from which

$$C_c = \sqrt{\frac{2\pi^2 E}{\sigma_y}} \tag{12–6}$$

The equation of the parabola or the J. B. Johnson formula is

$$\sigma_{cr} = \frac{P_{cr}}{A} = \left[1 - \frac{(kL/r)^2}{2C_c^2}\right]\sigma_y \tag{12–7}$$

The Euler formula applies when kL/r is greater than C_c, and the J. B. Johnson formula applies when kL/r is less than C_c. For $kL/r = C_c$, both formulas give the same result. Note that the Euler formula applies to all materials, whereas the J. B. Johnson formula applies mainly to ductile materials.

─────── **EXAMPLE 12–3** ───────────────────────────────────

Determine the allowable compressive load of a 4-in. standard steel pipe 25 ft long. The column is made of A36 steel with $\sigma_y = 36$ ksi and is welded to fixed supports at both ends. Use F.S. $= 3$ and $E = 29 \times 10^3$ ksi.

SOLUTION

From Table A–5, for a 4-in. standard steel pipe, $A = 3.17$ in.^{2} and $r = 1.51$ in. The slenderness ratio is

$$\frac{kL}{r} = \frac{0.5(25 \times 12 \text{ in.})}{1.51 \text{ in.}} = 99.3$$

From Eq. (12–7), the value of the transition slenderness ratio C_c is

$$C_c = \sqrt{\frac{2\pi^2 E}{\sigma_y}} = \sqrt{\frac{2\pi^2 (29\ 000 \text{ ksi})}{36 \text{ ksi}}} = 126.1$$

Since $kL/r < C_c$, the J. B. Johnson formula applies. From Eq. (12–6),

$$\sigma_{cr} = \left[1 - \frac{(kL/r)^2}{2C_c^2}\right]\sigma_y = \left[1 - \frac{(99.3)^2}{2(126.1)^2}\right](36 \text{ ksi}) = 24.8 \text{ ksi}$$

Thus

$$P_{cr} = \sigma_{cr}A = (24.8 \text{ kips/in.}^2)(3.17 \text{ in.}^2) = 78.6 \text{ kips}$$

$$P_{allow} = \frac{P_{cr}}{\text{F.S.}} = \frac{78.6 \text{ kips}}{3} = 26.2 \text{ kips}$$

PROBLEMS

In Problems **12–22** *to* **12–25**, *determine the critical buckling loads of the columns specified in the following table.*

	Material	E	Section	Length	End Restraints
12–22	A36 steel, σ_y = 250 MPa	200 GPa	100 mm × 200 mm rectangle	2.5 m	Both ends pinned
12–23	A36 steel, σ_y = 250 MPa	200 GPa	100-mm-diameter circle	3.5 m	One end pinned, one end fixed
12–24	A441 steel, σ_y = 50 ksi	29 × 10³ ksi	5-in. standard steel pipe	6 ft	One end fixed, one end free
12–25	A242 steel, σ_y = 50 ksi	29 × 10³ ksi	Wide-flange steel section W14 × 74	40 ft	Both ends fixed

12–26 In Fig. P12–26, compression member *BD* has a rectangular section 1 in. by 2 in. and is made of A441 steel with σ_y = 50 ksi. Determine the maximum weight *W* that can be supported by the assembly. Assume that the beam *AC* and the connections are all properly designed. Use a factor of safety of 2.5. (**HINT:** The maximum weight *W* depends on the allowable compressive load of member *BD*.)

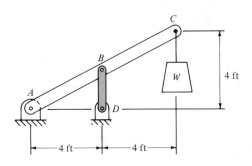

FIGURE P12–26

12–27 Compression member *AB* acts as a spreader bar between the cables shown in Fig. P12–27. The bar has a circular section of 100 mm diameter and is made of A36 steel with σ_y = 250 MPa and *E* = 200 GPa. Determine the maximum pulling force *F* that can be applied to the assembly based on the allowable compressive load of the bar for a factor of safety of 3.

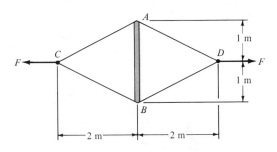

FIGURE P12–27

12–6
AISC FORMULAS FOR STEEL COLUMN DESIGN

The *AISC* (American Institute of Steel Construction) *Manual* (8th edition, 1980) gives the following formulas for computing the allowable compressive stresses for the steel column design.

1. For long columns, $kL/r \geq C_c = \sqrt{2\pi^2 E/\sigma_y}$
   (but in no case should kL/r be greater than 200)

   $$\sigma_{\text{allow}} = \frac{P_{\text{allow}}}{A} = \frac{\sigma_{\text{cr}} \text{ from the Euler formula}}{\text{F.S.}} = \frac{\pi^2 E/(kL/r)^2}{\text{F.S.}}$$

   where $E = 29 \times 10^3$ ksi and F.S. $= 1.92$; thus

   $$\sigma_{\text{allow}} = \frac{\pi^2 (29\ 000)/(kL/r)^2}{1.92} = \frac{149\ 100\ (\text{ksi})}{(kL/r)^2} \qquad (12\text{–}8)$$

2. For intermediate and short columns, $kL/r \leq C_c = \sqrt{2\pi^2 E/\sigma_y}$

   $$\sigma_{\text{allow}} = \frac{P_{\text{allow}}}{A} = \frac{\sigma_{\text{cr}} \text{ from the J. B. Johnson formula}}{\text{F.S.}}$$

   or

   $$\sigma_{\text{allow}} = \frac{\left[1 - \dfrac{(kL/r)^2}{2C_c^2} \right] \sigma_y\ (\text{ksi})}{\text{F.S.}} \qquad (12\text{–}9)$$

   where the factor of safety F.S. is computed from the equation

   $$\text{F.S.} = \frac{5}{3} + \frac{3(kL/r)}{8C_c} - \frac{(kL/r)^3}{8C_c^3} \qquad (12\text{–}10)$$

It is interesting to note that F.S. varies from $\frac{5}{3}$ ($= 1.67$) when $kL/r = 0$ to $\frac{23}{12}$ ($= 1.92$) when $kL/r = C_c$, being more conservative for longer columns.

As a design aid, values of the allowable compressive stress computed from the AISC formulas corresponding to $\sigma_y = 36$ ksi and $\sigma_y = 50$ ksi are tabulated for kL/r values from 1 to 200 in Table 12–1.

Table 12–2 shows the value of the AISC recommended effective length factor k for steel column design when the ideal end restraints are approximated.

───── **EXAMPLE 12–4** ─────────────────────────────────────

Determine the allowable axial compressive load for a 10-ft-long standard L6 × 4 × $\frac{1}{2}$ steel angle of A36 steel if the supporting conditions are (a) pinned at both ends, (b) fixed at both ends. Use the AISC formulas and the recommended k values.

SOLUTION
From Table A–4 of the Appendix Tables, for an L6 × 4 × $\frac{1}{2}$ steel angle, $A = 4.75$ in.^{2} and the least radius of gyration is $r_z = 0.870$ in.

TABLE 12–1(a)   AISC Allowable Compressive Stress for Steel Columns for σ_y = 36 ksi

$\dfrac{kL}{r}$	σ_{allow} (ksi)	$\dfrac{kL}{r}$	σ_{allow} (ksi)	$\dfrac{kL}{r}$	σ_{allow} (ksi)	$\dfrac{kL}{r}$	σ_{allow} (ksi)	$\dfrac{kL}{r}$	σ_{allow} (ksi)
1	21.56	41	19.11	81	15.24	121	10.14	161	5.76
2	21.52	42	19.03	82	15.13	122	9.99	162	5.69
3	21.48	43	18.95	83	15.02	123	9.85	163	5.62
4	21.44	44	18.86	84	14.90	124	9.70	164	5.55
5	21.39	45	18.78	85	14.79	125	9.55	165	5.49
6	21.35	46	18.70	86	14.67	126	9.41	166	5.42
7	21.30	47	18.61	87	14.56	127	9.26	167	5.35
8	21.25	48	18.53	88	14.44	128	9.11	168	5.29
9	21.21	49	18.44	89	14.32	129	8.97	169	5.23
10	21.16	50	18.35	90	14.20	130	8.84	170	5.17
11	21.10	51	18.26	91	14.09	131	8.70	171	5.11
12	21.05	52	18.17	92	13.97	132	8.57	172	5.05
13	21.00	53	18.08	93	13.84	133	8.44	173	4.99
14	20.95	54	17.99	94	13.72	134	8.32	174	4.93
15	20.89	55	17.90	95	13.60	135	8.19	175	4.88
16	20.83	56	17.81	96	13.48	136	8.07	176	4.82
17	20.78	57	17.71	97	13.35	137	7.96	177	4.77
18	20.72	58	17.62	98	13.23	138	7.84	178	4.71
19	20.66	59	17.53	99	13.10	139	7.73	179	4.66
20	20.60	60	17.43	100	12.98	140	7.62	180	4.61
21	20.54	61	17.33	101	12.85	141	7.51	181	4.56
22	20.48	62	17.24	102	12.72	142	7.41	182	4.51
23	20.41	63	17.14	103	12.59	143	7.30	183	4.46
24	20.35	64	17.04	104	12.47	144	7.20	184	4.41
25	20.28	65	16.94	105	12.33	145	7.10	185	4.36
26	20.22	66	16.84	106	12.20	146	7.01	186	4.32
27	20.15	67	16.74	107	12.07	147	6.91	187	4.27
28	20.08	68	16.64	108	11.94	148	6.82	188	4.23
29	20.01	69	16.53	109	11.81	149	6.73	189	4.18
30	19.94	70	16.43	110	11.67	150	6.64	190	4.14
31	19.87	71	16.33	111	11.54	151	6.55	191	4.09
32	19.80	72	16.22	112	11.40	152	6.46	192	4.05
33	19.73	73	16.12	113	11.26	153	6.38	193	4.01
34	19.65	74	16.01	114	11.13	154	6.30	194	3.97
35	19.58	75	15.90	115	10.99	155	6.22	195	3.93
36	19.50	76	15.79	116	10.85	156	6.14	196	3.89
37	19.42	77	15.69	117	10.71	157	6.06	197	3.85
38	19.35	78	15.58	118	10.57	158	5.98	198	3.81
39	19.27	79	15.47	119	10.43	159	5.91	199	3.77
40	19.19	80	15.36	120	10.28	160	5.83	200	3.73

Note: C_c = 126.1.

TABLE 12–1(b)   AISC Allowable Compressive Stress for Steel Columns for $\sigma_y = 50$ ksi

$\frac{kL}{r}$	σ_{allow} (ksi)	$\frac{kL}{r}$	σ_{allow} (ksi)	$\frac{kL}{r}$	σ_{allow} (ksi)	$\frac{kL}{r}$	σ_{allow} (ksi)	$\frac{kL}{r}$	σ_{allow} (ksi)
1	29.94	41	25.69	81	18.81	121	10.20	161	5.76
2	29.87	42	25.55	82	18.61	122	10.03	162	5.69
3	29.80	43	25.40	83	18.41	123	9.87	163	5.62
4	29.73	44	25.26	84	18.20	124	9.71	164	5.55
5	29.66	45	25.11	85	17.99	125	9.56	165	5.49
6	29.58	46	24.96	86	17.79	126	9.41	166	5.42
7	29.50	47	24.81	87	17.58	127	9.26	167	5.35
8	29.42	48	24.66	88	17.37	128	9.11	168	5.29
9	29.34	49	24.51	89	17.15	129	8.97	169	5.23
10	29.26	50	24.35	90	16.94	130	8.84	170	5.17
11	29.17	51	24.19	91	16.72	131	8.70	171	5.11
12	29.08	52	24.04	92	16.50	132	8.57	172	5.05
13	28.99	53	23.88	93	16.29	133	8.44	173	4.99
14	28.90	54	23.72	94	16.06	134	8.32	174	4.93
15	28.80	55	23.55	95	15.84	135	8.19	175	4.88
16	28.71	56	23.39	96	15.62	136	8.07	176	4.82
17	28.61	57	23.22	97	15.39	137	7.96	177	4.77
18	28.51	58	23.06	98	15.17	138	7.84	178	4.71
19	28.40	59	22.89	99	14.94	139	7.73	179	4.66
20	28.30	60	22.72	100	14.71	140	7.62	180	4.61
21	28.19	61	22.55	101	14.47	141	7.51	181	4.56
22	28.08	62	22.37	102	14.24	142	7.41	182	4.51
23	27.97	63	22.20	103	14.00	143	7.30	183	4.46
24	27.86	64	22.02	104	13.77	144	7.20	184	4.41
25	27.75	65	21.85	105	13.53	145	7.10	185	4.36
26	27.63	66	21.67	106	13.29	146	7.01	186	4.32
27	27.52	67	21.49	107	13.04	147	6.91	187	4.27
28	27.40	68	21.31	108	12.80	148	6.82	188	4.23
29	27.28	69	21.12	109	12.57	149	6.73	189	4.18
30	27.15	70	20.94	110	12.34	150	6.64	190	4.14
31	27.03	71	20.75	111	12.12	151	6.55	191	4.09
32	26.90	72	20.56	112	11.90	152	6.46	192	4.05
33	26.77	73	20.38	113	11.69	153	6.38	193	4.01
34	26.64	74	20.19	114	11.49	154	6.30	194	3.97
35	26.51	75	19.99	115	11.29	155	6.22	195	3.93
36	26.38	76	19.80	116	11.10	156	6.14	196	3.89
37	26.25	77	19.61	117	10.91	157	6.06	197	3.85
38	26.11	78	19.41	118	10.72	158	5.98	198	3.81
39	25.97	79	19.21	119	10.55	159	5.91	199	3.77
40	25.83	80	19.01	120	10.37	160	5.83	200	3.73

Note: $C_c = 107.0$.

TABLE 12–2 AISC Recommended k Values

End Restraints	Pinned Ends	Fixed Ends	Fixed–Pinned Ends	Fixed–Free Ends
Theoretical k value	1.0	0.5	0.7	2.0
AISC recommended k value	1.0	0.65	0.8	2.10

(a) For pinned ends, $k = 1$:

$$\frac{kL}{r} = \frac{(1)(10 \times 12 \text{ in.})}{0.870 \text{ in.}} = 137.9$$

For A36 steel, $\sigma_y = 36$ ksi; then

$$C_c = \sqrt{\frac{2\pi^2 E}{\sigma_y}} = \sqrt{\frac{2\pi^2(29\ 000 \text{ ksi})}{36 \text{ ksi}}} = 126.1$$

Since $kL/r > C_c$, Eq. (12–8) applies. Thus

$$\sigma_{\text{allow}} = \frac{149\ 100 \text{ ksi}}{(kL/r)^2} = \frac{149\ 100 \text{ ksi}}{(137.9)^2} = 7.84 \text{ ksi}$$

Or from Table 12–1(a), for $\sigma_y = 36$ ksi and $kL/r = 138$, the allowable compressive stress is $\sigma_{\text{allow}} = 7.84$ ksi, the same as calculated above. Thus

$$P_{\text{allow}} = \sigma_{\text{allow}} A = (7.84 \text{ kips/in.}^2)(4.75 \text{ in.}^2) = 37.3 \text{ kips}$$

(b) For fixed ends, the AISC recommended $k = 0.65$ (from Table 12–2). The slenderness ratio is

$$\frac{kL}{r} = \frac{(0.65)(10 \times 12 \text{ in.})}{0.870 \text{ in.}} = 89.7$$

which is less than $C_c = 126.1$; thus Eq. (12–9) applies. From Eq. (12–10),

$$\text{F.S.} = \frac{5}{3} + \frac{3(kL/r)}{8C_c} - \frac{(kL/r)^3}{8C_c^3}$$

$$= \frac{5}{3} + \frac{3(89.7)}{8(126.1)} - \frac{(89.7)^3}{8(126.1)^3} = 1.89$$

Substitution in Eq. (12–9) gives

$$\sigma_{\text{allow}} = \frac{\left[1 - \dfrac{(kL/r)^2}{2C_c^2}\right]\sigma_y}{\text{F.S.}} = \frac{\left[1 - \dfrac{(89.7)^2}{2(126.1)^2}\right](36 \text{ ksi})}{1.89} = 14.23 \text{ ksi}$$

Or from Table 12–1(a), for $\sigma_y = 36$ ksi and $kL/r = 90$, the allowable compressive stress is $\sigma_{allow} = 14.20$ ksi. Thus

$$P_{allow} = \sigma_{allow} A = (14.23 \text{ kips/in.}^2)(4.75 \text{ in.}^2) = 67.6 \text{ kips}$$

Hence we see that the allowable compressive load of a steel column is substantially increased by imposing stiffer constraints on the ends of the column. In general, it can be concluded that stiffer constraints usually mean higher allowable compressive loads.

EXAMPLE 12–5

Using AISC code, select a W shape for a 15-ft-long fixed-ended column to carry an axial compressive load of 500 kips. Use A441 steel, which has $\sigma_y = 50$ ksi.

SOLUTION
Since the size of the W shape is unknown, a trial-and-error procedure is needed. Assume the least radius of gyration to be $r = 3$ in.; then the slenderness ratio would be

$$\frac{kL}{r} = \frac{(0.65)(15 \times 12 \text{ in.})}{3 \text{ in.}} = 39$$

From Table 12–1(b), for $\sigma_y = 50$ ksi and $kL/r = 39$, the allowable compressive stress is $\sigma_{allow} = 25.97$ ksi. Thus the required cross-sectional area of the column is

$$A_{req} = \frac{P}{\sigma_{allow}} = \frac{500 \text{ kips}}{25.97 \text{ kips/in.}^2} = 19.3 \text{ in.}^2$$

From Table A–1 in the Appendix Tables it is found that a W14 × 68 comes nearest to having the required area. This section will be tentatively selected to determine if it meets design requirements:

[first trial]   W14 × 68 $A = 20.0$ in.^{2} least $r_y = 2.46$ in.

$$\frac{kL}{r} = \frac{(0.65)(15 \times 12 \text{ in.})}{2.46 \text{ in.}} = 48$$

From Table 12–1(b), $\sigma_{allow} = 24.66$ ksi. Thus

$$P_{allow} = \sigma_{allow} A = (24.66 \text{ kips/in.}^2)(20.0 \text{ in.}^2) = 493 \text{ kips}$$

which is only 1.4 percent under the required capacity; the section is therefore satisfactory. However, a more careful examination of Table A–1 shows that a lighter W12 × 65 section has approximately the same area, $A = 19.1$ in.^{2}, and has a considerably larger value of the least radius of gyration, $r_y = 3.02$ in. Hence another trial is made to determine if this section is satisfactory.

[second trial] W12 × 65 $A = 19.1$ in.^{2} least $r_y = 3.02$ in.

$$\frac{kL}{r} = \frac{(0.65)(15 \times 12 \text{ in.})}{3.02} = 39$$

From Table 12–1(b), $\sigma_{\text{allow}} = 25.97$ ksi; thus

$$P_{\text{allow}} = \sigma_{\text{allow}}A = (25.97 \text{ kips/in.}^2)(19.1 \text{ in.}^2) = 496 \text{ kips}$$

which is less than 1 percent under the required capacity; hence W12 × 65 is a satisfactory section for the column. ∎

PROBLEMS

In Problems **12–28** *to* **12–31**, *determine the allowable axial compressive loads of the steel columns specified in the following table. Use AISC formulas and the recommended k values. Use Table 12–1 to verify the computations.*

	Section	Steel	σ_y	Length	End Restraints
12–28	3-in. standard steel pipe	A36	36 ksi	(a) 5 ft (b) 8 ft	One end fixed, one end free
12–29	W8 × 40	A441	50 ksi	(a) 20 ft (b) 30 ft	Both ends fixed
12–30	W10 × 112	A36	36 ksi	(a) 30 ft (b) 40 ft	One end fixed, one end pinned
12–31	L5 × 5 × $\frac{1}{2}$	A242	50 ksi	(a) 10 ft (b) 15 ft	Both ends fixed

In Problems **12–32** *to* **12–35**, *assume that building columns are to be designed by using the AISC code. Select the lightest W sections for the columns specified in the following table.*

	Steel	σ_y	Allowable Load	Length	End Restraints
12–32	A36	36 ksi	400 kips	20 ft	Both ends pinned
12–33	A36	36 ksi	150 kips	35 ft	Fixed–pinned ends
12–34	A441	50 ksi	200 kips	30 ft	Both ends fixed
12–35	A242	50 ksi	500 kips	40 ft	Both ends fixed

12–36 Two standard steel C10 × 20 channels form a 25-ft-long square compression member. As shown in Fig. P12–36, the channels are tied together by end tie plates and lacing bars to make the two channels act together as one unit. The tie plates and the lacings are not effective in resisting compression. The channels are made of A36 steel with $\sigma_y = 36$ ksi, and the ends of the columns are considered hinged. Determine the allowable axial force of the member according to AISC code.

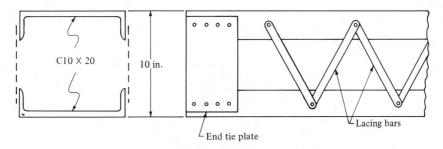

FIGURE P12–36

12–37 A 40-ft compression member made of two C12 × 30 channels is arranged as shown in Fig. P12–37. The channels are properly laced together to act as one unit, they are made of A441 steel with $\sigma_y = 50$ ksi. The ends of the column are considered hinged. Determine **(a)** the value of the distance b so that the section will have equal moments of inertia about the x- and y-axes, **(b)** the allowable axial force of the member according to the AISC code.

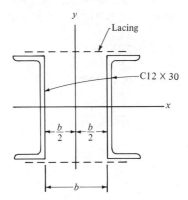

FIGURE P12–37

*12–7
COMPUTER PROGRAM ASSIGNMENTS

In the following assignments, write the computer programs either in FORTRAN language or in BASIC language for column analysis.

C12–1 Develop a computer program to compute the critical stresses from the Euler formula [Eq. (12–3)] or the J. B. Johnson formula [Eq. (12–6)]. These formulas are shown in the following for convenience:

$$\text{Euler formula:} \qquad \sigma_{cr} = \frac{\pi^2 E}{(SR)^2} \qquad \text{(for } SR \geq CC\text{)}$$

$$\text{J. B. Johnson formula:} \quad \sigma_{cr} = \left(1 - \frac{1}{2}Q^2\right)\sigma_y \quad \text{(for } SR < CC\text{)}$$

where

$$SR = \text{slenderness ratio}$$

$$E = \text{modulus of elasticity}$$

$$\sigma_y = \text{yield stress}$$

$$CC = \sqrt{\frac{2\pi^2 E}{\sigma_y}}$$

$$Q = \frac{SR}{CC}$$

Do not use a specific unit system in the program. When consistent units are used in the input data, the computed results are in the same system of units. Input data for E and σ_y. Output a table listing the values of critical stresses corresponding to the values of the slenderness ratio varying from 0 to 200 at an increment of 10. Run the program with $E = 29\,000$ ksi and $\sigma_y = 50$ ksi and plot a curve similar to Fig. 12–6 showing the variation of critical stresses of steel column with $\sigma_y = 50$ ksi versus slenderness ratio from 0 to 200.

C12–2 Develop a computer program to compute the AISC allowable compressive stresses for the steel column design. Use Eqs. (12–8) and (12–9). These equations are shown in the following for convenience:

$$\sigma_{\text{allow}} = \frac{\pi^2 E}{\text{F.S. (SL)}^2)} \qquad \text{(for SL} \geq \text{CC)}$$

$$\sigma_{\text{allow}} = \frac{(1 - \frac{1}{2}Q^2)\sigma_y}{\text{F.S.}} \ \text{(ksi)} \qquad \text{(for SL} < \text{CC)}$$

where

$$SL = \text{slenderness ratio}$$

$$E = \text{modulus of elasticity} = 29\,000 \text{ ksi}$$

$$CC = \sqrt{\frac{2\pi^2 E}{\sigma_y}}$$

$$Q = \frac{SL}{CC}$$

$$\text{F.S.} = \text{factor of safety} \begin{cases} = 1.92 & \text{(for SR} \geq \text{CC)} \\ = \frac{5}{3} + \frac{3}{8}Q - \frac{1}{8}Q^3 & \text{(for SR} < \text{CC)} \end{cases}$$

Input the value for σ_y in ksi. Output a table listing the values of σ_{allow} corresponding to values of the slenderness ratio (SL) varying from 1 to 200 at an increment of 1. Run the program inputting $\sigma_y = 36$ ksi and compare the results with those in Table 12–1(a). Set up a new table for $\sigma_y = 60$ ksi. (**HINT:** For best results, treat the allowable stresses as elements in a one-dimensional array. Compute the allowable stresses in a DO-loop. Use another DO-loop to print out the array in the format as shown in Table 12–1.)

CHAPTER 13

Structural Connections

13–1
INTRODUCTION

This chapter considers the analysis and design of several types of connections for structural members. The design of structural connections is partly empirical, based mainly on past experience and experimental research.

The major connectors used in steel structures are rivets, bolts, and welding. Riveted connections were extensively used for many years in the past, but the current trend has been more and more toward welding and high-strength steel bolting.

The riveted connections are discussed first, followed by a study of high-strength steel bolts. Welded connections are treated at the end of the chapter.

13–2
RIVETED CONNECTIONS

Plates and platelike parts such as legs of angles, webs and flanges of beams, can be connected by rivets. In general, the design of a connection is concerned with the transfer of forces from one component to another through connection. For a riveted connection, the forces are transmitted through shear forces in the rivets and the bearing force between the rivets and the connected plates. A rivet can be in single or double shear, depending on whether one or two sections of the rivet are subjected to shear forces, which was discussed in Section 2–3. The bearing stress between a rivet and a plate was discussed in Section 2–4. The actual distribution of bearing stress is rather complicated (see Fig. 2–9 on p. 35). In practice, the bearing stress distribution is approximated on the basis of an average bearing stress acting over the projected area of the rivet's shank onto the cross section of a plate, that is, of a rectangular area td, as shown in Fig. 2–9(c).

Holes $\frac{1}{16}$ in. larger than the rivet diameter are punched or drilled for the insertion of red-hot rivets. One end of a rivet has a head, the rivet is heated, and the projecting shank at the other end is driven by a power riveter to form another head. Upon cooling, the rivet contracts so that the plates being connected are tightly pressed together.

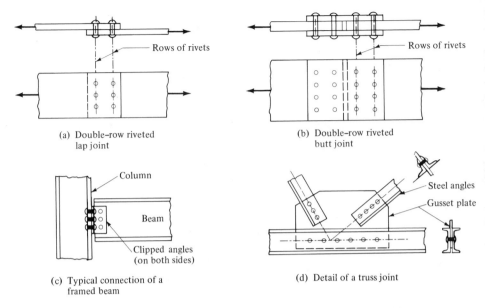

(a) Double–row riveted
lap joint

(b) Double–row riveted
butt joint

(c) Typical connection of a
framed beam

(d) Detail of a truss joint

FIGURE 13–1

The following assumptions are made for rivet connections:

1. The rivets completely fill the holes.
2. The friction forces between the connected plates are ignored.
3. A load applied to the member without any eccentricity with respect to the centroid of the rivets is assumed to be shared equally by all the rivets.
4. The shear stress is assumed to be uniformly distributed over a section (or two sections in double shear) of a rivet. The bearing stress is assumed to be uniformly distributed over the projected area td, where t is the thickness of the plate and d is the diameter of the rivet. Stress concentrations (see Section 3–12) at rivet holes in the plate are ignored, and tensile stress is assumed to be uniformly distributed across the net section of a plate.

For ordinary construction, rivets $\frac{3}{4}$ in. and $\frac{7}{8}$ in. in diameter are most commonly used, but rivets are available in standard sizes from $\frac{1}{2}$ to $1\frac{1}{2}$ in., in $\frac{1}{8}$-in. increments. It is preferable that all rivets on the same structure be of one size.

Steel rivets used for structural purposes are classified as ASTM* A502, grades 1 and 2; these are designated as A502-1 and A502-2, respectively.

Several typical arrangements of riveted connections are shown in Fig. 13–1.

13–3
STRENGTH OF RIVETED CONNECTIONS

Riveted joints may fail in one of the following ways:

1. Failure in shear of rivets, as shown in Fig. 13–2(a) and (b).
2. Failure in bearing when the rivets crush the material of the plate against which the rivets bear, as shown in Fig. 13–2(c).

* ASTM stands for the American Society for Testing and Materials.

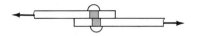

(a) Failure due to single shear of rivets

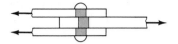

(b) Failure due to double shear of rivets

(c) Failure due to crushing of plate

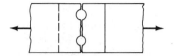

(d) Failure due to tension through the net section

FIGURE 13–2

3. Failure in tension when the connected plate is torn apart at the riveted cross section, which is weakened by the rivet holes, as shown in Fig. 13–2(d).

It is presumed that the joint will ultimately fail in one of the three ways listed above. The allowable loads based on these three methods of failure are outlined below.

1. Allowable load in shear:

$$(P_s)_{\text{allow}} = nA_s\tau_{\text{allow}} \qquad (13\text{--}1)$$

where

n = total number of shear planes of the rivets in the joint
A_s = cross-sectional area of rivet = $\frac{1}{4}\pi d^2$
d = diameter of rivets
τ_{allow} = allowable shear stress of rivets

The allowable shear stresses for rivets, specified in the *AISC Manual*, 8th edition, 1980, for both single shear and double shear are

$$\tau_{\text{allow}} = 15 \text{ ksi for A502--1}$$

$$\tau_{\text{allow}} = 20 \text{ ksi for A502--2}$$

2. Allowable load in bearing:

$$(P_b)_{\text{allow}} = n(td)(\sigma_b)_{\text{allow}} \qquad (13\text{--}2)$$

where

n = total number of bearing surfaces on plate
t = thickness of plate
d = diameter of rivet
$(\sigma_b)_{\text{allow}}$ = allowable bearing stress of the connected plate

The AISC specification gives the allowable bearing stress on the projected area as

$$(\sigma_b)_{\text{allow}} = 1.35\sigma_y$$

where σ_y is the yield stress of the connected plate.

3. Allowable load in tension:

$$(P_t)_{allow} = A_{net} \, (\sigma_t)_{allow} = (b_{net}t) \, (\sigma_t)_{allow} \qquad (13\text{--}3)$$

where

b_{net} = net width of the connected plate through the critical section
$\qquad$ = $b - n(d + \frac{1}{8})$
$\quad b$ = width of plate
$\quad d$ = rivet diameter
$\quad n$ = number of rivet holes in the critical section
$\quad t$ = thickness of plate
$(\sigma_t)_{allow}$ = allowable tensile stress of plate

Note that in calculating the net width, $\frac{1}{8}$ in. is added to the rivet diameter to obtain the diameter of the rivet holes. This is done to account for the fact that the holes are $\frac{1}{16}$ in. larger than the rivet diameters, plus an additional $\frac{1}{16}$ in. for possible damage to the rim of the hole when it is punched. According to the AISC Manual, the allowable tensile stress in the net section of plate is

$$(\sigma_t)_{allow} = 0.60 \, \sigma_y$$

where σ_y is the yield stress of the connected plate.

$\quad$ The smallest of the three allowable loads is the strength of the joint. The ratio of this strength divided by the strength of a solid plate or member (without holes), expressed in percent, is called the *joint efficiency:*

$$\text{joint efficiency} = \frac{\text{strength of the joint}}{\text{strength of the solid member}} \times 100\% \qquad (13\text{--}4)$$

──────── **EXAMPLE 13–1** ────────────────────────────────

Determine the strength and efficiency of the lap joint shown. The $\frac{7}{8}$-in.-diameter rivets are made of A502–1 steel, and the plates are made of A36 steel with $\sigma_y = 36$ ksi.

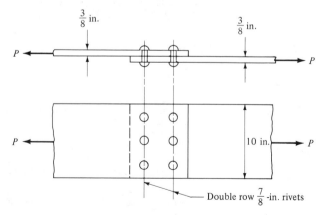

SOLUTION
The cross-sectional area of the rivet is

$$A_s = \tfrac{1}{4}\pi d^2 = \tfrac{1}{4}\pi(\tfrac{7}{8} \text{ in.})^2 = 0.601 \text{ in.}^2$$

With six rivets in single shear, the allowable load in shear calculated from Eq. (13–1) is

$$(P_s)_{allow} = nA_s\tau_{allow} = 6(0.601 \text{ in.}^2)(15 \text{ kips/in.}^2) = 54.1 \text{ kips}$$

With six bearing surfaces on each plate, the allowable load in bearing calculated from Eq. (13–2) is

$$(P_b)_{allow} = n(td)(\sigma_b)_{allow} = 6(\tfrac{3}{8} \times \tfrac{7}{8} \text{ in.}^2)(1.35 \times 36 \text{ kips/in.}^2) = 95.7 \text{ kips}$$

The net width through the critical section is

$$b_{net} = b - n(d + \tfrac{1}{8}) = 10 - 3(\tfrac{7}{8} + \tfrac{1}{8}) = 7.00 \text{ in.}$$

From Eq. (13–3), the allowable load in tension is

$$(P_t)_{allow} = b_{net}t(\sigma_t)_{allow} = (7.00 \text{ in.})(\tfrac{3}{8} \text{ in.})(0.60 \times 36 \text{ kips/in.}^2) = 56.7 \text{ kips}$$

The allowable load in shear of rivets is the smallest of the three; hence the strength of the joint is

$$P = (P_s)_{allow} = 54.1 \text{ kips}$$

From Eq. (13–4), the joint efficiency is

$$\text{joint efficiency} = \frac{\text{strength of the joint}}{\text{strength of the solid plate}} \times 100\%$$

$$= \frac{54.1 \text{ kips}}{(\tfrac{3}{8} \times 10 \text{ in.}^2)(0.60 \times 36 \text{ kips/in.}^2)} \times 100\% = 66.8\%$$ ∎

EXAMPLE 13–2

Determine the strength and efficiency of the butt joint shown in Figs. (a) and (b). The $\tfrac{3}{4}$-in.-diameter rivets are made of A502–2 steel, and the plates are made of A441 steel with $\sigma_y = 50$ ksi.

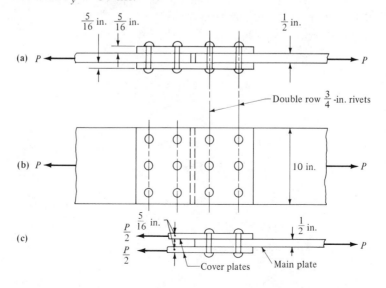

$\tfrac{5}{16}$ in. $\tfrac{5}{16}$ in. $\tfrac{1}{2}$ in.

(a) P P

Double row $\tfrac{3}{4}$-in. rivets

(b) P 10 in. P

$\tfrac{5}{16}$ in. $\tfrac{1}{2}$ in.

(c) $\dfrac{P}{2}$ $\dfrac{P}{2}$ P

Cover plates Main plate

SOLUTION

The load is transmitted from the main plate at each side by six rivets to the cover plates, as shown in Fig. (c). The six rivets are in double shear; this makes the total number of rivet sections under shear stress equal to 12. The bearing on the main plate is more critical than the bearing on the cover plates, since the combined thickness of the cover plate is greater than that of the main plate. Thus

$$A_s = \tfrac{1}{4}\pi d^2 = \tfrac{1}{4}\pi(\tfrac{3}{4} \text{ in.})^2 = 0.442 \text{ in.}^2$$

$$(P_s)_{\text{allow}} = 6(2A_s)\tau_{\text{allow}} = 12(0.442 \text{ in.}^2)(20 \text{ kips/in.}^2) = 106.1 \text{ kips}$$

$$(P_b)_{\text{allow}} = 6(td)(\sigma_b)_{\text{allow}} = 6(\tfrac{1}{2} \times \tfrac{3}{4} \text{ in.})(1.35 \times 50 \text{ kips/in.}^2) = 151.9 \text{ kips}$$

$$b_{\text{net}} = b - n(d + \tfrac{1}{8}) = 10 - 3(\tfrac{3}{4} + \tfrac{1}{8}) = 7.38 \text{ in.}$$

$$(P_t)_{\text{allow}} = b_{\text{net}}t(\sigma_t)_{\text{allow}} = (7.38 \text{ in.})(\tfrac{1}{2} \text{ in.})(0.60 \times 50 \text{ kips/in.}^2) = 110.6 \text{ kips}$$

The strength of the joint is controlled by the allowable shear load of the rivets; thus

$$P = 106.1 \text{ kips}$$

and, by definition,

$$\text{joint efficiency} = \frac{P}{P_{\text{solid plate}}} \times 100\%$$

$$= \frac{106.1 \text{ kips}}{(\tfrac{1}{2} \times 10 \text{ in.}^2)(0.60 \times 50 \text{ kips/in.}^2)} \times 100\% = 70.6\% \quad\blacksquare$$

EXAMPLE 13–3

A tension member composed of a pair of L4 × 3 × $\tfrac{5}{16}$ angles arranged back to back is connected to the $\tfrac{1}{2}$-in.-thick gusset plate at the joint shown. Both the angle and the gusset plate are made of A36 steel. The four $\tfrac{3}{4}$-in.-diameter rivets are made of A502–2 steel. Determine the capacity and the efficiency of the connection.

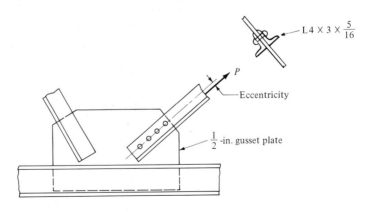

SOLUTION

The net cross-sectional area of the angles is obtained by deducting the area of the rivet holes from the gross area of the angles. Thus for two angles

$$A_{net} = 2[A_{angle} - t(d + \tfrac{1}{8})] = 2[2.09 - \tfrac{5}{16}(\tfrac{3}{4} + \tfrac{1}{8})] = 3.63 \text{ in.}^2$$

The allowable tensile force on the net section is

$$(P_t)_{allow} = A_{net}(\sigma_t)_{allow} = (3.63 \text{ in.}^2)(0.60 \times 36 \text{ kips/in.}^2) = 78.5 \text{ kips}$$

The tensile force in the member is transmitted to the gusset plate through four rivets in double shear; thus

$$A_s = \tfrac{1}{4}\pi d^2 = \tfrac{1}{4}\pi(\tfrac{3}{4} \text{ in.})^2 = 0.442 \text{ in.}^2$$

$$(P_s)_{allow} = 4(2A_s)\,\tau_{allow} = 8(0.442 \text{ in.}^2)(20 \text{ kips/in.}^2) = 70.7 \text{ kips}$$

Each angle has four bearing surfaces; there are eight bearing surfaces in both angles; thus for bearing on the angles,

$$(P_b)_{allow} = 8(td)(\sigma_b)_{allow} = 8(\tfrac{5}{16} \times \tfrac{3}{4} \text{ in.}^2)(1.35 \times 36 \text{ kips/in.}^2) = 91.1 \text{ kips}$$

The rivets bear on the gusset plate at four surfaces; thus for bearing on the gusset plate,

$$(P_b)_{allow} = 4(t_g d)(\sigma_b)_{allow} = 4(\tfrac{1}{2} \times \tfrac{3}{4} \text{ in.}^2)(1.35 \times 36 \text{ kips/in.}^2) = 72.9 \text{ kips}$$

It is seen that the capacity of the connection is governed by the shear of the rivets, which gives

$$P = 70.7 \text{ kips}$$

The efficiency of the connection is

$$\text{joint efficiency} = \frac{P}{2A_{angle}(\sigma_t)_{allow}} \times 100\%$$

$$= \frac{70.7 \text{ kips}}{(2 \times 2.09 \text{ in.}^2)(0.60 \times 36 \text{ kips/in.}^2)} \times 100\% = 78.3\%$$

As indicated in the figure, there is a small eccentricity of the line of action of tensile force P (which acts through the centroid of the angles) from the centerline of the rivets, but this eccentricity is small and is usually ignored. ∎

EXAMPLE 13–4

Find the capacity of a standard AISC connection for joining a W16 × 57 beam to a W12 × 87 column, as shown in the figure. The connection consists of two 9-in.-long clipped L3½ × 3½ × ¼ angles jointed to the web of the beam and the flange of the column by $\frac{3}{4}$-in. rivets made of A502–2 steel. The beam, the column, and the clipped angles are all made of A441 steel with $\sigma_y = 50$ ksi.

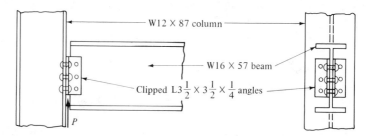

SOLUTION

This connection between a beam and a column is considered to be a simple support. The loads on the beam are transmitted to the column via shear forces in the rivets and bearing forces between the rivets and the bearing plates.

For three rivets in double shear (where the angles are connected to the beam) or six rivets in single shear (where the angles are connected to the column):

$$A_s = \tfrac{1}{4}\pi d^2 = \tfrac{1}{4}\pi(\tfrac{7}{8} \text{ in.})^2 = 0.601 \text{ in.}^2$$

$$(P_s)_{\text{allow}} = 3(2A_s)\tau_{\text{allow}} = 6(0.601 \text{ in.}^2)(20 \text{ kips/in.}^2) = 72.2 \text{ kips}$$

For bearing of three rivets on the web of the W16 × 57 beam,

$$(P_b)_{\text{allow}} = 3(t_w d)(\sigma_b)_{\text{allow}}$$
$$= 3(0.430 \times \tfrac{7}{8} \text{ in.}^2)(1.35 \times 50 \text{ kips/in.}^2) = 76.2 \text{ kips}$$

For bearing of three rivets on two L$3\frac{1}{2}$ × $3\frac{1}{2}$ × $\frac{1}{4}$ angles,

$$(P_b)_{\text{allow}} = 3 \times 2(td)(\sigma_b)_{\text{allow}}$$
$$= 6(\tfrac{1}{4} \times \tfrac{7}{8} \text{ in.}^2)(1.35 \times 50 \text{ kips/in.}^2) = 88.6 \text{ kips}$$

For bearing of six rivets on the flange of the W12 × 87 column,

$$(P_b)_{\text{allow}} = 6(t_f d)(\sigma_b)_{\text{allow}}$$
$$= 6(0.810 \times \tfrac{7}{8} \text{ in.}^2)(1.35 \times 50 \text{ kips/in.}^2) = 287 \text{ kips}$$

The capacity is governed by the shear of rivets; thus the capacity of the connection is

$$P = 72.2 \text{ kips}$$

■

PROBLEMS

13–1 Determine the strength and efficiency of the lap joint shown in Fig. P13–1. The $\frac{3}{4}$-in.-diameter rivets are made of A502–1 steel, and the plates are made of A36 steel with $\sigma_y = 36$ ksi.

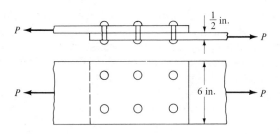

FIGURE P13–1

13–2 Determine the strength and efficiency of the butt joint shown in Fig. P13–2. The $\frac{7}{8}$-in.-diameter rivets are made of A502–2 steel, and the plates are made of A36 steel with $\sigma_y = 36$ ksi.

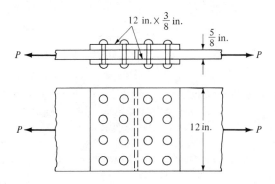

FIGURE P13–2

13–3 A tension member in a roof truss consists of a single L5 $\times$ 3 $\times$ $\frac{3}{8}$ connected to the $\frac{1}{2}$-in. gusset plate shown in Fig. P13–3. The angle and the gusset plate are both of A36 steel. The connection is made by $\frac{7}{8}$-in.-diameter A502–2 rivets. Using the AISC specifications, determine the number of rivets required so that the maximum strength of the member can be developed.

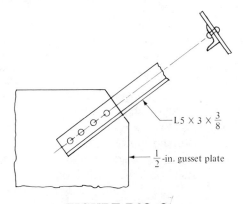

FIGURE P13–3

13–4 A tension member in a bridge truss consisting of a pair of L5 × 3 × $\frac{3}{8}$ angles arranged back to back is connected to the $\frac{5}{8}$-in. gusset plate shown in Fig. P13–4. The angles and gusset plate are made of A36 steel with σ_y = 36 ksi. The four 1-in.-diameter rivets are made of A502–1 steel. Determine the capacity and the efficiency of the connection.

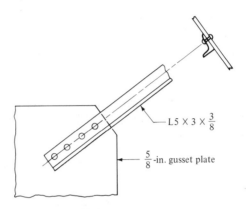

L5 × 3 × $\frac{3}{8}$

$\frac{5}{8}$-in. gusset plate

FIGURE P13–4

13–5 The standard connection of a W18 × 60 beam to a W12 × 87 column is shown in Fig. P13–5. The connection consists of two clipped L4 × 4 × $\frac{3}{8}$ angles. Four $\frac{7}{8}$-in. rivets connect the angles to the web of the beam, and eight rivets of the same size connect the angles to the flange of the column. Determine the capacity of the connection if the beam, column, and the angles are made of A36 steel and the rivets are of A502–1 steel.

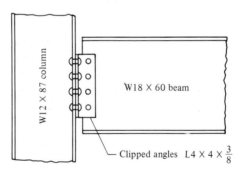

W12 × 87 column

W18 × 60 beam

Clipped angles L4 × 4 × $\frac{3}{8}$

FIGURE P13–5

13–6 A W18 × 50 beam is connected to two W12 × 65 columns by means of the standard beam connection shown in Fig. P13–6. The connection on each side of the beam consists of two clipped L4 × $3\frac{1}{2}$ × $\frac{5}{16}$ angles and twelve $\frac{3}{4}$-in. rivets. Determine the maximum uniform load that the beam can carry if the beam, the column, and the angles are made of A36 steel and the rivets are made of A502–1 steel. (**HINT:** Consider both the strength of the connection and the flexural strength of the beam using the allowable flexural stress of $0.66\sigma_y$, and consider the beam to be simply supported.)

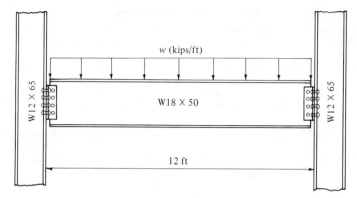

FIGURE P13–6

13–4
HIGH-STRENGTH STEEL BOLTS

Although the use of high-strength steel bolts is a relatively new development, these bolts have already become the leading fastener for connections done in the field. Generally speaking, any joint that can be connected by rivets can be connected by bolts. The design considerations of a bolted joint is similar to the design of a riveted joint.

The high-strength steel bolts used in structural joints are classified as ASTM A325 and A490. They have tensile strengths several times those of ordinary bolts. All high-strength bolts in a joint must be tightened to a specified minimum initial tension equal to 70 percent of the bolt tensile strength. The resulting tension in the bolt develops a reliable clamping force and the load is transmitted essentially by friction between the surfaces of the connected plates, which are pressed tightly together.

Joints connected by high-strength steel bolts are of two types: the friction type and the bearing type. In a friction-type joint the bolt is tightened until it is stressed to the tensile load specified, so that the clamping force is large enough to allow the full load on the joint to be transmitted by friction. This type of connection is used for structures subjected to impact and vibration, where a high factor of safety against slippage is necessary.

In a bearing-type joint, the load is transmitted by the bearing of the bolts against the joined parts. Friction resistance probably shares the load, but is not considered in the design analysis. This type of connection is usually used for structures subjected primarily to a static load; the allowable shear stress for a bearing-type connection is considerably higher because a smaller factor of safety against slippage is used.

For a friction-type connection, no bearing stresses need to be considered, since the joint is not supposed to slip. For a bearing-type connection, the allowable bearing stress on the projected area of the bolts is the same as that for rivets:

$$(\sigma_b)_{\text{allow}} = 1.35\sigma_y$$

where σ_y is the yield stress of the connected part.

The allowable shear stresses specified by the AISC code for A325 and A490 bolts are given in Table 13–1.

**TABLE 13–1   Allowable Shear Stresses
in High-Strength Bolts**

Bolt Material	Friction Type	Bearing Type
A325	15 ksi	22 ksi
A490	20 ksi	32 ksi

━━━━━ **EXAMPLE 13–5** ━━━━━━━━━━━━━━━━━━━━━━━━━━━━━━━━━━━━━━

Rework Example 13–1 if A325 bearing-type high-strength steel bolts $\frac{7}{8}$ in. in diameter are used instead of rivets.

SOLUTION

From Table 13–1, the allowable shear stress for A325 bearing type is 22 ksi. Thus

$$A_s = \tfrac{1}{4}\pi d^2 = \tfrac{1}{4}\pi(\tfrac{7}{8}\text{-in.})^2 = 0.601 \text{ in.}^2$$

$$(P_s)_{\text{allow}} = nA_s\tau_{\text{allow}} = 6(0.601 \text{ in.}^2)(22 \text{ kips/in.}^2) = 79.4 \text{ kips}$$

The allowable tensile force and the allowable bearing force are the same as those calculated in Example 13–1, which gives

$$(P_t)_{\text{allow}} = 56.7 \text{ kips}$$

$$(P_b)_{\text{allow}} = 95.7 \text{ kips}$$

Thus the capacity of the joint is governed by tension of the plate; it is

$$P = 56.7 \text{ kips}$$

and the efficiency of the joint is

$$\text{efficiency} = \frac{P}{bt(\sigma_t)_{\text{allow}}} \times 100\%$$

$$= \frac{56.7 \text{ kips}}{(\tfrac{3}{8} \times 10 \text{ in.}^2)(0.60 \times 36 \text{ kips/in.}^2)} \times 100\% = 70.0\%$$

■

━━━━━ **EXAMPLE 13–6** ━━━━━━━━━━━━━━━━━━━━━━━━━━━━━━━━━━━━━━

Rework Example 13–3 if A490 friction-type high-strength steel bolts $\frac{3}{4}$ in. in diameter are used instead of rivets.

SOLUTION

From Table 13–1, the allowable shear stress for A490 friction-type is 20 ksi. Thus

$$A_s = \tfrac{1}{4}\pi d^2 = \tfrac{1}{4}\pi(\tfrac{3}{4} \text{ in.})^2 = 0.442 \text{ in.}^2$$

$$(P_s)_{\text{allow}} = 4(2A_s)\tau_{\text{allow}} = 8(0.442 \text{ in.}^2)(20 \text{ kips/in.}^2) = 70.7 \text{ kips}$$

The allowable force in tension is the same as that calculated in Example 13–3, which gives

$$(P_t)_{\text{allow}} = 78.5 \text{ kips}$$

Since the joint is not supposed to slip in the friction-type connection, failure due to bearing will not occur, and bearing strength need not be considered.

Thus the capacity of the joint is

$$P = 70.7 \text{ kips}$$

which is the same as that of the riveted connection in Example 13–3. The joint efficiency should also be the same; that is,

$$\text{joint efficiency} = 78.3\%$$

Because the friction-type connection has a high factor of safety against slippage, the connection by high-strength bolts in this example will prove to have a higher fatigue strength and a better resistance to the dynamic loads to which a bridge truss is usually subjected.

PROBLEMS

13–7   Rework Problem 13–1 (on p. 378) if A325 bearing-type high-strength bolts of $\frac{3}{4}$-in. diameter are used instead of rivets.

13–8   Rework Problem 13–2 (on p. 379) if A490 bearing-type high-strength bolts of $\frac{7}{8}$-in. diameter are used instead of rivets.

13–9   Rework Problem 13–3 (on p. 379) if A490 bearing-type high-strength bolts of $\frac{7}{8}$-in. diameter are used instead of rivets.

13–10   Rework Problem 13–4 (on p. 380) if A490 friction-type high-strength bolts of 1-in. diameter are used instead of rivets.

13–11   Rework Problem 13–5 (on p. 380) if A325 bearing-type high-strength bolts of $\frac{7}{8}$-in. diameter are used instead of rivets.

13–12   Rework Problem 13–6 (on p. 380) if A490 friction-type high-strength bolts of $\frac{3}{4}$-in. diameter are used instead of rivets.

*13–5
ECCENTRICALLY LOADED RIVETED
AND BOLTED CONNECTIONS

In the previous sections, the line of action of the applied load passes through the centroid of the connectors; thus the load is assumed to be shared equally by all the connectors. When the load is applied with eccentricity from the centroid of

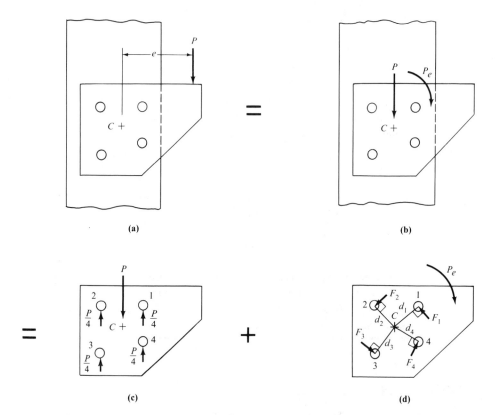

FIGURE 13–3

the connectors, the eccentric load produces direct shear as well as torsion. The rivets are no longer subjected to equal forces. Figure 13–3(a) shows a connection subjected to a load P applied at an eccentricity e from the centroid of the connectors. The eccentric load is equivalent to a direct shear force P and a torque Pe, as shown in Fig. 13–3(b). The direct shear force P is resisted equally by the four connectors; each carries a load of $P/4$, as shown in Fig. 13–3(c). To resist the torque Pe, the resisting force on each connector is proportional to its distance to the centroid of the connectors, and acts in the direction perpendicular to the line joining the centroid and the connector, as shown in Fig. 13–3(d). We have

$$\frac{F_1}{d_1} = \frac{F_2}{d_2} = \frac{F_3}{d_3} = \frac{F_4}{d_4}$$

or

$$F_2 = \frac{F_1 d_2}{d_1} \qquad F_3 = \frac{F_1 d_3}{d_1} \qquad F_4 = \frac{F_1 d_4}{d_1} \tag{a}$$

The sum of the moments of the resisting forces F_1, F_2, F_3, and F_4 must balance the couple Pe. Thus

$$F_1 d_1 + F_2 d_2 + F_3 d_3 + F_4 d_4 = Pe \tag{b}$$

Substituting Eq. (a) in Eq. (b), we get

$$F_1 d_1 + \frac{F_1 d_2^2}{d_1} + \frac{F_1 d_3^2}{d_1} + \frac{F_1 d_4^2}{d_1} = Pe$$

or

$$\frac{F_1}{d_1}(d_1^2 + d_2^2 + d_3^2 + d_4^2) = \frac{F_1}{d_1}\Sigma d^2 = Pe$$

from which

$$F_1 = \frac{Pe}{\Sigma d^2}d_1 = Kd_1$$

Substituting in Eq. (a), we have

$$F_2 = Kd_2 \qquad F_3 = Kd_3 \qquad F_4 = Kd_4$$

or in general, we have

$$F_i = Kd_i \tag{13–5}$$

where

$$K = \frac{Pe}{\Sigma d^2} \tag{13–6}$$

Let the horizontal and vertical components of the distance d_i be represented by x_i and y_i, respectively. Then

$$d_i^2 = x_i^2 + y_i^2$$
$$\Sigma d^2 = \Sigma x^2 + \Sigma y^2$$

and

$$K = \frac{Pe}{\Sigma x^2 + \Sigma y^2} \tag{13–7}$$

Let the horizontal and vertical components of the force F_i on the ith connector be represented by H_i and V_i as shown in Fig. 13–4. The force F_i is perpendicular to the line d_i; hence if the angle between d_i and x_i is θ, then the angle between F_i and V_i is also θ. We have

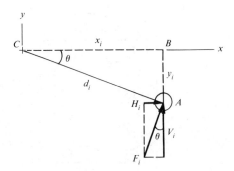

FIGURE 13–4

$$H_i = F_i \sin \theta = Kd_i \sin \theta$$
$$V_i = F_i \cos \theta = Kd_i \cos \theta$$

From triangle ABC we have

$$d_i \sin \theta = y_i \qquad d_i \cos \theta = x_i$$

Therefore,

$$H_i = Ky_i \qquad V_i = Kx_i \tag{13–8}$$

───── **EXAMPLE 13–7** ─────────────────────────────────

Determine the required size of the A502–1 rivets and the required thickness of the gusset plate in the connection shown, which connects a beam seat to a W14 × 90 column. The column and the gusset plate are of A36 steel.

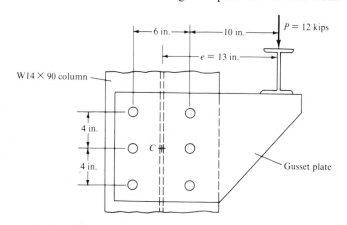

SOLUTION

The eccentric load P is equivalent to a direct shear force $P = 12$ kips and a torque $Pe = (12 \text{ kips})(13 \text{ in.}) = 156$ kip-in. Due to the direct shear load, the resisting force on each rivet is

$$\frac{P}{6} = \frac{12 \text{ kips}}{6} = 2 \text{ kips}$$

acting upward, as shown in Fig. (a).

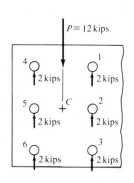

(a)

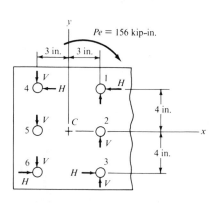

(b)

Due to the torque Pe, the horizontal and vertical components of the resisting force on each rivet are shown in Fig. (b). To find these components, we first compute

$$\Sigma x^2 + \Sigma y^2 = 6(3 \text{ in.})^2 + 4(4 \text{ in.})^2 = 118 \text{ in.}^2$$

$$K = \frac{Pe}{\Sigma x^2 + \Sigma y^2} = \frac{156 \text{ kip-in.}}{118 \text{ in.}^2} = 1.322 \text{ kips/in.}$$

The y-coordinates of rivets 2 and 5 are zero; hence the horizontal components of the two rivets due to the torque is zero. The horizontal components of other rivets are equal, because these rivets have the same absolute value of y. The absolute value of H is

$$H = Ky = (1.322 \text{ kips/in.})(4 \text{ in.}) = 5.29 \text{ kips}$$

The vertical components of all the rivets are equal, because they all have the same absolute value of x. The absolute value of V is

$$V = Kx = (1.322 \text{ kips/in.})(3 \text{ in.}) = 3.97 \text{ kips}$$

The vertical components of rivets 1 and 3 due to direct shear force and due to torque are both upward, as shown in Figs. (a) and (b); hence these two rivets are most critically loaded. Resisting forces of the two rivets are, respectively,

$$F_1 = F_3 = \sqrt{(5.29 \text{ kips})^2 + (2 \text{ kips} + 3.97 \text{ kips})^2} = 7.98 \text{ kips}$$

The required diameter of the rivets is determined by limiting the shear stress in the rivets to the allowable shear stress of 15 ksi for A501-1 steel. Thus

$$\tau = \frac{F}{A} = \frac{7.98 \text{ kips}}{\frac{1}{4}\pi d^2} = \frac{10.2 \text{ kips}}{d^2} \le 15 \text{ ksi}$$

$$d \ge \sqrt{\frac{10.2 \text{ kips}}{15 \text{ kips/in.}^2}} = 0.823 \text{ in.}$$

$$d_{\text{req}} = 0.823 \text{ in.}$$

Rivets of $\frac{7}{8}$-in. (= 0.875 in.) diameter may be used.

The required thickness of the gusset plate is determined by limiting the bearing stress in the gusset plate to the allowable value of

$$(\sigma_b)_{\text{allow}} = 1.35\sigma_y = 1.35(36 \text{ ksi}) = 48.6 \text{ ksi}$$

Thus

$$\sigma_b = \frac{F}{td} = \frac{7.98 \text{ kips}}{t(0.875 \text{ in.})} \leq 48.6 \text{ ksi}$$

$$t \geq \frac{7.98 \text{ kips}}{(0.875 \text{ in.})(48.6 \text{ kips/in.}^2)} = 0.188 \text{ in.}$$

$$t_{\text{req}} = 0.188 \text{ in.}$$

A $\frac{1}{4}$-in.-thick gusset plate may be used. ∎

PROBLEMS

13–13 A steel plate is attached to a machine with five $\frac{7}{8}$-in. rivets as shown in Fig. P13–13. If the allowable shear stress is 15 ksi, determine the maximum load P that can be applied to the plate.

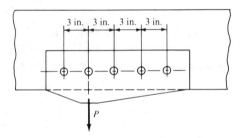

FIGURE P13–13

13–14 Determine the required size of the bolts in Fig. P13–14 if the allowable shear stress is 20 ksi.

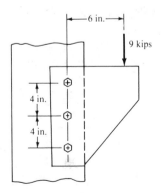

FIGURE P13–14

13–15 The bracket shown in Fig. P13–15 is connected to the flange of a steel column by four $\frac{3}{4}$-in. A490 friction-type bolts with an allowable shear stress of 20 ksi. Determine the maximum allowable load P that can be applied to the bracket.

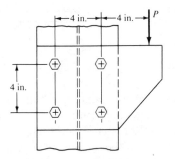

FIGURE P13–15

13–16 The bracket is connected to the column by six $\frac{3}{4}$-in. rivets arranged as shown in Fig. P13–16. Determine the maximum shear stress in the most critically loaded rivet.

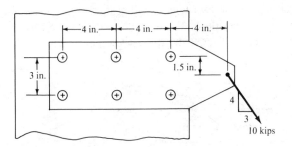

FIGURE P13–16

13–17 In Fig. P13–17, the W12 × 65 column and the $\frac{1}{2}$-in. gusset plate are both made of A36 steel. The connection is made of eight $\frac{7}{8}$-in. A325 bearing-type bolts. Determine the maximum load P that can be applied to the bracket.

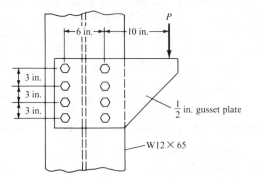

FIGURE P13–17

13–6
WELDED CONNECTIONS

Welding is the process of connecting metallic parts by heating the surfaces to a plastic or fluid state and allowing the melted parts to join together. Welding is a very widely used method for joining metallic parts, both in the shop and in the field.

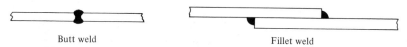

Butt weld   Fillet weld

FIGURE 13–5

The two main type of welds are butt welds and fillet welds, as shown in Fig. 13–5. Most structural connections are made with fillet welds. The main problem in a butt weld is that it is difficult to get the pieces to fit together in the field. For the fillet weld the amount of overlap can be freely adjusted.

The AISC specifications allow the same tensile stress in the butt weld as in the base metal if the member is subjected to static loads. Thus the strength of a butt weld is simply the strength of the weakest component being connected.

Fillet welds are designated by the size of the legs, as shown in Fig. 13–6. Although the weld surface is usually curved, the smallest inscribed triangle, shown by dashed lines, is considered to be the theoretical dimensions of the fillet weld. The corner of the two legs is called the *root*. The smallest distance from the root to the opposite side of the triangle is called the *throat* of a fillet weld. For fillet welds with equal legs, the throat is equal to leg $\times$ sin 45° = 0.707 (size).

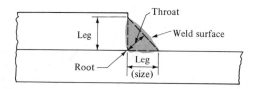

FIGURE 13–6

Tests have shown that failure of a fillet weld commonly occurs through the throat due to shear stress. Therefore, the strength of a fillet weld, regardless of the direction of the applied load, is equal to the cross-sectional area at the throat multiplied by the allowable shear stress for the weld metal. According to the AISC specifications, the allowable shear stress is 0.3 times the electrode tensile strength. For example, the allowable shear stress of a fillet weld with E70 electrodes, which have a tensile strength of 70 ksi, is 0.3 $\times$ 70 = 21 ksi. The allowable force q per inch of length of a fillet weld with E70 electrodes is then

$$q = 0.707(\text{size, in.})(21 \text{ kips/in.}^2) = 14.8(\text{size})(\text{kips/in.}) \qquad (13\text{–}9)$$

───────── **EXAMPLE 13–8** ─────────────────────────────────

Determine the total length required for a $\frac{5}{16}$-in. fillet weld used to connect two plates shown. The joint is to develop the full strength of the plate 6 in. by $\frac{3}{8}$ in. made of A36 steel ($\sigma_y = 36$ ksi). Use the AISC specification and E70 electrodes.

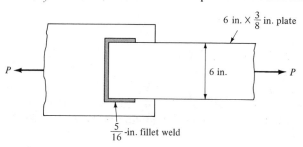

SOLUTION

The capacity of a plate 6 in. by $\frac{3}{8}$ in. is

$$P = A(\sigma_t)_{\text{allow}} = (6 \times \tfrac{3}{8} \text{ in.}^2)(0.6 \times 36 \text{ kips/in.}^2) = 48.6 \text{ kips}$$

From Eq. (13–9), the allowable load per inch of $\frac{5}{16}$-in. fillet weld is

$$q = 14.8(\text{size}) = 14.8(\tfrac{5}{16}) = 4.63 \text{ kips/in.}$$

The length of weld required is

$$L = \frac{P}{q} = \frac{48.6 \text{ kips}}{4.63 \text{ kips/in.}} = 10.5 \text{ in.}$$

This required length can be provided by either one of the following arrangements:

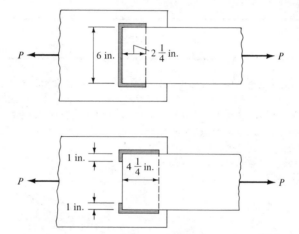

The end returns in the second arrangement are required by the AISC to reduce the effect of stress concentration. The end returns are included as part of the effective length of the weld.

■

EXAMPLE 13–9

Determine the size of the fillet weld necessary for the lap joint connecting two 10 in. by $\frac{3}{4}$ in. plates, as shown. The plates are A36 steel, and the joint is to develop the full strength of the plates. The fillet weld is along the full width of the plates. Use the AISC specification and E70 electrodes.

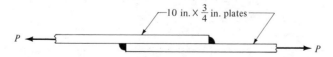

SOLUTION

Strength of plate:

$$P = (10 \times \tfrac{3}{4} \text{ in.}^2)(0.6 \times 36 \text{ kips/in.}^2) = 162 \text{ kips}$$

Allowable load of weld:

$$q = 14.8(\text{size})(\text{kips/in.})$$

Length of weld:

$$L = 2 \times 10 = 20 \text{ in.}$$

To develop the full strength of plate, Lq must be equal to or greater than P. Thus

$$Lq = (20 \text{ in.})[14.8(\text{size}) (\text{kips/in.})] \geq 162 \text{ kips}$$

from which

$$\text{size} \geq 0.547 \text{ in.}$$

Use $\frac{5}{8}$-in. (0.625-in.) fillet weld.

███

━━━━ EXAMPLE 13–10 ━━━━

The long leg of a steel angle L4 $\times$ $3\frac{1}{2}$ $\times$ $\frac{1}{2}$ is connected by a $\frac{7}{16}$ in. fillet weld to the bottom chord of a truss at the joint shown. Determine (a) the total length of weld required if the full strength of the angle is developed, (b) the lengths L_1 and L_2 so that the centroid of the weld is coincident with the centroid of the angle and thus avoids eccentric loading of the member. Use the AISC specification, A36 steel, and E70 electrodes.

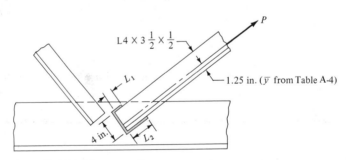

SOLUTION

Strength of the angle:

$$P = A(0.60\sigma_y) = (3.50 \text{ in.}^2)(0.60 \times 36 \text{ kips/in.}^2) = 75.6 \text{ kips}$$

Allowable load of weld:

$$q = 14.8(\text{size}) = 14.8(\tfrac{7}{16}) = 6.48 \text{ kips/in.}$$

(a) The total required length of weld is

$$L = \frac{P}{q} = \frac{75.6}{6.48} = 11.7 \text{ in.} \qquad \text{say 12 in.}$$

(b) If the centroid of the fillet weld is coincident with the centroid of the angle, the sum of the first moment of the weld segments about a reference line must be equal to the first moment of the total length L located at the centroid of the angle about the same line. Taking the reference line along L_2, we have

$$L_1(4) + 4(2) + L_2(0) = L(1.25)$$

from which

$$L_1 = \frac{1.25L - 4(2)}{4} = \frac{1.25 \times 12 - 8}{4} = 1.75 \text{ in.} = 1\tfrac{3}{4} \text{ in.}$$

Then

$$L_2 = L - L_1 - 4 = 12 - 1.75 - 4 = 6.25 \text{ in.} = 6\tfrac{1}{4} \text{ in.}$$

PROBLEMS

For the following problems, use the AISC specifications, A36 steel, and E70 electrodes.

13–18 Determine the strength and efficiency of the welded connection shown in Fig. P13–18. The $\tfrac{1}{2}$-in. fillet weld is along the entire width of the plates.

FIGURE P13–18

13–19 In Problem 13–18, determine the size of the fillet weld (to the nearest $\tfrac{1}{16}$ in.) if the joint is to develop full strength of the plates.

13–20 Determine the strength and efficiency of the welded connection shown in Fig. P13–20. The $\tfrac{1}{2}$-in. fillet weld is along the entire width of the plates.

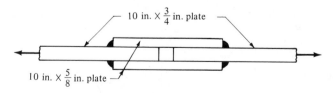

FIGURE P13–20

13–21 In Fig. P13–20, determine the required size of the fillet weld (to the nearest $\tfrac{1}{16}$ in.) if the joint is to develop full strength of the plates.

13–22 Determine the length L_1 required for a $\frac{5}{16}$-in. fillet weld in the connection shown in Fig. P13–22. The joint is to develop the full strength of the plate.

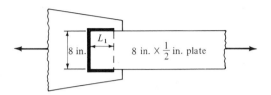

FIGURE P13–22

13–23 Rework Example 13–10 for an L6 × 4 × $\frac{3}{4}$ angle using $\frac{1}{2}$-in. fillet welds.

13–24 The structural joint shown in Fig. P13–24 is fillet-welded and it has a strength equal to the full strength of the angle. Determine the proper length L_1 and L_2 if the centroid of the weld must be coincident with the centroid of the angle to avoid eccentric loading. The end returns are required by the AISC code to reduce the effect of stress concentrations, and the end returns can be included as part of the effective length of the weld.

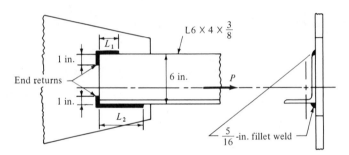

FIGURE P13–24

Appendix Tables ────────────

Acknowledgment: Data for Tables A–1 through A–5 are taken from the 8th edition (copyright 1980) of AISC Manual of Steel Construction and are reproduced by permission of the American Institute of Steel Construction, Inc.

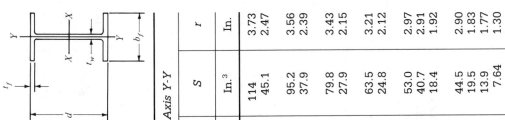

TABLE A–1 American Wide-Flange Steel Beams (W Shapes): Design Properties (Abridged List)

Designation	Area, A	Depth, d	Web Thickness, t_w	Flange Width, b_f	Flange Thickness, t_f	Axis X-X I	Axis X-X S	Axis X-X r	Axis Y-Y I	Axis Y-Y S	Axis Y-Y r
	In.²	In.	In.	In.	In.	In.⁴	In.³	In.	In.⁴	In.³	In.
W36 × 230	67.6	35.90	0.760	16.470	1.260	15000	837	14.9	940	114	3.73
× 150	44.2	35.85	0.625	11.975	0.940	9040	504	14.3	270	45.1	2.47
W33 × 201	59.1	33.68	0.715	15.745	1.150	11500	684	14.0	749	95.2	3.56
× 130	38.3	33.09	0.580	11.510	0.855	6710	406	13.2	218	37.9	2.39
W30 × 173	50.8	30.44	0.655	14.985	1.065	8200	539	12.7	598	79.8	3.43
× 108	31.7	29.83	0.545	10.475	0.760	4470	299	11.9	146	27.9	2.15
W27 × 146	42.9	27.38	0.605	13.965	0.975	5630	411	11.4	443	63.5	3.21
× 94	27.7	26.92	0.490	9.990	0.745	3270	243	10.9	124	24.8	2.12
W24 × 131	38.5	24.48	0.605	12.855	0.960	4020	329	10.2	340	53.0	2.97
× 104	30.6	24.06	0.500	12.750	0.750	3100	258	10.1	259	40.7	2.91
× 76	22.4	23.92	0.440	8.990	0.680	2100	176	9.69	82.5	18.4	1.92
W21 × 111	32.7	21.51	0.550	12.340	0.875	2670	249	9.05	274	44.5	2.90
× 83	24.3	21.43	0.515	8.355	0.835	1830	171	8.67	81.4	19.5	1.83
× 62	18.3	20.99	0.400	8.240	0.615	1330	127	8.54	57.5	13.9	1.77
× 50	14.7	20.83	0.380	6.530	0.535	984	94.5	8.18	24.9	7.64	1.30

TABLE A-1 (continued)

Designation	Area, A In.²	Depth, d In.	Web Thickness, t_w In.	Flange Width, b_f In.	Flange Thickness t_f In.	Axis X-X I In.⁴	Axis X-X S In.³	Axis X-X r In.	Axis Y-Y I In.⁴	Axis Y-Y S In.³	Axis Y-Y r In.
W18×97	28.5	18.59	0.535	11.145	0.870	1750	188	7.82	201	36.1	2.65
×60	17.6	18.24	0.415	7.555	0.695	984	108	7.47	50.1	13.3	1.69
×50	14.7	17.99	0.355	7.495	0.570	800	88.9	7.38	40.1	10.7	1.65
×46	13.5	18.06	0.360	6.060	0.605	712	78.8	7.25	22.5	7.43	1.29
×35	10.3	17.70	0.300	6.000	0.425	510	57.6	7.04	15.3	5.12	1.22
W16×100	29.4	16.97	0.585	10.425	0.985	1490	175	7.10	186	35.7	2.51
×89	26.2	16.75	0.525	10.365	0.875	1300	155	7.05	163	31.4	2.49
×57	16.8	16.43	0.430	7.120	0.715	758	92.2	6.72	43.1	12.1	1.60
×50	14.7	16.26	0.380	7.070	0.630	659	81.0	6.68	37.2	10.5	1.59
×36	10.6	15.86	0.295	6.985	0.430	448	56.5	6.51	24.5	7.00	1.52
×26	7.68	15.69	0.250	5.500	0.345	301	38.4	6.26	9.59	3.49	1.12
W14×342	101.0	17.54	1.540	16.360	2.470	4900	559	6.98	1810	221	4.24
×132	38.8	14.66	0.645	14.725	1.030	1530	209	6.28	548	74.5	3.76
×90	26.5	14.02	0.440	14.520	0.710	999	143	6.14	362	49.9	3.70
×82	24.1	14.31	0.510	10.130	0.855	882	123	6.05	148	29.3	2.48
×74	21.8	14.17	0.450	10.070	0.785	796	112	6.04	134	26.6	2.48
×68	20.0	14.04	0.415	10.035	0.720	723	103	6.01	121	24.2	2.46
×61	17.9	13.89	0.375	9.995	0.645	640	92.2	5.98	107	21.5	2.45
×53	15.6	13.92	0.370	8.060	0.660	541	77.8	5.89	57.7	14.3	1.92
×43	12.6	13.66	0.305	7.995	0.530	428	62.7	5.82	45.2	11.3	1.89
×38	11.2	14.10	0.310	6.770	0.515	385	54.6	5.87	26.7	7.88	1.55
×34	10.0	13.98	0.285	6.745	0.455	340	48.6	5.83	23.3	6.91	1.53
×30	8.85	13.84	0.270	6.730	0.385	291	42.0	5.73	19.6	5.82	1.49

Designation	Area	d	t_w	b_f	t_f	I_x	S_x	r_x	I_y	S_y	r_y
W12 × 87	25.6	12.53	0.515	12.125	0.810	740	118	5.38	241	39.7	3.07
× 65	19.1	12.12	0.390	12.000	0.605	533	87.9	5.28	174	29.1	3.02
× 53	15.6	12.06	0.345	9.995	0.575	425	70.6	5.23	95.8	19.2	2.48
× 40	11.8	11.94	0.295	8.005	0.515	310	51.9	5.13	44.1	11.0	1.93
× 35	10.3	12.50	0.300	6.560	0.520	285	45.6	5.25	24.5	7.47	1.54
× 30	8.79	12.34	0.260	6.520	0.440	238	38.6	5.21	20.3	6.24	1.52
× 25	7.65	12.22	0.230	6.490	0.380	204	33.4	5.17	17.3	5.34	1.51
W10 ×112	32.9	11.36	0.755	10.415	1.250	716	126	4.66	236	45.3	2.68
×100	29.4	11.10	0.680	10.340	1.120	623	112	4.60	207	40.0	2.65
× 88	25.9	10.84	0.605	10.265	0.990	534	98.5	4.54	179	34.8	2.63
× 77	22.6	10.60	0.530	10.190	0.870	455	85.9	4.49	154	30.1	2.60
× 60	17.6	10.22	0.420	10.080	0.680	341	66.7	4.39	116	23.0	2.57
× 49	14.4	9.98	0.340	10.000	0.560	272	54.6	4.35	93.4	18.7	2.54
× 45	13.3	10.10	0.350	8.020	0.620	248	49.1	4.32	53.4	13.3	2.01
× 39	11.5	9.92	0.315	7.985	0.530	209	42.1	4.27	45.0	11.3	1.98
× 33	9.71	9.73	0.290	7.960	0.435	170	35.0	4.19	36.6	9.20	1.94
× 30	8.84	10.47	0.300	5.810	0.510	170	32.4	4.38	16.7	5.75	1.37
× 22	6.49	10.17	0.240	5.750	0.360	118	23.2	4.27	11.4	3.97	1.33
W 8 × 67	19.7	9.00	0.570	8.280	0.935	272	60.4	3.72	88.6	21.4	2.12
× 58	17.1	8.75	0.510	8.220	0.810	228	52.0	3.65	75.1	18.3	2.10
× 48	14.1	8.50	0.400	8.110	0.685	184	43.3	3.61	60.9	15.0	2.08
× 40	11.7	8.25	0.360	8.070	0.560	146	35.5	3.53	49.1	12.2	2.04
× 35	10.3	8.12	0.310	8.020	0.495	127	31.2	3.51	42.6	10.6	2.03
× 31	9.13	8.00	0.285	7.995	0.435	110	27.5	3.47	37.1	9.27	2.02
× 28	8.25	8.06	0.285	6.535	0.465	98.0	24.3	3.45	21.7	6.63	1.62
× 24	7.08	7.93	0.245	6.495	0.400	82.8	20.9	3.42	18.3	5.63	1.61
× 21	6.16	8.28	0.250	5.270	0.400	75.3	18.2	3.49	9.77	3.71	1.26
× 18	5.26	8.14	0.230	5.250	0.330	61.9	15.2	3.43	7.97	3.04	1.23

TABLE A–2 American Standard Steel I-Beams (S Shapes): Design Properties

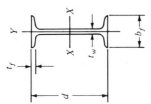

Designation	Area, A	Depth, d	Web Thickness, t_w	Flange Width, b_f	Flange Thickness, t_f	Axis X-X I	Axis X-X S	Axis X-X r	Axis Y-Y I	Axis Y-Y S	Axis Y-Y r
	In.²	In.	In.	In.	In.	In.⁴	In.³	In.	In.⁴	In.³	In.
S24 × 121	35.6	24.50	0.800	8.050	1.090	3160	258	9.43	83.3	20.7	1.53
× 106	31.2	24.50	0.620	7.870	1.090	2940	240	9.71	77.1	19.6	1.57
S24 × 100	29.3	24.00	0.745	7.245	0.870	2390	199	9.02	47.7	13.2	1.27
× 90	26.5	24.00	0.625	7.125	0.870	2250	187	9.21	44.9	12.6	1.30
× 80	23.5	24.00	0.500	7.000	0.870	2100	175	9.47	42.2	12.1	1.34
S20 × 96	28.2	20.30	0.800	7.200	0.920	1670	165	7.71	50.2	13.9	1.33
× 86	25.3	20.30	0.660	7.060	0.920	1580	155	7.89	46.8	13.3	1.36
S20 × 75	22.0	20.00	0.635	6.385	0.795	1280	128	7.62	29.8	9.32	1.16
× 66	19.4	20.00	0.505	6.255	0.795	1190	119	7.83	27.7	8.85	1.19
S18 × 70	20.6	18.00	0.711	6.251	0.691	926	103	6.71	24.1	7.72	1.08
× 54.7	16.1	18.00	0.461	6.001	0.691	804	89.4	7.07	20.8	6.94	1.14

Designation											
S15×50	14.7	15.00	0.550	5.640	0.622	486	64.8	5.75	15.7	5.57	1.03
×42.9	12.6	15.00	0.411	5.501	0.622	447	59.6	5.95	14.4	5.23	1.07
S12×50	14.7	12.00	0.687	5.477	0.659	305	50.8	4.55	15.7	5.74	1.03
×40.8	12.0	12.00	0.462	5.252	0.659	272	45.4	4.77	13.6	5.16	1.06
S12×35	10.3	12.00	0.428	5.078	0.544	229	38.2	4.72	9.87	3.89	0.980
×31.8	9.35	12.00	0.350	5.000	0.544	218	36.4	4.83	9.36	3.74	1.00
S10×35	10.3	10.00	0.594	4.944	0.491	147	29.4	3.78	8.36	3.38	0.901
×25.4	7.46	10.00	0.311	4.661	0.491	124	24.7	4.07	6.79	2.91	0.954
S 8×23	6.77	8.00	0.441	4.171	0.426	64.9	16.2	3.10	4.31	2.07	0.798
×18.4	5.41	8.00	0.271	4.001	0.426	57.6	14.4	3.26	3.73	1.86	0.831
S 7×20	5.88	7.00	0.450	3.860	0.392	42.4	12.1	2.69	3.17	1.64	0.734
×15.3	4.50	7.00	0.252	3.662	0.392	36.7	10.5	2.86	2.64	1.44	0.766
S 6×17.25	5.07	6.00	0.465	3.565	0.359	26.3	8.77	2.28	2.31	1.30	0.675
×12.5	3.67	6.00	0.232	3.332	0.359	22.1	7.37	2.45	1.82	1.09	0.705
S 5×14.75	4.34	5.00	0.494	3.284	0.326	15.2	6.09	1.87	1.67	1.01	0.620
×10	2.94	5.00	0.214	3.004	0.326	12.3	4.92	2.05	1.22	0.809	0.643
S 4×9.5	2.79	4.00	0.326	2.796	0.293	5.79	3.39	1.56	0.903	0.646	0.569
×7.7	2.26	4.00	0.193	2.663	0.293	6.08	3.04	1.64	0.764	0.574	0.581
S 3×7.5	2.21	3.00	0.349	2.509	0.260	2.93	1.95	1.15	0.586	0.468	0.516
×5.7	1.67	3.00	0.170	2.330	0.260	2.52	1.68	1.23	0.455	0.390	0.522

TABLE A-3   American Standard Steel Channels: Design Properties

Designation	Area, A In.^{2}	Depth, d In.	Web Thickness, t_w In.	Flange Width, b_f In.	Flange Average Thickness, t_f In.	Axis X-X I In.^{4}	S In.^{3}	r In.	Axis Y-Y I In.^{4}	S In.^{3}	r In.	$\bar{x}$ In.
C15×50	14.7	15.00	0.716	3.716	0.650	404	53.8	5.24	11.0	3.78	0.867	0.798
×40	11.8	15.00	0.520	3.520	0.650	349	46.5	5.44	9.23	3.37	0.886	0.777
×33.9	9.96	15.00	0.400	3.400	0.650	315	42.0	5.62	8.13	3.11	0.904	0.787
C12×30	8.82	12.00	0.510	3.170	0.501	162	27.0	4.29	5.14	2.06	0.763	0.674
×25	7.35	12.00	0.387	3.047	0.501	144	24.1	4.43	4.47	1.88	0.780	0.674
×20.7	6.09	12.00	0.282	2.942	0.501	129	21.5	4.61	3.88	1.73	0.799	0.698
C10×30	8.82	10.00	0.673	3.033	0.436	103	20.7	3.42	3.94	1.65	0.669	0.649
×25	7.35	10.00	0.526	2.886	0.436	91.2	18.2	3.52	3.36	1.48	0.676	0.617
×20	5.88	10.00	0.379	2.739	0.436	78.9	15.8	3.66	2.81	1.32	0.692	0.606
×15.3	4.49	10.00	0.240	2.600	0.436	67.4	13.5	3.87	2.28	1.16	0.713	0.634

C 9 ×20	5.88	9.00	0.448	2.648	0.413	60.9	13.5	3.22	2.42	1.17	0.642	0.583
×15	4.41	9.00	0.285	2.485	0.413	51.0	11.3	3.40	1.93	1.01	0.661	0.586
×13.4	3.94	9.00	0.233	2.433	0.413	47.9	10.6	3.48	1.76	0.962	0.669	0.601
C 8 ×18.75	5.51	8.00	0.487	2.527	0.390	44.0	11.0	2.82	1.98	1.01	0.599	0.565
×13.75	4.04	8.00	0.303	2.343	0.390	36.1	9.03	2.99	1.53	0.854	0.615	0.553
×11.5	3.38	8.00	0.220	2.260	0.390	32.6	8.14	3.11	1.32	0.781	0.625	0.571
C 7 ×14.75	4.33	7.00	0.419	2.299	0.366	27.2	7.78	2.51	1.38	0.779	0.564	0.532
×12.25	3.60	7.00	0.314	2.194	0.366	24.2	6.93	2.60	1.17	0.703	0.571	0.525
×9.8	2.87	7.00	0.210	2.090	0.366	21.3	6.08	2.72	0.968	0.625	0.581	0.540
C 6 ×13	3.83	6.00	0.437	2.157	0.343	17.4	5.80	2.13	1.05	0.642	0.525	0.514
×10.5	3.09	6.00	0.314	2.034	0.343	15.2	5.06	2.22	0.866	0.564	0.529	0.499
×8.2	2.40	6.00	0.200	1.920	0.343	13.1	4.38	2.34	0.693	0.492	0.537	0.511
C 5 ×9	2.64	5.00	0.325	1.885	0.320	8.90	3.56	1.83	0.632	0.450	0.489	0.478
×6.7	1.97	5.00	0.190	1.750	0.320	7.49	3.00	1.95	0.479	0.378	0.493	0.484
C 4 ×7.25	2.13	4.00	0.321	1.721	0.296	4.59	2.29	1.47	0.433	0.343	0.450	0.459
×5.4	1.59	4.00	0.184	1.584	0.296	3.85	1.93	1.56	0.319	0.283	0.449	0.457
C 3 ×6	1.76	3.00	0.356	1.596	0.273	2.07	1.38	1.08	0.305	0.268	0.416	0.455
×5	1.47	3.00	0.258	1.498	0.273	1.85	1.24	1.12	0.247	0.233	0.410	0.438
×4.1	1.21	3.00	0.170	1.410	0.273	1.66	1.10	1.17	0.197	0.202	0.404	0.436

TABLE A–4 Steel Angles with Equal Legs and Unequal Legs: Design Properties

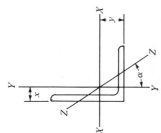

Size and Thickness		Weight per Foot	Area	Axis X-X				Axis Y-Y					Axis Z-Z	
				I	*S*	*r*	*y*	*I*	*S*	*r*	*x*	*r*	Tan *α*
In.		lb	In.²	In.⁴	In.³	In.	In.	In.⁴	In.³	In.	In.	In.	
L 8 ×8 ×1⅛		56.9	16.7	98.0	17.5	2.42	2.41	98.0	17.5	2.42	2.41	1.56	1.000
	1	51.0	15.0	89.0	15.8	2.44	2.37	89.0	15.8	2.44	2.37	1.56	1.000
	⅞	45.0	13.2	79.6	14.0	2.45	2.32	79.6	14.0	2.45	2.32	1.57	1.000
	¾	38.9	11.4	69.7	12.2	2.47	2.28	69.7	12.2	2.47	2.28	1.58	1.000
	⅝	32.7	9.61	59.4	10.3	2.49	2.23	59.4	10.3	2.49	2.23	1.58	1.000
	½	26.4	7.75	48.6	8.36	2.50	2.19	48.6	8.36	2.50	2.19	1.59	1.000
L 8 ×6 ×1		44.2	13.0	80.8	15.1	2.49	2.65	38.8	8.92	1.73	1.65	1.28	0.543
	¾	33.8	9.94	63.4	11.7	2.53	2.56	30.7	6.92	1.76	1.56	1.29	0.551
	½	23.0	6.75	44.3	8.02	2.56	2.47	21.7	4.79	1.79	1.47	1.30	0.558
L 8 ×4 ×1		37.4	11.0	69.6	14.1	2.52	3.05	11.6	3.94	1.03	1.05	0.846	0.247
	¾	28.7	8.44	54.9	10.9	2.55	2.95	9.36	3.07	1.05	0.953	0.852	0.258
	½	19.6	5.75	38.5	7.49	2.59	2.86	6.74	2.15	1.08	0.859	0.865	0.267
L 7 ×4 ×	¾	26.2	7.69	37.8	8.42	2.22	2.51	9.05	3.03	1.09	1.01	0.860	0.324
	½	17.9	5.25	26.7	5.81	2.25	2.42	6.53	2.12	1.11	0.917	0.872	0.335
	⅜	13.6	3.98	20.6	4.44	2.27	2.37	5.10	1.63	1.13	0.870	0.880	0.340

Size and Thickness	Wt per ft (lb)	Area (in²)	I x-x (in⁴)	S x-x (in³)	r x-x (in)	y (in)	I y-y (in⁴)	S y-y (in³)	r y-y (in)	x (in)	r z-z (in)	tan α
L 6 ×6 ×1	37.4	11.0	35.5	8.57	1.80	1.86	35.5	8.57	1.80	1.86	1.17	1.000
7/8	33.1	9.73	31.9	7.63	1.81	1.82	31.9	7.63	1.81	1.82	1.17	1.000
3/4	28.7	8.44	28.2	6.66	1.83	1.78	28.2	6.66	1.83	1.78	1.17	1.000
5/8	24.2	7.11	24.2	5.66	1.84	1.73	24.2	5.66	1.84	1.73	1.18	1.000
1/2	19.6	5.75	19.9	4.61	1.86	1.68	19.9	4.61	1.86	1.68	1.18	1.000
3/8	14.9	4.36	15.4	3.53	1.88	1.64	15.4	3.53	1.88	1.64	1.19	1.000
L 6 ×4 × 3/4	23.6	6.94	24.5	6.25	1.88	2.08	8.68	2.97	1.12	1.08	0.860	0.428
5/8	20.0	5.86	21.1	5.31	1.90	2.03	7.52	2.54	1.13	1.03	0.864	0.435
1/2	16.2	4.75	17.4	4.33	1.91	1.99	6.27	2.08	1.15	0.987	0.870	0.440
3/8	12.3	3.61	13.5	3.32	1.93	1.94	4.90	1.60	1.17	0.941	0.877	0.446
L 6 ×3½× 3/8	11.7	3.42	12.9	3.24	1.94	2.04	3.34	1.23	0.988	0.787	0.767	0.350
5/16	9.8	2.87	10.9	2.73	1.95	2.01	2.85	1.04	0.996	0.763	0.772	0.352
L 5 ×5 × 7/8	27.2	7.98	17.8	5.17	1.49	1.57	17.8	5.17	1.49	1.57	0.973	1.000
3/4	23.6	6.94	15.7	4.53	1.51	1.52	15.7	4.53	1.51	1.52	0.975	1.000
1/2	16.2	4.75	11.3	3.16	1.54	1.43	11.3	3.16	1.54	1.43	0.983	1.000
3/8	12.3	3.61	8.74	2.42	1.56	1.39	8.74	2.42	1.56	1.39	0.990	1.000
5/16	10.3	3.03	7.42	2.04	1.57	1.37	7.42	2.04	1.57	1.37	0.994	1.000
L 5 ×3½× 3/4	19.8	5.81	13.9	4.28	1.55	1.75	5.55	2.22	0.977	0.996	0.748	0.464
1/2	13.6	4.00	9.99	2.99	1.58	1.66	4.05	1.56	1.01	0.906	0.755	0.479
3/8	10.4	3.05	7.78	2.29	1.60	1.61	3.18	1.21	1.02	0.861	0.762	0.486
5/16	8.7	2.56	6.60	1.94	1.61	1.59	2.72	1.02	1.03	0.838	0.766	0.489
L 5 ×3 × 1/2	12.8	3.75	9.45	2.91	1.59	1.75	2.58	1.15	0.829	0.750	0.648	0.357
3/8	9.8	2.86	7.37	2.24	1.61	1.70	2.04	0.888	0.845	0.704	0.654	0.364
5/16	8.2	2.40	6.26	1.89	1.61	1.68	1.75	0.753	0.853	0.681	0.658	0.368
1/4	6.6	1.94	5.11	1.53	1.62	1.66	1.44	0.614	0.861	0.657	0.663	0.371
L 4 ×4 × 3/4	18.5	5.44	7.67	2.81	1.19	1.27	7.67	2.81	1.19	1.27	0.778	1.000
5/8	15.7	4.61	6.66	2.40	1.20	1.23	6.66	2.40	1.20	1.23	0.779	1.000
1/2	12.8	3.75	5.56	1.97	1.22	1.18	5.56	1.97	1.22	1.18	0.782	1.000
3/8	9.8	2.86	4.36	1.52	1.23	1.14	4.36	1.52	1.23	1.14	0.788	1.000
5/16	8.2	2.40	3.71	1.29	1.24	1.12	3.71	1.29	1.24	1.12	0.791	1.000
1/4	6.6	1.94	3.04	1.05	1.25	1.09	3.04	1.05	1.25	1.09	0.795	1.000

TABLE A–4 (continued)

Size and Thickness	Weight per Foot	Area	Axis X-X				Axis Y-Y				Axis Z-Z	
			I	S	r	y	I	S	r	x	r	Tan α
In.	lb	In.²	In.⁴	In.³	In.	In.	In.⁴	In.³	In.	In.	In.	
L 4 × 3½ × ½	11.9	3.50	5.32	1.94	1.23	1.25	3.79	1.52	1.04	1.00	0.722	0.750
⅜	9.1	2.67	4.18	1.49	1.25	1.21	2.95	1.17	1.06	0.955	0.727	0.755
5⁄16	7.7	2.25	3.56	1.26	1.26	1.18	2.55	0.994	1.07	0.932	0.730	0.757
¼	6.2	1.81	2.91	1.03	1.27	1.16	2.09	0.808	1.07	0.909	0.734	0.759
L 4 × 3 × ½	11.1	3.25	5.05	1.89	1.25	1.33	2.42	1.12	0.864	0.827	0.639	0.543
⅜	8.5	2.48	3.96	1.46	1.26	1.28	1.92	0.866	0.879	0.782	0.644	0.551
5⁄16	7.2	2.09	3.38	1.23	1.27	1.26	1.65	0.734	0.887	0.759	0.647	0.554
¼	5.8	1.69	2.77	1.00	1.28	1.24	1.36	0.599	0.896	0.736	0.651	0.558
L 3½ × 3½ × ⅜	8.5	2.48	2.87	1.15	1.07	1.01	2.87	1.15	1.07	1.01	0.687	1.000
5⁄16	7.2	2.09	2.45	0.976	1.08	0.990	2.45	0.976	1.08	0.990	0.690	1.000
¼	5.8	1.69	2.01	0.794	1.09	0.968	2.01	0.794	1.09	0.968	0.694	1.000
L 3½ × 3 × ⅜	7.9	2.30	2.72	1.13	1.09	1.08	1.85	0.851	0.897	0.830	0.625	0.721
5⁄16	6.6	1.93	2.33	0.954	1.10	1.06	1.58	0.722	0.905	0.808	0.627	0.724
¼	5.4	1.56	1.91	0.776	1.11	1.04	1.30	0.589	0.914	0.785	0.631	0.727
L 3½ × 2½ × ⅜	7.2	2.11	2.56	1.09	1.10	1.16	1.09	0.592	0.719	0.660	0.537	0.496
5⁄16	6.1	1.78	2.19	0.927	1.11	1.14	0.939	0.504	0.727	0.637	0.540	0.501
¼	4.9	1.44	1.80	0.755	1.12	1.11	0.777	0.412	0.735	0.614	0.544	0.506
L 3 × 3 × ½	9.4	2.75	2.22	1.07	0.898	0.932	2.22	1.07	0.898	0.932	0.584	1.000
⅜	7.2	2.11	1.76	0.833	0.913	0.888	1.76	0.833	0.913	0.888	0.587	1.000
5⁄16	6.1	1.78	1.51	0.707	0.922	0.865	1.51	0.707	0.922	0.865	0.589	1.000
¼	4.9	1.44	1.24	0.577	0.930	0.842	1.24	0.577	0.930	0.842	0.592	1.000
3⁄16	3.71	1.09	0.962	0.441	0.939	0.820	0.962	0.441	0.939	0.820	0.596	1.000

Size	Thickness												
L 3 × 2½ ×	3/8	6.6	1.92	1.66	0.810	0.928	0.956	1.04	0.581	0.736	0.706	0.522	0.676
	1/4	4.5	1.31	1.17	0.561	0.945	0.911	0.743	0.404	0.753	0.661	0.528	0.684
	3/16	3.39	0.996	0.907	0.430	0.954	0.888	0.577	0.310	0.761	0.638	0.533	0.688
L 3 × 2 ×	3/8	5.9	1.73	1.53	0.781	0.940	1.04	0.543	0.371	0.559	0.539	0.430	0.428
	5/16	5.0	1.46	1.32	0.664	0.948	1.02	0.470	0.317	0.567	0.516	0.432	0.435
	1/4	4.1	1.19	1.09	0.542	0.957	0.993	0.392	0.260	0.574	0.493	0.435	0.440
	3/16	3.07	0.902	0.842	0.415	0.966	0.970	0.307	0.200	0.583	0.470	0.439	0.446
L 2½ × 2½ ×	3/8	5.9	1.73	0.984	0.566	0.753	0.762	0.984	0.566	0.753	0.762	0.487	1.000
	5/16	5.0	1.46	0.849	0.482	0.761	0.740	0.849	0.482	0.761	0.740	0.489	1.000
	1/4	4.1	1.19	0.703	0.394	0.769	0.717	0.703	0.394	0.769	0.717	0.491	1.000
	3/16	3.07	0.902	0.547	0.303	0.778	0.694	0.547	0.303	0.778	0.694	0.495	1.000
L 2½ × 2 ×	3/8	5.3	1.55	0.912	0.547	0.768	0.831	0.514	0.363	0.577	0.581	0.420	0.614
	5/16	4.5	1.31	0.788	0.466	0.776	0.809	0.446	0.310	0.584	0.559	0.422	0.620
	1/4	3.62	1.06	0.654	0.381	0.784	0.787	0.372	0.254	0.592	0.537	0.424	0.626
	3/16	2.75	0.809	0.509	0.293	0.793	0.764	0.291	0.196	0.600	0.514	0.427	0.631
L 2 × 2 ×	3/8	4.7	1.36	0.479	0.351	0.594	0.636	0.479	0.351	0.594	0.636	0.389	1.000
	5/16	3.92	1.15	0.416	0.300	0.601	0.614	0.416	0.300	0.601	0.614	0.390	1.000
	1/4	3.19	0.938	0.348	0.247	0.609	0.592	0.348	0.247	0.609	0.592	0.391	1.000
	3/16	2.44	0.715	0.272	0.190	0.617	0.569	0.272	0.190	0.617	0.569	0.394	1.000
	1/8	1.65	0.484	0.190	0.131	0.626	0.546	0.190	0.131	0.626	0.546	0.398	1.000

TABLE A–5   Properties of Standard Steel Pipes

Dimensions				Weight per Foot- Lbs. Plain ends	Properties			
Nominal Diameter (in.)	Outside Diameter (in.)	Inside Diameter (in.)	Wall Thickness (in.)		A (in.^{2})	I (in.^{4})	S (in.^{3})	r (in.)
Standard Weight								
$\frac{1}{2}$	0.840	0.622	0.109	0.85	0.250	0.017	0.041	0.261
$\frac{3}{4}$	1.050	0.824	0.113	1.13	0.333	0.037	0.071	0.334
1	1.315	1.049	0.133	1.68	0.494	0.087	0.133	0.421
$1\frac{1}{4}$	1.660	1.380	0.140	2.27	0.669	0.195	0.235	0.540
$1\frac{1}{2}$	1.900	1.610	0.145	2.72	0.799	0.310	0.326	0.623
2	2.375	2.067	0.154	3.65	1.07	0.666	0.561	0.787
$2\frac{1}{2}$	2.875	2.469	0.203	5.79	1.70	1.53	1.06	0.947
3	3.500	3.068	0.216	7.58	2.23	3.02	1.72	1.16
$3\frac{1}{2}$	4.000	3.548	0.226	9.11	2.68	4.79	2.39	1.34
4	4.500	4.026	0.237	10.79	3.17	7.23	3.21	1.51
5	5.563	5.047	0.258	14.62	4.30	15.2	5.45	1.88
6	6.625	6.065	0.280	18.97	5.58	28.1	8.50	2.25
8	8.625	7.981	0.322	28.55	8.40	72.5	16.8	2.94
10	10.750	10.020	0.365	40.48	11.9	161	29.9	3.67
12	12.750	12.000	0.375	49.56	14.6	279	43.8	4.38

TABLE A–6 Properties of American Standard Timber Sections

Nominal Size	American Standard Dressed Size	Area of Section	Weight per Foot	Moment of Inertia	Section Modulus
in.	in.	in.^{2}	lb/ft	in.^{4}	in.^{3}
2×4	$1\frac{5}{8} \times 3\frac{5}{8}$	5.89	1.64	6.45	3.56
6	$5\frac{5}{8}$	9.14	2.54	24.1	8.57
8	$7\frac{1}{2}$	12.2	3.39	57.1	15.3
10	$9\frac{1}{2}$	15.4	4.29	116	24.4
12	$11\frac{1}{2}$	18.7	5.19	206	35.8
14	$13\frac{1}{2}$	21.9	6.09	333	49.4
16	$15\frac{1}{2}$	25.2	6.99	504	65.1
18	$17\frac{1}{2}$	28.4	7.90	726	82.9
3×4	$2\frac{5}{8} \times 3\frac{5}{8}$	9.52	2.64	10.4	5.75
6	$5\frac{5}{8}$	14.8	4.10	38.9	13.8
8	$7\frac{1}{2}$	19.7	5.47	92.3	24.6
10	$9\frac{1}{2}$	24.9	6.93	188	39.5
12	$11\frac{1}{2}$	30.2	8.39	333	57.9
14	$13\frac{1}{2}$	35.4	9.84	538	79.7
16	$15\frac{1}{2}$	40.7	11.3	815	105
18	$17\frac{1}{2}$	45.9	12.8	1172	134
4×4	$3\frac{5}{8} \times 3\frac{5}{8}$	13.1	3.65	14.4	7.94
6	$5\frac{5}{8}$	20.4	5.66	53.8	19.1
8	$7\frac{1}{2}$	27.2	7.55	127	34.0
10	$9\frac{1}{2}$	34.4	9.57	259	54.5
12	$11\frac{1}{2}$	41.7	11.6	459	79.9
14	$13\frac{1}{2}$	48.9	13.6	743	110
16	$15\frac{1}{2}$	56.2	15.6	1125	145
18	$17\frac{1}{2}$	63.4	17.6	1619	185
6×6	$5\frac{1}{2} \times 5\frac{1}{2}$	30.3	8.40	76.3	27.7
8	$7\frac{1}{2}$	41.3	11.4	193	51.6
10	$9\frac{1}{2}$	52.3	14.5	393	82.7
12	$11\frac{1}{2}$	63.3	17.5	697	121
14	$13\frac{1}{2}$	74.3	20.6	1128	167
16	$15\frac{1}{2}$	85.3	23.6	1707	220
18	$17\frac{1}{2}$	96.3	26.7	2456	281
20	$19\frac{1}{2}$	107.3	29.8	3398	349
8×8	$7\frac{1}{2} \times 7\frac{1}{2}$	56.3	15.6	264	70.3
10	$9\frac{1}{2}$	71.3	19.8	536	113
12	$11\frac{1}{2}$	86.3	23.9	951	165
14	$13\frac{1}{2}$	101.3	28.0	1538	228
16	$15\frac{1}{2}$	116.3	32.0	2327	300
18	$17\frac{1}{2}$	131.3	36.4	3350	383
20	$19\frac{1}{2}$	146.3	40.6	4643	475
22	$21\frac{1}{2}$	161.3	44.8	6211	578

All properties and weights given are for dressed size only. The weights given above are based on an assumed average weight of 40 lb per cubic foot. Table compiled by the National Lumber Manufacturers Association.

TABLE A–6 (continued)

Nominal Size	American Standard Dressed Size	Area of Section	Weight per Foot	Moment of Inertia	Section Modulus
in.	in.	in.^{2}	lb/ft	in.^{4}	in.^{3}
10 × 10	9½ × 9½	90.3	25.0	679	143
12	11½	109	30.3	1204	209
14	13½	128	35.6	1948	289
16	15½	147	40.9	2948	380
18	17½	166	46.1	4243	485
20	19½	185	51.4	5870	602
22	21½	204	56.7	7868	732
24	23½	223	62.0	10274	874
12 × 12	11½ × 11½	132	36.7	1458	253
14	13½	155	43.1	2358	349
16	15½	178	49.5	3569	460
18	17½	201	55.9	5136	587
20	19½	224	62.3	7106	729
22	21½	247	68.7	9524	886
24	23½	270	75.0	12437	1058
14 × 14	13½ × 13½	182	50.6	2768	410
16	15½	209	58.1	4189	541
18	17½	236	65.6	6029	689
20	19½	263	73.1	8342	856
22	21½	290	80.6	11181	1040
24	23½	317	88.1	14600	1243
16 × 16	15½ × 15½	240	66.7	4810	621
18	17½	271	75.3	6923	791
20	19½	302	83.9	9578	982
22	21½	333	92.5	12837	1194
24	23½	364	101	16763	1427
18 × 18	17½ × 17½	306	85.0	7816	893
20	19½	341	94.8	10813	1109
22	21½	376	105	14493	1348
24	23½	411	114	18926	1611
26	25½	446	124	24181	1897
20 × 20	19½ × 19½	380	106	12049	1236
22	21½	419	116	16150	1502
24	23½	458	127	21089	1795
26	25½	497	138	26945	2113
28	27½	536	149	33795	2458
24 × 24	23½ × 23½	552	153	25415	2163
26	25½	599	166	32472	2547
28	27½	646	180	40727	2962
30	29½	693	193	50275	3408

TABLE A–7 Typical Mechanical Properties of Common Engineering Materials

Material	Unit Weight, γ		Elastic Moduli				Ultimate Strength						Yield Strength				Coefficient of Thermal Expansion, α	
			E		G (Shear)		Tension, $(\sigma_u)_t$		Compr.,[a] $(\sigma_u)_c$		Shear,[b] τ_u		Tension, σ_y		Shear, τ_y			
	lb/ft³	kN/m³	×10³ksi	GPa	×10³ksi	GPa	ksi	MPa	ksi	MPa	ksi	MPa	ksi	MPa	ksi	MPa	×10⁻⁶/°F	×10⁻⁶/°C
Aluminum alloy																		
2024-T4[c]	173	27.2	10.6	73	4.0	28	60	410	—	—	32	220	44	300	25	170	12.9	23.2
6061-T6[c]	173	27.2	10.0	70	3.75	26	38	260	—	—	24	170	35	240	20	140	13.0	23.4
Cast iron																		
Gray	470	74	13	90	6	41	30	210	120	830	—	—	—	—	—	—	5.8	10
Malleable	470	74	25	172	12	83	54	370	—	—	48	330	36	250	24	170	6.7	12
Concrete																		
Low strength	150	24	3	21	—	—	—	—	3	21	—	—	—	—	—	—	6.0	11
High strength	150	24	5	34	—	—	—	—	5	34	—	—	—	—	—	—	6.0	11
Copper, cold-drawn	556	87.3	17	117	6.4	44	45	310	—	—	—	—	40	280	—	—	9.3	17
Magnesium alloy, AM100A[d]	110	17	6.5	45	2.4	17	40	280			21	140	22	150	—	—	14.0	25.2
Steel																		
0.2% C, Hot rolled	490	77	30	210	12	83	65	450	—	—	48	330	36	250	24	170	6.5	12
0.2% C, Cold rolled	490	77	30	210	12	83	80	550	—	—	60	410	60	410	36	250	6.5	12
0.6% C, Hot rolled	490	77	30	210	12	83	100	690	—	—	80	550	60	410	36	250	6.5	12
0.6% C, Quenched	490	77	30	210	12	83	120	830	—	—	100	690	75	520	45	310	6.5	12
Wood																		
Douglas Fir	31	4.9	1.8	12	—	—	—	—	7.4	51	1.1	8	—	—	—	—	—	—
Southern Pine	36	5.7	1.8	12	—	—	—	—	8.4	58	1.5	10	—	—	—	—	—	—

[a] For ductile materials the ultimate strength in compression is indefinite; may be assumed to be the same as that in tension. The compressive strength of wood is parallel to the grain on short blocks.

[b] The shear strength of wood is parallel to the grain.

[c] Aluminum Association designation.

[d] American Society of Testing Materials designation.

411

TABLE A–8   Abbreviations and Symbols

Abbreviations

al	aluminum (used in subscript)
allow	allowable (used in subscript)
avg	average (used in subscript)
b	bearing (used in subscript)
br	brass or bronze (used in subscript)
cn	concrete (used in subscript)
cr	critical (used in subscript)
cu	copper (used in subscript)
F.S.	factor of safety
ft	feet
hp	horsepower
in.	inches
kg	kilograms
kips	kilopounds (1000 lb)
ksi	kips per square inch
lb	pounds
m	meters
max	maximum (used in subscript)
min	minimum (used in subscript), minute
N	newtons
N.A.	neutral axis
Pa	pascal (N/m^2)
psi	pounds per square inch
rad	radians
req	required (used in subscript)
rpm	revolutions per minute
s	seconds, shear (used in subscript)
st	steel (used in subscript)
ult	ultimate (used in subscript)
wd	wood (used in subscript)
yp or y	yield point (used in subscript)

Italic Letter Symbols

A	area, area of cross section
A'	partial area of beam section, area of inclined section
b	breadth, width
C	constant of integration
c	distance from neutral axis or from center of twist to extreme fiber
D	diameter
d	diameter, distance between parallel axes, depth
E	modulus of elasticity in tension or compression
e	eccentricity
F	force
G	modulus of elasticity in shear
g	acceleration of gravity
h	height, depth of beam
I	moment of inertia of cross-sectional area
J	polar moment of inertia of cross-sectional area
K	stress concentration factor
k	effective length factor (for columns)
L	length, span length
M	moment, bending moment
m	mass
N	number of revolutions per minute
n	number, ratio of moduli of elasticity
P	force, axial force, concentrated load, power

TABLE A–8   (continued)

p	intensity of pressure
Q	first moment of area A' about the neutral axis of a beam
q	shear flow (shear force per unit length), allowable force of fillet weld per unit length
R	reaction, radius
r	radius of gyration, radius, radial distance
S	section modulus ($S = I/c$)
T	torque, temperature, tension
t	thickness, width, tangential deviation
V	shear force, volume
W	total weight
w	weight per unit length, intensity of distributed load
x	distance along the x-axis, distance from left end of a beam to a general section
y	distance along the y-axis, distance from neutral axis of a beam, beam deflection

Greek Letter Symbols

α	(alpha)	coefficient of thermal expansion, general angle
β	(beta)	general angle
γ	(gamma)	shear strain, unit weight (weight per unit volume)
Δ	(delta)	change of any designated function
δ	(delta)	axial deformation of axially loaded member
ε	(epsilon)	linear strain
θ	(theta)	slope angle of deflection curve, angle between an inclined plane and the vertical plane, general angle
μ	(mu)	Poisson's ratio
π	(pi)	ratio of the circumference of a circle to its diameter
ρ	(rho)	radius, radius of curvature
Σ	(sigma)	summation of any designated quantities
σ	(sigma)	normal stress
τ	(tau)	shear stress
ϕ	(phi)	total angle of twist, general angle
ω	(omega)	angular velocity in rad/s

Answers to Selected Problems*

Chapter 1

1–1 (a) 6.38×10^6 kg (b) 9×10^5 m (c) 37.6 Mg (d) 0.070 (e) 23.4 kN

1–2 23.6 kN/m^3

1–4 $R_{Ax} = 6$ kips $\rightarrow$, $R_{Ay} = 4$ kips $\uparrow$, $R_B = 4$ kips $\uparrow$

1–5 $R_{Ax} = 34.6$ kN $\rightarrow$, $R_{Ay} = 60$ kN, $R_B = 40$ kN $\downarrow$

1–7 $R_{Ax} = 5$ kN $\rightarrow$, $R_{Ay} = 27$ kN $\uparrow$, $M_A = 79$ kN $\circlearrowleft$

1–8 $R_{Ax} = 1.73$ kips $\rightarrow$, $R_{Ay} = 1.87$ kips $\uparrow$, $R_B = 3.13$ kips $\uparrow$

1–10 $R_A = 866$ lb $\rightarrow$, $F_{BD} = 1000$ lb(C)

1–11 $R_{Ax} = 0$, $R_{Ay} = 2$ kN $\uparrow$, $M_A = 3$ kN $\cdot$ m $\circlearrowright$, $F_{BD} = 4.24$ kN(C), $C_x = 3$ kN, $C_y = 1$ kN

1–13 $dy/dx = 14x + 4$

1–14 $dy/dx = 2x + 4$

1–16 $dw/du = 3u^2 - 5$

1–17 slope $= +4$

1–19 slope $= +6$

1–20 slope $= -\frac{1}{4}$

1–22 $t = \frac{1}{6}s$

1–23 $x^3 + C$

1–25 $2x^3 - \dfrac{1}{x} + C$

1–26 $\frac{2}{3}x^{3/2} + C$

1–28 42

1–29 6.75

1–31 18

1–32 12

1–34 40.8

1–35 10.67

Chapter 2

2–1 $\sigma_{1-1} = 15.92$ ksi(T), $\sigma_{2-2} = 3.18$ ksi(T), $\sigma_{3-3} = 6.37$ ksi(C)

2–2 $\sigma_{1-1} = 51.4$ MPa(T), $\sigma_{2-2} = 0$, $\sigma_{3-3} = 51.4$ MPa(T)

2–4 $\sigma_{AB} = 8.97$ ksi(C), $\sigma_{BC} = 6.31$ ksi(C),

2–5 $\sigma_{AB} = 318$ MPa(T), $\sigma_{BC} = 5.33$ MPa(C)

* Note: Answers to every third problem and all the questions are omitted.

2–7 $\sigma_A = \sigma_B = 59.7$ MPa(T)

2–8 $\sigma_{BC} = 2040$ psi

2–10 (a) $\tau = 95.5$ MPa   (b) $\sigma_b = 125$ Mpa

2–11 $\tau = 11.3$ ksi, $\sigma_b = 26.7$ ksi

2–13 (a) $\tau = 14.4$ ksi   (b) $\sigma_b = 28.8$ ksi

2–14 590 kN

2–16 $\tau = 3.70$ MPa, $\sigma_b = 8.33$ MPa

2–17 $(A_{BD})_{req} = (A_{BE})_{req} = 383$ mm^2, $(A_{CE})_{req} = 343$ mm^2

2–19 $(d_A)_{req} = 6.61$ mm, $(d_B)_{req} = 7.29$ mm

2–20 $d_{req} = 2.06$ in.

2–22 (a) $A_{req} = 0.5$ in.^{2} (b) $d_{req} = 0.728$ in.

2–23 $d_{req} = 2.12$ in., $D_{req} = 4.15$ in.

2–25 $P_{min} = 12.8$ kips

2–26 $P = 57.8$ kN

2–28 $P_{max} = 3.55$ kN

Chapter 3

3–10 (a) $\sigma_p = 30$ ksi   (b) $E = 15 \times 10^3$ ksi   (c) $\sigma_y = 42$ ksi
(d) $\sigma_u = 56$ ksi   (e) Percent elongation = 39%
(f) Percent reduction in area = 32.6%

3–11 (a) $\sigma_p = 60$ ksi   (b) $E = 30 \times 10^3$ ksi   (c) $\sigma_y = 70$ ksi
(d) $\sigma_u = 127$ ksi   (e) Percent elongation = 15.5%
(f) Percent reduction in area = 20.6%

3–13 $L = 3.08$ m

3–14 $\sigma = 15.3$ ksi, $\varepsilon = 0.000\ 527$, $L' = 20.011$ ft

3–16 Stretched length = 100.003 ft

3–17 $\delta = 0.758$ mm (elongation)

3–19 $\delta_C = 1.82$ mm

3–20 $\delta_{AD} = 0.0529$ in. (elongation)

3–22 $\delta_{AD} = 0.782$ mm (elongation)

3–23 $\delta_{AC} = 0.005\ 43$ in. (elongation), $\delta_{BD} = 0.0109$ in. (elongation)

3–25 $d_{req} = 9.21$ mm

3–28 $E = 117$ GPa, $\mu = 0.358$, $G = 43$ GPa

3–29 $\delta_D = 2.07 \times 10^{-4}$ in. (contraction)

3–34 (a) $\sigma_{max} = 11.9$ ksi   (b) $\sigma_{max} = 12.5$ ksi   (c) $\sigma_{max} = 17.0$ ksi

3–35 $\sigma_{max} = 71$ MPa

3–37 $P_{max} = 48.7$ kN

3–38 $P/P' = 1.6$

3–40 (a) $P = 52.6$ kN   (b) $P = 100$ kN

Chapter 4

4–1 $\bar{x} = 3.5$ ft, $\bar{y} = 1.33$ ft

4–2 $\bar{x} = 762$ mm, $\bar{y} = 308$ mm

4–4 $\bar{x} = 783$ mm, $\bar{y} = 307$ mm

4–5 $\bar{x} = 0$, $\bar{y} = 4.30$ in.

4–7 $\bar{x} = 0$, $\bar{y} = 4.84$ in.

4–8 $\bar{x} = 0$, $\bar{y} = 140$ mm

4–11 $I_x = 8340$ in.^{4}

4–13 $I_{x'} = 13\ 440$ in.^{4}

4–14 $\bar{J} = 8.33 \times 10^{-5}$ m^4

4–16 $r_x = 40$ mm, $r_y = 89.4$ mm
4–17 $r_x = 3.00$ in., $r_y = 3.80$ in.
4–19 $\bar{I}_x = 820$ in.^{4}
4–20 $\bar{I}_x = 7.39 \times 10^{-4}$ m^4
4–22 $\bar{I}_x = 8.94 \times 10^{-5}$ m^4

4–23 $J_0 = \dfrac{\pi}{2}(R_0^4 - R_i^4)$

4–25 $\bar{J}_0 = 5.11 \times 10^{-3}$ m^4
4–26 $I_x = 2990$ in.^{4}
4–28 $\bar{I}_x = 1974$ in.^{4}, $\bar{r}_x = 6.86$ in.
4–29 $\bar{I}_x = 727$ in.^{4}, $\bar{r}_x = 4.52$ in.
4–31 $\bar{I}_x = 5590$ in.^{4}, $\bar{r}_x = 8.69$ in.
4–32 $\bar{I}_x = 670$ in.^{4}, $\bar{r}_x = 6.34$ in.
4–34 $\bar{r}_x = 5.78$ in.

Chapter 5
5–1 $T_{AB} = +2$ kN $\cdot$ m, $T_{BC} = -3$ kN $\cdot$ m
5–2 $T_{AB} = +3$ kip-in., $T_{BC} = +8$ kip-in., $T_{CD} = -6$ kip-in.
5–4 $T_{AB} = +4500$ lb-in., $T_{BC} = +6500$ lb-in., $T_{CD} = +1500$ lb-in.
5–5 $T_{AB} = -5$ kN $\cdot$ m, $T_{BC} = +6$ kN $\cdot$ m, $T_{CD} = +3$ kN $\cdot$ m
5–8 $\tau_A = 59.7$ MPa, $\tau_B = 29.8$ MPa
5–10 $\tau_{max} = 4660$ psi, $\tau_{min} = 3490$ psi
5–11 $\tau_{max} = 6.52$ MPa
5–13 $T_{allow} = 5940$ lb-in.
5–16 25 hp
5–17 At 100 rpm: $\tau_{max} = 10.04$ ksi, At 300 rpm: $\tau_{max} = 3.35$ ksi
5–19 $\tau_{max} = 38.5$ MPa
5–20 $\tau_{max} = 46.4$ MPa
5–22 (a) $T_{AB} = +3150$ lb-in., $T_{BC} = -12\ 600$ lb-in., $T_{CD} = -6300$ lb-in.
(b) $\tau_{max} = 8210$ psi
5–23 $0.267°$
5–25 $\phi_C = \phi_{C/A} = -0.489°$
5–26 $\phi_{D/A} = +1.14°$
5–28 $\phi_{D/A} = -0.573°$
5–29 $d_{req} = 77.7$ mm
5–31 $d_{req} = 72.4$ mm
5–32 $(d_0)_{req} = 3.19$ in., $W_{hollow}/W_{solid} = 0.47$
5–34 six bolts
5–35 $P = 1220$ hp
5–37 $\tau_{max} = 80.9$ MPa

Chapter 6
6–1 $R_A = 7.5$ kips $\uparrow$, $R_B = 10.5$ kips $\uparrow$
6–2 $R_A = 15$ kN $\uparrow$, $R_B = 45$ kN $\uparrow$
6–4 $R_B = 12$ kN $\uparrow$, $M_B = 56$ kip-ft $\circlearrowright$
6–5 $R_B = 15$ kN $\uparrow$, $M_B = 19$ kN $\cdot$ m $\circlearrowright$

6–7 $V_{1-1} = -10$ kN, $V_{2-2} = -20$ kN, $V_{3-3} = -20$ kN, $M_{1-1} = -5$ kN $\cdot$ m,
$M_{2-2} = -20$ kN $\cdot$ m, $M_{3-3} = -40$ kN $\cdot$ m

6–8 $V_{1-1} = +0.5$ kip, $V_{2-2} = -2.5$ kip, $V_{3-3} = +2$ kips,
$M_{1-1} = +1.0$ kip-ft, $M_{2-2} = +1.0$ kip-ft, $M_{3-3} = -2$ kip-ft

6–10 $V_{1-1} = +14$ kips, $V_{2-2} = +2$ kips, $V_{3-3} = +2$ kips,
$M_{1-1} = M_{2-2} = +14$ kip-ft, $M_{3-3} = +16$ kip-ft

6–11 $V_{1-1} = +4800$ lb, $V_{2-2} = +1800$ lb, $V_{3-3} = +450$ lb,
$M_{1-1} = -20\,100$ lb-ft, $M_{2-2} = -3600$ lb-ft, $M_{3-3} = -450$ lb-ft

6–13 $V_{A-} = 0$, $V_{A+} = V_{B-} = +6.8$ kips, $V_{B+} = V_C = V_D = -1.2$ kips,
$V_E = -2.2$ kips, $V_{F-} = -3.2$ kips, $V_{F+} = 0$, $M_A = 0$, $M_B = +6.8$ kip-ft,
$M_C = +5.6$ kip-ft, $M_D = +4.4$ kip-ft, $M_E = +2.7$ kip-ft, $M_F = 0$

6–14 $V_{A-} = 0$, $V_{A+} = +5$ kN, $V_B = +1$ kN, $V_C = -3$ kN, $V_D = -7$ kN,
$V_{E-} = -11$ kN, $V_{E+} = V_{F-} = +12$ kN, $V_{F+} = 0$, $M_A = 0$,
$M_B = +3$ kN $\cdot$ m, $M_C = +2$ kN $\cdot$ m, $M_D = -3$ kN $\cdot$ m,
$M_E = -12$ kN $\cdot$ m, $M_F = 0$

6–16 $V_{max}^{(+)} = +6.8$ kips, $V_{max}^{(-)} = -3.2$ kips, $M_{max} = +6.8$ kip-ft

6–17 $V_{max}^{(+)} = +12$ kN, $V_{max}^{(-)} = -11$ kN, $M_{max}^{(+)} = +3.125$ kN $\cdot$ m, $M_{max}^{(-)} = -12$ kN $\cdot$ m

6–19 $V_{max}^{(+)} = +25$ kN, $V_{max}^{(-)} = -55$ kN, $M_{max}^{(+)} = +37.8$ kN $\cdot$ m,
$M_{max}^{(-)} = -20$ kN $\cdot$ m

6–20 $V_{max}^{(+)} = +16$ kips, $V_{max}^{(-)} = -8$ kips, $M_{max}^{(-)} = -16$ kip-ft

6–22 $V = -6x$, $M = -3x^2$

6–23 $V = 6 - 2x$, $M = 6x - x^2$

6–25 $V_{CA} = -20$, $V_{AD} = +25$, $V_{DB} = 145 - 40x$
$M_{CA} = -20x$, $M_{AD} = 25x - 45$, $M_{DB} = -20x^2 + 145x - 225$

6–26 $V_{AB} = -2x$, $V_{BC} = +16$, $M_{AB} = -x^2$, $M_{BC} = 16x - 80$

6–28 $V_{max}^{(+)} = +2$ kips, $V_{max}^{(-)} = -6$ kip, $M_{max}^{(+)} = +4.5$ kip-ft

6–29 $V_{max}^{(+)} = +10$ kN, $V_{max}^{(-)} = -8.4$ kN $\cdot$ m, $M_{max}^{(+)} = +7.22$ kN $\cdot$ m,
$M_{max}^{(-)} = -10$ kN $\cdot$ m

6–31 $V_{max}^{(+)} = |V_{max}^{(-)}| = 26$ kN, $M_{max}^{(+)} = +24.5$ kN $\cdot$ m

6–32 $V_{max}^{(-)} = -wL$, $M_{max}^{(-)} = -\frac{1}{2}wL^2$

6–34 $V_{max}^{(+)} = +1200$ lb, $V_{max}^{(-)} = -1360$ lb, $M_{max}^{(+)} = +1610$ lb-ft,
$M_{max}^{(-)} = -1280$ lb-ft

6–35 $V_{max}^{(+)} = +40$ kN, $V_{max}^{(-)} = -32$ kN, $M_{max}^{(+)} = +20$ kN $\cdot$ m,
$M_{max}^{(-)} = -40$ kN $\cdot$ m

6–37 $V_{max}^{(-)} = -54$ kN, $M_{max}^{(-)} = 102$ kN $\cdot$ m

6–38 $V_{max}^{(+)} = +11$ kips, $V_{max}^{(-)} = -13$ kips, $M_{max}^{(+)} = +44$ kip-ft

6–40 $V_{max}^{(+)} = +5$ kips, $V_{max}^{(-)} = -4$ kips, $M_{max}^{(+)} = +4.5$ kip-ft, $M_{max}^{(-)} = -12$ kip-ft

6–41 $V_{max}^{(+)} = +110$ kN, $V_{max}^{(-)} = -90$ kN,
$M_{max}^{(+)} = +142.5$ kN $\cdot$ m, $M_{max}^{(-)} = -160$ kN $\cdot$ m

6–43 $V_{max}^{(+)} = +90$ kN, $V_{max}^{(-)} = -50$ kN, $M_{max}^{(+)} = +150$ kN $\cdot$ m,
$M_{max}^{(-)} = -15$ kN $\cdot$ m

6–44 $V_{max}^{(+)} = +22$ kips, $M_{max}^{(-)} = -162$ kip-ft

6–46 $V_{max}^{(+)} = +10$ kips, $V_{max}^{(-)} = -20$ kips, $M_{max}^{(+)} = +50$ kip-ft,
$M_{max}^{(-)} = -100$ kip-ft

6–47 $V_{max}^{(+)} = +3$ kN, $V_{max}^{(-)} = -8$ kN, $M_{max}^{(+)} = +18$ kN $\cdot$ m,
$M_{max}^{(-)} = -6$ kN $\cdot$ m

Chapter 7

7–1 $\sigma_A = 750$ psi(C), $\sigma_B = 0$, $\sigma_C = 500$ psi(T)

7–2 $\sigma_A = 50.9$ MPa(T), $\sigma_B = 0$, $\sigma_C = -101.8$ MPa

7–4 $\sigma_A = 6.62$ ksi(T), $\sigma_B = 2.21$ ksi(T), $\sigma_C = 11.03$ ksi(C)

7–5 $\sigma_A = 24.9$ MPa(C), $\sigma_B = 1.55$ MPa(C), $\sigma_C = 21.8$ MPa(T)

7–8 $M_{max} = 833$ N · m

7–10 $M_{max} = 162$ kip-ft

7–11 $M_{max} = 101.6$ kip-ft

7–13 $M_{max}^{(+)} = 18.2$ kN · m

7–14 $M_{max}^{(-)} = 9.39$ kN · m

7–16 $w_{max} = 4000$ lb/ft

7–17 $\sigma_{max}^{(T)} = 67.4$ MPa, $\sigma_{max}^{(C)} = 101$ MPa

7–19 $\tau_A = 0$, $\tau_B = 66.0$ psi, $\tau_C = 118.8$ psi

7–20 $\tau_A = 0$, $\tau_B = \tau_C = 2.55$ MPa

7–22 $P_{max} = 8$ kN

7–23 $w_{max} = 696$ lb/ft

7–25 $P_{max} = 5.00$ kN

7–26 $\tau_{max} = 6.24$ MPa

7–28 $\tau_1 = 0$, $\tau_2 = 29.4$ psi, $\tau_{2'} = 88.2$ psi, $\tau_3 = 91.9$ psi, $\tau_4 = 68.9$ psi, $\tau_5 = 0$

7–29 $\tau_{max} = 2.15$ MPa, $\sigma_{max} = 13.0$ MPa

7–31 $\tau_{max} = 12.0$ MPa, $\sigma_{max}^{(T)} = 75.0$ MPa, $\sigma_{max}^{(C)} = 150$ MPa

7–32 $\sigma_A = 0$, $\tau_A = 1.52$ ksi, $\sigma_B = 8.42$ ksi(C), $\tau_B = 1.45$ ksi

7–34 8×16 timber section

7–35 6×12 timber section

7–37 6×8 timber section

7–38 6×12 timber section

7–40 S20 $\times$ 66 steel section

7–41 W18 $\times$ 60 steel section

7–43 W14 $\times$ 30 steel section

7–44 W10 $\times$ 22 steel section

7–46 $w_{max} = 341$ lb/ft

7–47 $P_{max} = 3.67$ kN

7–49 $p_{max} = 3.90$ in.

7–50 $(p_{AB})_{req} = 6.75$ in., $(p_{BC})_{req} = 3.38$ in.

7–52 $(\sigma_{wd})_{max} = 10.3$ MPa (C), $(\sigma_{st})_{max} = 95.4$ MPa(T)

7–53 $(\sigma_{st})_{max} = 116$ MPa, $(\sigma_{wd})_{max} = 5.38$ MPa

7–55 $M_{allow} = 44.3$ kN · m

7–56 $M_{allow} = 27.8$ kN · m

7–58 $x = 5.80$ in.

7–59 $(\sigma_{cn})_{max} = 1.30$ ksi(C), $(\sigma_{st})_{max} = 20.2$ ksi(T)

7–61 (a) $x = 6$ in., (b) $A_{st} = 1.8$ in.^{2}, (c) $w = 1480$ lb/ft

Chapter 8

8–5 $y_{max} = 0.385$ in. ↓

8–7 $y_{max} = 0.462$ in. ↓

8–8 $y_{max} = 8.54$ mm ↓

8–10 $y_C = 7.57$ mm ↓, $y_D = 11.11$ mm ↓, $y_E = 8.26$ mm ↓

8–11 $y_C = y_E = 0.728$ in. ↓, $y_D = 1.04$ in. ↓

8–13 $y_C = 0.142$ in. ↑, $y_D = 0.338$ in. ↓

8–14 $y_C = 3.63$ mm $\uparrow$, $y_D = 13.34$ mm $\downarrow$

8–16 $y = -\dfrac{Px^2}{6EI}(3L - x)$

8–17 $y = -\dfrac{wx^2}{24EI}(6L^2 - 4Lx + x^2)$

8–19 $y_{max} = 0.692$ in. $\downarrow$

8–20 $y_{max} = 17.2$ mm $\downarrow$

8–22 $y_{max} = 8.54$ mm $\downarrow$

8–23 $y_C = 34.9$ mm $\downarrow$

8–25 $y_C = 0.320$ in. $\downarrow$

8–26 $y_C = 9.35$ mm $\downarrow$

8–28 $y_C = 0.142$ in. $\uparrow$, $y_D = 0.337$ in. $\downarrow$

8–29 $y_C = 3.64$ mm $\uparrow$, $y_D = 13.3$ mm $\downarrow$

8–31 $y_C = 3.92$ mm $\uparrow$, $y_D = 10.5$ mm $\downarrow$

8–32 $y_C = 0.161$ in. $\downarrow$, $y_D = 0.0940$ in. $\uparrow$

8–34 $M_A = -230$ kN $\cdot$ m

8–35 $M_C = +20$ KN $\cdot$ m

8–37 $M_C = +400$ lb-ft

8–38 $M_A = -10$ kN $\cdot$ m

8–40 $M_A = -30$ kip-ft

8–41 $\theta_{max} = \dfrac{Pa^2}{2EI}$, $y_{max} = \dfrac{Pa^2}{6EI}(3L - a)$

8–43 $\theta_{max} = \dfrac{wL^3}{24EI}$, $y_{max} = \dfrac{wL^4}{30EI}$

8–44 $y_{max} = 0.692$ in. $\downarrow$

8–46 $y_{max} = 0.476$ in. $\downarrow$

8–47 $y_{max} = 8.54$ mm $\downarrow$

8–49 $y_{max} = \dfrac{PL^3}{48EI}$

8–50 $y_{max} = 34.9$ mm $\downarrow$

8–52 $y_{max} = 10.0$ mm $\downarrow$

8–53 $y_{max} = 0.596$ in. $\downarrow$

8–55 $y_C = 9.35$ mm $\downarrow$

8–56 $y_C = 6.63$ mm $\downarrow$

8–58 $y_D = 3.57$ mm $\downarrow$

8–59 $y_C = 3.92$ mm $\uparrow$, $y_D = 10.5$ mm $\downarrow$

8–61 $y_C = 0.142$ in. $\uparrow$, $y_D = 0.337$ in. $\downarrow$

8–62 $y_C = 3.64$ mm $\uparrow$, $y_D = 13.3$ mm $\downarrow$

Chapter 9

9–1 $\sigma_{AC} = 10.25$ ksi(C), $\sigma_{BC} = 12.83$ ksi(T)

9–2 $P_{AC} = +21.8$ kN, $P_{CD} = -58.2$ kN, $P_{DB} = +101.8$ kN

9–4 $P_{allow} = 181$ kips

9–5 13 ksi, 7 ksi

9–7 $T_{CD} = 24$ kN, $T_{EF} = 48$ kN

9–8 $\sigma_{st} = 13\ 080$ psi(T), $\sigma_{al} = 3980$ psi(C)

9–10 $\sigma = 3890$ psi

9–11 $T = 49.7°C$

9–13 (a) $\sigma_{al} = 19.2$ MPa(C), $\sigma_{st} = 5.77$ MPa(C)

(b) $\Delta T = 32.1°C$ (decrease)

9–14 $\sigma_{st} = 5160$ psi(T), $\sigma_{al} = 4130$ psi(C)

9–16 $\tau_{max} = 97.8$ MPa

9–17 $\tau_{max} = 15.04$ ksi

9–19 $T_{AC} = -2300$ N $\cdot$ m, $T_{CD} = +700$ N $\cdot$ m, $T_{DB} = 2200$ N $\cdot$ m

9–20 $M_A = \frac{3}{16}PL \circlearrowleft$

9–22 $R_B = \frac{5}{4}wL \uparrow$

9–23 $R_B = 7.33$ kips $\uparrow$

9–25 $M_A = M_B = \frac{1}{12}wL^2$, $R_A = R_B = \frac{1}{2}wL$

9–26 $R_B = 1400$ lb $\uparrow$, $R_A = 11\,400$ lb $\uparrow$, $M_A = 57\,600$ lb-ft $\circlearrowleft$

9–28 $R_A = 1120$ lb $\uparrow$, $R_B = 1040$ lb $\uparrow$, $M_B = 1920$ lb-ft $\circlearrowright$, $V_{max}^{(+)} = +1120$ lb, $V_{max}^{(-)} = -1040$ lb, $M_{max}^{(+)} = +2240$ lb, $M_{max}^{(-)} = -1920$ lb-ft

9–29 $M_A = M_B = 24$ kN $\cdot$ m, $R_A = R_B = 24$ kN, $V_{max}^{(+)} = |V_{max}^{(-)}| = 24$ kN, $M_{max}^{(+)} = |M_{max}^{(-)}| = 24$ kN $\cdot$ m

9–31 $R_A = R_D = 8$ kN, $R_B = R_C = 22$ kN, $V_{max}^{(+)} = |V_{max}^{(-)}| = 12$ kN, $M_{max}^{(+)} = +6.4$ kN $\cdot$ m, $M_{max}^{(-)} = -8$ kN $\cdot$ m

Chapter 10

10–1 $\sigma_A = 7.46$ MPa(T), $\sigma_B = 5.89$ MPa(C)

10–2 $\sigma_A = 20.7$ ksi(T), $\sigma_B = 17.5$ ksi(C),
Zero stress occurs at 0.916 m above point B.

10–4 $\sigma_A = 7.35$ ksi(T), $\sigma_B = 17.35$ ksi(C),
Zero stress occurs at 4.16 in. to the right of point A.

10–5 $\sigma_A = 89.1$ MPa(C), $\sigma_B = 114.5$ MPa(T)

10–7 $P_{max} = 5050$ lb

10–8 $\sigma_{max}^{(T)} = 8.31$ ksi, $\sigma_{max}^{(C)} = 10.81$ ksi

10–10 $\sigma_{max}^{(T)} = 574$ psi, $\sigma_{max}^{(C)} = 722$ psi

10–11 The line of zero stress is $0.1875y + 0.25x = 0$, $\sigma_B = \sigma_D = 0$, $\sigma_A = 1.50$ ksi(C), $\sigma_C = 1.50$ ksi(T)

10–13 (a) $\sigma_Q = 133.3y + 426.7x$
(b) The line of zero stress is $133.3y + 426.7x = 0$
(c) $\sigma_A = 240$ psi(C), $\sigma_B = 1040$ psi(T), $\sigma_C = 240$ psi(T), $\sigma_D = 1040$ psi(C)

10–14 $a_{req} = 91.5$ mm

10–16 W12 $\times$ 26 steel section

10–17 $\sigma_A = 600$ psi(C), $\sigma_B = 1200$ psi(T)

10–20 $e_{max} = 5.40$ in.

10–22 $P = 49.4$ kips

10–23 $\sigma_A = 5.27$ MPa(T), $\sigma_B = 3.27$ MPa(C), $\sigma_C = 9.27$ MPa(C), $\sigma_D = 0.73$ MPa(C), The line of zero stress is $28.5x + 30y + 1 = 0$

10–25 $\sigma = 0$, $\tau = 2.36$ ksi

10–26 $\sigma = 0$, $\tau = 10.6$ MPa

10–28 At A: $\sigma = 96.6$ MPa(T), $\tau = 4.83$ MPa
At B: $\sigma = 0$, $\tau = 7.85$ MPa

10–29 At A: $\sigma = 19.15$ ksi(T), $\tau = 1.56$ ksi
At B: $\sigma = 0$, $\tau = 1.638$ ksi

10–31 $t_{min} = 7.5$ mm

10–32 $p_{max} = 750$ psi

10–34 $p_{max} = 4.33$ ksi

10–35 $\sigma_x = 48$ MPa(T), $\sigma_y = 32.8$ MPa(T)

Chapter 11

11–1 $\sigma_\theta = +5$ ksi, $\tau_\theta = -5$ ksi

11–2 $\sigma_\theta = 0$, $\tau_\theta = +50$ MPa

11–4 $\sigma_\theta = +2.83$ ksi, $\tau_\theta = -6.83$ ksi

11–5 $\sigma_\theta = -99.5$ MPa, $\tau_\theta = -5.21$ MPa

11–7 $\sigma_\theta = +5$ ksi, $\tau_\theta = -5$ ksi

11–8 $\sigma_\theta = 0$, $\tau_\theta = +50$ MPa

11–10 $\sigma_\theta = +2.9$ ksi, $\tau_\theta = -6.8$ ksi

11–11 $\sigma_\theta = -100$ MPa, $\tau_\theta = -5$ MPa

11–13 **(a)** $\sigma_x = +75$ MPa, $\sigma_y = +150$ MPa, $\tau_x = 0$
(b) $\sigma_\theta = +131.3$ MPa, $\tau_\theta = -32.5$ MPa

11–14 **(a)** $\sigma_x = +450$ psi, $\sigma_y = 0$, $\tau_x = -14.1$ psi
(b) $\sigma_\theta = +100.3$ psi, $\tau_\theta = -187.8$ psi

11–16 $\sigma_1 = +4$ ksi, $\sigma_2 = -6$ ksi, $\theta_1 = -18.4°$, $\tau_{max} = |\tau_{min}| = 5$ ksi,
$\sigma' = -1$ ksi, $\theta_2 = -63.4°$

11–17 $\sigma_1 = +12$ MPa, $\sigma_2 = -8$ MPa, $\theta_1 = -63.4°$,
$\tau_{max} = |\tau_{min}| = 10$ MPa, $\sigma' = 2$ MPa, $\theta_2 = +71.6°$

11–19 $\sigma_1 = +20$ ksi, $\sigma_2 = 0$, $\theta_1 = +63.4°$, $\tau_{max} = |\tau_{min}| = 10$ ksi,
$\sigma' = +10$ ksi, $\theta_2 = +18.4°$

11–20 $\sigma_1 = +52.4$ MPa, $\sigma_2 = -32.4$ MPa, $\theta_1 = -67.5°$
$\tau_{max} = |\tau_{min}| = 42.4$ MPa, $\sigma' = +10$ MPa, $\theta_2 = +67.5°$

11–22 $\sigma_1 = +24$ ksi, $\sigma_2 = -6$ ksi, $\theta_1 = -26.6°$, $\tau_{max} = |\tau_{min}| = 15$ ksi,
$\sigma' = +9$ ksi, $\theta_2 = -71.6°$

11–23 At A: $\sigma_1 = +1.5$ psi, $\sigma_2 = -668$ psi, $\theta_1 = -88.5°$
At B: $\sigma_1 = +31.3$ psi, $\sigma_2 = -31.3$ psi, $\theta_1 = -45°$
At C: $\sigma_1 = +1000$ psi, $\sigma_2 = 0$, $\theta_1 = 0°$

11–25 $P = 806$ lb

11–26 $\sigma_1 = +26.5$ MPa, $\sigma_2 = -0.7$ MPa,
$\tau_{max} = |\tau_{min}| = 13.6$ MPa, $\sigma' = +12.9$ MPa

11–28 At A: $\sigma_1 = +96.8$ MPa, $\sigma_2 = -0.2$ MPa, $\theta_1 = -2.4°$
At B: $\sigma_1 = +7.85$ MPa, $\sigma_2 = -7.85$ MPa, $\theta_1 = -45°$

11–29 $\sigma_1 = +19.08$ ksi, $\sigma_2 = -0.12$ ksi, $\theta_1 = +85.4°$
$\tau_{max} = |\tau_{min}| = 9.70$ ksi, $\sigma' = 9.58$ ksi, $\theta_2 = +40.4°$

11–31 $P_{max} = 7.70$ kips

Chapter 12

12–1 166

12–2 120

12–4 109

12–5 Square section is better.

12–7 53.1

12–8 57.4

12–10 $P_{cr} = 158$ kN

12–11 $P_{cr} = 28.8$ kips

12–13 $P_{cr} = 20.5$ kips

12–14 $b = 24.3$ mm

12–16 $W_{max} = 39.5$ kN

12–17 $I_{req} = 1.31$ in.^{4}; use $2\frac{1}{2}$-in. standard steel pipe.

12–19 $d_{req} = 1.34$ in.

12–20 $F_{allow} = 4.4$ kips

12–22 $P_{cr} = 160$ kips

12–23 $P_{cr} = 1.37$ MN
12–25 $P_{cr} = 644$ kips
12–26 $W_{max} = 14$ kips
12–28 (a) $P_{allow} = 26.3$ kips (b) $P_{allow} = 11.0$ kips
12–29 (a) $P_{allow} = 230$ kips (b) $P_{allow} = 132$ kips
12–31 (a) $P_{allow} = 91.2$ kips (b) $P_{allow} = 50.0$ kips
12–32 W12 × 87 steel section
12–34 W10 × 49 steel section
12–35 W24 × 131 steel section
12–37 (a) $b = 7.09$ in. (b) $P_{allow} = 210$ kips

Chapter 13

13–1 39.8 kips, 61.4%
13–2 108 kips, 66.7%
13–4 94.2 kips, 76.2%
13–5 70.6 kips
13–7 45.9 kips, 70.8%
13–8 108 kips, 66.7%
13–10 105 kips, 85.0%
13–11 70.6 kips
13–13 $P_{max} = 22.5$ kips
13–14 Use $\frac{3}{4}$-in. bolts
13–16 $\tau_{max} = 11.6$ ksi
13–17 $P_{max} = 25.7$ kips
13–19 $\frac{9}{16}$-in. fillet weld
13–20 strength of weld = 148 kips, efficiency = 91.4%
13–22 $L_1 = 5\frac{1}{2}$ in.
13–23 $L_2 = 10\frac{3}{8}$ in.

Index

Beam Deflection Formulas

Case	Beam	General Deflection Formula (y is positive downward)	Slope at Ends	Maximum Deflection
1		$y = \dfrac{Px^2}{6EI}(3L - x)$	$\theta_{max} = \dfrac{PL^2}{2EI}$	$y_{max} = \dfrac{PL^3}{3EI}$
2		$y_{AB} = \dfrac{Px^2}{6EI}(3a - x)$ $y_{BC} = \dfrac{Pa^2}{6EI}(3x - a)$	$\theta_{max} = \dfrac{Pa^2}{2EI}$	$y_{max} = \dfrac{Pa^2}{6EI}(3L - a)$
3		$y = \dfrac{wx^2}{24EI}(x^2 + 6L^2 - 4Lx)$	$\theta_{max} = \dfrac{wL^3}{6EI}$	$y_{max} = \dfrac{wL^4}{8EI}$
4		$y = \dfrac{wx^2}{120EIL}(10L^3 - 10L^2x + 5Lx^2 - x^3)$	$\theta_{max} = \dfrac{wL^3}{24EI}$	$y_{max} = \dfrac{wL^4}{30EI}$
5		$y = \dfrac{Mx^2}{2EI}$	$\theta_{max} = \dfrac{ML}{EI}$	$y_{max} = \dfrac{ML^2}{2EI}$
6		$y_{AB} = \dfrac{Px}{12EI}\left(\dfrac{3L^2}{4} - x^2\right)$	$\theta_{max} = \dfrac{PL^2}{16EI}$	$y_{max} = \dfrac{PL^3}{48EI}$